Topics in Engineering
Volume 8

Rapidly Assembled Structures

Proceedings of the International conference held at Southampton, England, April 1991.

Editor: P.S. Bulson

Computational Mechanics Publications
Southampton UK and Boston USA

Dr. P.S. Bulson
Advanced Mechanics and Engineering Ltd.
1, Huxley Road
Surrey Research Park
Guildford
Surrey
(Mott MacDonald Group, U.K.)

British Library Cataloguing in Publication Data
Rapidly assembled structures.
1. Structures
I. Bulson, P.S. (Philip Stanley) *1925*- II. Computational Mechanics Institute III. Series
624
ISBN 1-85312-136-3

ISBN 1-85312-136-3 Computational Mechanics Publications, Southampton, UK
ISBN 1-56252-063-6 Computational Mechanics Publications, Boston, USA
ISBN 0952-5300 Series

Library of Congress Catalog Card Number 91-70382

see also p32 and p37

Printed in Great Britain by Antony Rowe, Ltd, Chippenham, Wilts.

PREFACE

There are many types of structure that must be rapidly assembled or unfolded before they are ready to use, and very often the requirement for fast assembley is associated with a need for rapid and efficient transportation to the site. Expertise in the design of such structures, often backed by research and analysis, has been accumulated over a long time and in many countries. Notable fields of the subject are military bridges, trackways, communication equipment and battlefield protective shelters. In the civil area demountable grandstands, scaffolding, circus tents, air structures and foldable seating come to mind.

The absence of books or conference proceedings on the subject suggests that the mechanical and structural science of the subject as a whole has rarely been treated as a separate field of technology, in the same way that use, form or material is often used as a way of classifying structures. It also suggests that workers in the field rarely meet to discuss common problems or to exchange philosophic views. There have been publications, conferences, or lectures dealing generally with moveable structures such as swing bridges or traversing radio telescopes, or the relatively recent concept of active structures that change their geometry to cater for changes in load; but these publications and reports are thought to be few in number.

It therefore seemed appropriate to try to bring together engineers from a number of areas of the subject, and the result was the two - day conference MARAS 91, Mobile and Rapidly Assembled Structures, held at Southampton,U.K., in April 1991. This book has been prepared from the collected papers, which were presented in two main sessions: Military and Civil Structures. The edited papers provide a 'state - of - the - art' review of the subject, which newcomers to the rapid - assembly field should find useful as an introduction to a fascinating area of structural engineering.

P.S. Bulson
April 1991

CONTENTS

INTRODUCTION

SECTION 1: RAPIDLY ASSEMBLED MILITARY STRUCTURES

SECTION 2: RAPIDLY ASSEMBLED AND ACTIVE CIVIL STRUCTURES

INTRODUCTION

The Efficiency of Rapidly Assembled Structures

P.S. Bulson

Advanced Mechanics and Engineering Ltd, Mott MacDonald Group

ABSTRACT

There are many types of rapidly assembled structure, and the paper starts by describing a proposed classification into hinged, pinned, clamped, sliding and fabric structures. The penalties of each type are discussed, and following this an attempt is made to compare the efficiencies of all types by means of structural coefficients. The numerical coefficients, or indices, deal with the following characteristics: load/volume, joint efficiency and vulnerability. A detailed survey is recommended that would set ranges and limits to these coefficients.

INTRODUCTION

We can select a few examples of rapid assembly from the distant and recent past. The nomadic tribes of ancient times carried their shelters around with them, perhaps just sticks and skins to form tents and awnings, but there must have been logistic decisions to be made: the best way of bundling sticks or which animals produced minimum weight hides. Noah designed and built what must have been one of the first mobile warehouses. William the Conqueror brought with him from Normandy, on 28 September 1066, a complete prefabricated, dismantled and pre-assembled wooden castle. It was in position in Hastings by the evening of that day as his headquarters. The instant house was often built throughout history by simply turning over a boat. The art of scaffold construction is an ancient one, and over the centuries there has been a strong military interest in rapidly assembled lifting and gap crossing equipment, originally using wooden poles and rope lashings, but more recently employing metal panels and girders. Nowadays we are all familiar with rapidly-stowed aircraft components such as retractable undercarriages, folding wings and escape chutes.

The diversity of the areas leads to the thought that there might be a common technical thread running through the subject, a common philosophy that would help to unite this diversity into a discrete subject in structural engineering. We are all familiar with the classification of structures by their use: bridge, ship or aircraft structures for example; or by their form: thin-walled or shell structures; or by their material: steel, concrete or aluminium structures. Is there a case for thinking of all rapidly assembled structures as part of a family, and can this family be rationalised and interlinked?

To explore this we must first look at the anatomy of rapid assembly and examine the engineering fundamentals that stem from it. The author proposes the following classification (see Fig 1):

a) Hinged structures: the elements are permanently connected to each other by pins or joints that allow articulation. Structural integrity after unfolding is achieved by the cunning use of linkages or sliding connections. Simple examples are the umbrella, the deck chair or the foldable music stand. More complex examples include the aircraft undercarriage, folding wings or variable geometry aircraft. The unfolding can be operated manually for small-scale structures or hydraulically for large-scale structures. In all cases the joint design is a key feature.

b) Pinned structures: the elements are transported separately and connected at the site by pins or bolts. Bolting does not always mean rapid assembly, so the emphasis is on the quick insertion of pins in jaws or shaped heads into slots. Well known examples are the hand-built military bridge girders, where the elements are structural panels; and space frameworks, where the elements are single bars in tension or compression.

c) Clamped structures: the elements are single bars or tubes, transported separately and connected to form space frames by separate clamping elements. The most obvious example is scaffolding, which is usually tubular and in which the clamping action is direct onto the tube wall. Sometimes the clamping action brings together circumferential flanges, as in the connection of lengths of pipeline by victaulic joints.

d) Sliding structures: the elements are deployed from the pre-assembled packed form by sliding actions. Telescopic structures of various designs come into this category, particularly telescopic structures in which the integrity of the deployed shape is achieved by oil hydraulic pressure. The pins, bolts or clamps are replaced by hydraulic or air pressure supplied by mechanical pumps.

e) Fabric structures: dismantling consists of folding flexible materials of various kinds so that they can be stowed and transported easily. Structural integrity requires the addition of pressure - often air pressure as in parachutes, airhouses or inflatable dinghies.

Using this first attempt at a classification as a basis, can we draw some unifying design philosophy described by simple parameters or coefficients? Can we classify solutions in terms of structural efficiency and economy? Let us first look at the penalties of rapid assembly.

PENALTIES

All innovation tends to be accompanied by drawbacks. In pinned rapidly assembled structures of the framework type, for example, there is clearly a penalty in that the weight of joints is higher than in a similar fabrication in which rapid assembly is not a design requirement. It has been shown that in a simple welded steel framed girder having a span of about 5 metres, the extra material required to ensure proper integrity of the welds adds about 4% to the basic weight of the structure. If the girder is made by bolts or rivets attached in groups through gusset plates, then the extra material adds between 8% and 11% to the basic weight, depending on the geometry. Groups of bolts are not capable of rapid assembly, so it is preferable to replace them by single, large pins or bolts. These concentrate stresses into small zones and lead to heavier lugs and jaws on the ends of components. These heavy joints often result in higher secondary bending stresses in the members of the structure, and this leads to a further increase in structural weight. A detailed study of these effects has not been made, but it is hard to imagine that the total weight of extra material adds less than 15% to the basic weight. So in broad terms the transformation of a welded framework into a rapidly assembled framework can add at least 10% to the basic weight of the structure.

It is possible to offset this increase by using higher strength materials to reduce the basic weight, but the possibility of structural fatigue then becomes an important consideration. Stresses must be reduced in key areas of the tension joints to eliminate premature cracking, and this again can add considerably to the basic weight. This problem is sometimes attacked by using multi-jaw joints, but the geometry of the framework does not always provide enough space to accommodate multi-jaws.

The 10% figure relates to a simple truss girder of the Pratt or Warren type, for example, designed to carry a total applied load of about 12 tonnes, but as the scale of the girder increases, and the applied load increases, individual components become too large to handle and the structure must be divided into more components and more joints. For some military structures, carrying heavy loads and having components that must be manhandled, the joint weight can rise to more than 30% of the total girder weight. This suggests that joint weight is influenced by the load to be carried, the span, the largest allowable weight of a single component, and the largest allowable depth of the girder.

The penalties in a clamped structure are the size and weight of the clamps, and the unused length of member that falls outside each clamp. The penalty in a sliding structure is the overlap between sliding units, and in telescopic beams that have to take bending and shear the overlap has to be quite substantial, to ensure that the local loads in the overlap area do not become too large. When telescopic structures are used as columns in compression, the overlap has an important influence on the stability of the structure, a fact well known to undercarriage designers.

Fabric or membrane structures have very different penalties. They are very efficient with regard to the ratio deployed volume/stowed volume, but there are great difficulties when they are subjected to localised loading. Local loads have to be taken into the membrane via large numbers of separate tensile cords, as in a parachute for example. They are very vulnerable to damage, and in the case of air structures, they rely on the output of a compressor to maintain integrity. Fabric or membrane structures tend to combine high efficiency with short life, so any parameter used to compare their performance must include life, both in service and on the shelf.

One of the penalties of the hinged structure is that the stowed and packed geometry is one unit having the same weight as the deployed structure. The same is true of telescopic structures. This limits the range of size and weights that are suitable, or else it means a high level of mechanical assistance in the construction phase. Thus we find most folding structures limited to a size and weight that can be handled by one individual (chairs, tables etc), whereas the larger telescopic structures are often permanently mounted on load-carrying vehicles, or have their own mobile support system (like the telescopic gangways at airports).

LOAD CAPACITY AND VOLUMETRIC ADVANTAGE

In conventional structures structural efficiency is thought of in terms of strength and weight, and the strength/weight ratio provides one method of comparing structures designed for similar purposes. Compressive structures, such as centrally loaded columns, are often classified according to a compressive structural loading coefficient P_c/L^2, where P_c is the compressive load and L the distance over which it is to be transmitted. Bending structures, such as beams with thin webs, are classified according to a shear loading coefficient V_s/d^2, where V_s is the shear load and d is the depth of the web plate. These efficiencies and coefficients are not appropriate to rapidly assembled structures, and as was remarked earlier the component weight and the number of joints to be brought together on assembly must be included in any discussion of efficiency.

Consider the simplest pin jointed structure, a single equilateral triangle formed from three identical bars, pinned together in assembly. If the length between joint centres = L, and the bars have a square section of side b, then the total stowed volume is approximately $3Lb^2$. The deployed volume is $(bL/2)(\sqrt{3}L/2) = 0{\cdot}43L^2b$ and the volumetric advantage (= deployed volume/stowed volume) is $0{\cdot}14L/b$. For a steel structure an efficient slenderness ratio for the compression members in the assembled structure is, say, 50, and the radius of gyration of a square section of side b is $b/\sqrt{12}$. This means that $b \simeq 0{\cdot}07L$, and the volumetric advantage, f_v = 2.

With these proportions it should be possible for the collapse stress of the compression member to reach $0{\cdot}9\sigma_Y$, so the carrying capacity of the triangular frame, supported at two joints and having a load P_D applied at the third joint is given by $P_D = 2(0{\cdot}9\sigma_Y)b^2\sqrt{3}/2 \simeq 1{\cdot}6\sigma_Y b^2$. The carrying capacity of the three identical bars, laid side by side and used as a beam is a central load, P_S, given by $P_S = 3\sigma_Y b^3/L$. The ratio of the carrying capacity is $P_D/P_S = 0{\cdot}5L/b$, and taking $b = 0{\cdot}07L$, as before, the ratio becomes $7{\cdot}6$. Thus the capacity to carry a central load P over a span L is increased by a factor of $7{\cdot}6$ by rapid assembly of three identical bars into a simple triangular framework.

The volumetric advantage changes with the form of the truss, as shown in Fig 2. For an efficiently designed warren truss composed of equilateral triangular elements, the advantage is in general $3(N-1)N$, where N is the number of identical bars. When $N=3$, as in the example above, the advantage is 2. Note that if we add more and more triangular elements, so that $N \to \infty$, the volumetric advantage cannot exceed 3. Of course, if the span is increased in this way the load carrying capacity of the framework must be reduced to ensure that the tensile and compressive strength of the bars is not exceeded. The load carrying capacity will also be reduced if the compression members have high slenderness ratios.

The characteristics of telescopic structures are quite different. Suppose there are N telescoping elements, with an overlap of C and a basic element length of L; then, the volumetric advantage = $1+(N-1)(1-C/L)$ if it assumed that all elements have the same cross section. If, as is more usual, each successive element has a smaller cross section, the above equation must be adjusted. If the ratio of successive cross sections is 0.9 and $c = 0.1L$, then successive values of the advantage as N is increased from 2 to 4 change from 1.9, 2.8 and 3.7 to 1.81, 2.53 and 3.16. So with sufficient elements the volumetric advantage has a greater potential than for simple trusses formed from separate bars. However, the problem is that the ratio between the total deployed structure weight, and the weight of a single stowed element is 1, because the elements cannot be disconnected. If the elements can be disconnected the structure classification changes from a telescopic to a pinned structure.

The structural efficiency of foldable membrane structures is very high, but is counterbalanced by vulnerability and a short life. An inflated bridge researched by the author some years ago had a volumetric advantage of about 25, but this does not take into account the size and weight of the air compressor. In a large air house the ratio between the enclosed volume and the stowed volume of the fabric could well be 400.

In military support bridges it was customary for a long time to design the main girders to be built without crane assistance, and to build them from basic triangular or rectangular panel units, pinned together to form girders. The number of panel units needed to form a single girder at maximum load class influenced the percentage of the total weight of the structure contained in the joints, and there was evidence of a linear relationship between the two, as indicated in Fig 3. It was also found that the weight of the structure calculated from the preliminary design layout grew by about 25% by the time the fully detailed and engineered drawings had been made. It was also found that if the panel units were short lengths of plate girder construction there was a 'rule of thumb' relationship between the percentage decrease in weight of the structure and the percentage increase in the ultimate strength of the material of construction, according to the equation: % in weight/%increase in ultimate strength $\triangleq$ 0.4.

We have already seen that a good volumetric advantage f_v can be offset by a relatively poor load carrying capacity, so in considering the overall load/weight/volume efficiency, it is appropriate to introduce a second factor, the maximum load (or the load producing the maximum stresses in the structure, P) divided by the total structural weight, W. If we denote this as f_W, then we can introduce a strength/volume index $S = f_V \cdot f_W$. This allows for a high value of f_V to be offset by a low f_W, and is therefore one way of comparing the qualities of rapidly assembled structures of varying anatomy.

THE WEIGHT AND LIFE OF JOINTS

In all forms of rapidly assembled structure the efficiency of the joints is important, particularly when joint weight is 20% or 30% of the total structural weight. The author suggests that a way of measuring this is to use a joint weight factor, f_j, which is the ratio between the weight of continuous structural detail when no joint is required and the same detail when a joint of equal strength is inserted in or clamped to the detail. To take a simple example the factor f_j for a telescopic column or girder is the ratio between the weight of a single tube equal in length to the overlap, and the total weight of the overlap. In this case the value of f_j is about 0.5, as indicated in Fig 4(a). A tension member consisting of two elements, welded together to form one continuous element, would have $f_j = 1.0$. A single pin joint of the type illustrated in part (b) of Figure 4 might well have a value of f_j less than 0.5.

A potential problem in a rapidly assembled structure occurs when the life of the joint does not match the life of the structural elements it joins. In a welded metal structure, for example, the insertion of a joint may well introduce areas of stress concentration associated with short lengths of weld set transversely to the general direction of tensile stress. Under fluctuating tensile stress this could lead to fatigue cracking and premature failure of the joint. So the efficiency of a joint is linked not only to its geometry, but to the possibility of reduction in life. An obvious index that enables this to be set in numerical form is f_l = life of joint/life of the structure without the joint. In a welded high strength aluminium structure, for example, the life under a given cyclic design stress for unwelded material may be three times greater than the life of a transversely welded detail under the same general stress pattern. This can be accounted for by treating the Joint Efficiency as the product of a weight factor and a life factor, ie $J = f_j f_l$, where J is the Joint Efficiency Index.

VULNERABILITY

We have already noted that certain rapidly assembled structures with relatively high values of the load/volume index can easily suffer damage. This applies particularly to thin-walled metal structures and to inflated fabric structures. So any comparison of the efficiency of this type of structure should take vulnerability to damage into account. High impact stresses under local point loading, associated with material having low ductility or poor fracture toughness, can lead to premature failure by tearing. Structural joints with poor fatigue characteristics can have limited lives, as we have already seen, but a further hazard is the vulnerability of certain types of joint to localised damage. Thin jaws can be easily deformed, inadequate clamps can suffer loss of grip, long thin tension links are often bent or twisted, and the surface protection of pins can be damaged.

Is it possible to codify vulnerability in a way that gives a quick numerical comparison between different types of structure? Percentage ductility, as given in material specifications, coupled with a measure of the ease with which material can be locally deformed, is one possible approach. The relationship between local pressure under the head of a circular rod pressed transversely into the structural element and the local tensile strength in the element caused by this action in the vicinity of the pressure area will give a measure of the vulnerability as indicated in Fig 5. Thin sheet material or fabrics will have a relatively low pressure/stress ratio; thicker sections and chunky details will have a relatively higher ratio. If this ratio f_s is combined with the percentage ductility, f_d, we have a possible numerical damage vulnerability index, V, where $V = f_s \cdot f_d$. Once an index of this type has been accepted, experiments can define the limits of survivability for various types of structure, particularly in the areas of joints or large unsupported sheets.

THE INFLUENCE OF SIZE AND COST

The indices discussed so far do not address the problem of practical limitations to the size of both the deployed and the stowed structure. A music stand that folds into a length that fits into a music case is acceptable, but a stand that folds into a package 3 metres in length, however brilliant the folding concept, would not be practical. Think of the problems of carrying it on the Metro. Similarly, the panels of a pinned military bridge truss, or the box girders of a hinged military bridge girder, must not be so heavy, long or wide that they cannot readily be packed on to the bed of a truck or carried on a trailer.

Thus the weight of the components of the disassembled structure, which we discussed earlier, is not enough to define its adequacy, and the main dimensions, length, breadth and depth of the components must be related to the method of transportation. The concept of an umbrella is that it must be carried and deployed by a single person, and its folded geometry must be such that the person can easily carry it. An umbrella that was so heavy that it took two people to carry and open it would probably not be acceptable. The folded size of a parachute must fit a pack that can be readily carried by one person. There seem to be two obvious methods of transport that impose size limitations, the human body, and the capacity of trucks, vans or trailers. In the first method, a length of 1 metre would seem to be a useful limitation, anything longer is difficult to transport by hand. Of course certain sporting structures like the javelin or the billiard cue represent a problem because disassembly into manageable sections is not generally possible.

In the case of mechanically assisted transportation, the bed length and width of trucks, not usually greater than 10 x 3 metres, is the limiting area, and this is generally linked with a maximum allowable depth of about 2 metres. A load of greater depth than 2 metres on a standard height of truck bed would give problems when negotiating most railway bridges. Deeper loads would require special low level trailers and could reduce transportation speed.

In general, smaller components ease transportation and handling problems, but can increase the assembly time. In military structures assembly time is linked to the tactical requirements governing the use of the system. In civil structures assembly time is often linked to the cost of the labour force and equipment needed to assemble the structure.

The remaining factor is cost. The design, development and manufacture of structures that employ rapid-assembly joints is likely to be more expensive than conventional structures of equivalent size and capacity. Complex joints often involve research and testing well beyond normally accepted levels. There are so many factors to be considered that it is not easy to see how this aspect can be put in terms of a single numerical coefficient, but the individual influence of key factors can be discussed. High initial cost is associated with: expensive materials, complex joint design, mechanical assistance to fast assembly, highly trained labour. High operating costs are associated with vulnerability to damage or misuse, low material life, the need for protection, the frequency of assembly and disassembly, and the running costs of transportation. Certain of these factors can be linked directly to the load/volume index, the joint efficiency index and the vulnerability index, but others are linked to more abstract decisions involving the operational requirements of the user and the company policy of the manufacturer.

CONCLUSIONS

It is suggested that the efficiencies of a range of rapidly assembled structures can be compared by means of numerical indices. The first of these is the strength/volume index, S , which is the product of a volumetric advantage, f_V , and a strength/weight factor, f_w . The second is the Joint Efficiency Index, J , which is the product of a joint weight factor, f_j , and a joint life factor, f_l . The third is a Vulnerability Index, V , which is the product of a pressure/stress ratio, f_s , and percentage ductility, f_d .

These indices do not take into account all aspects of size and cost, but represent a first attempt to put the efficiency of rapidly assembled structures on a more rational basis that hitherto. It is recommended that future research projects in this field examine many types of structure and establish the ranges and numerical limits for the coefficients.

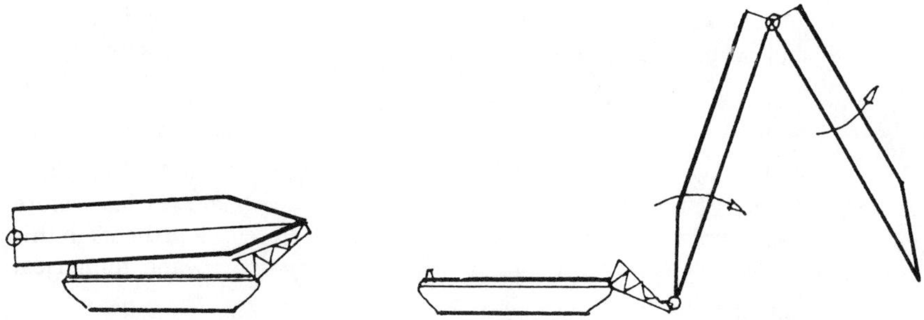

(a) Hinged structure - military tank bridge

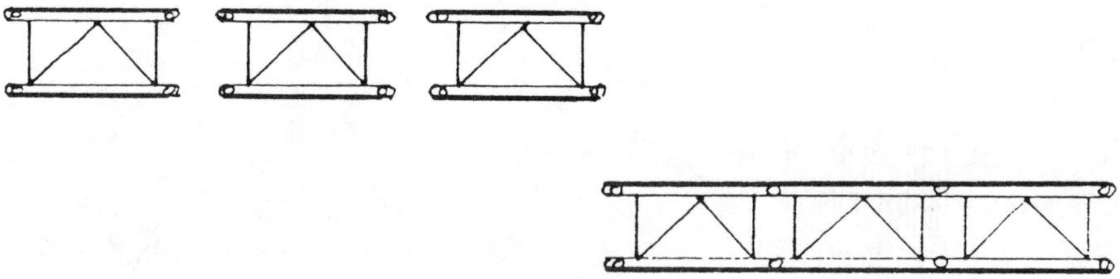

(b) Pinned structure - girder formed from panels

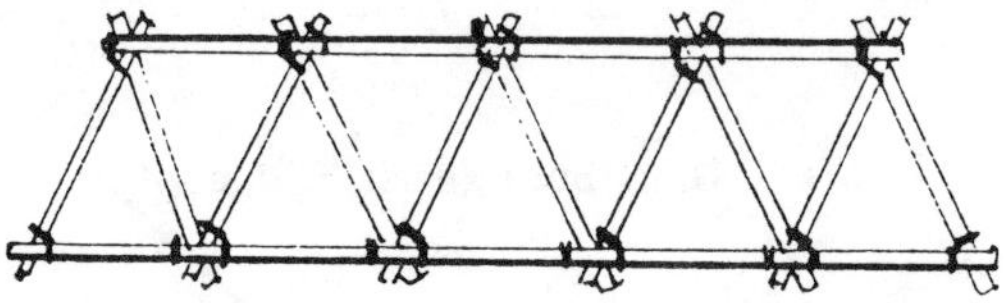

(c) Clamped structure - scaffolding

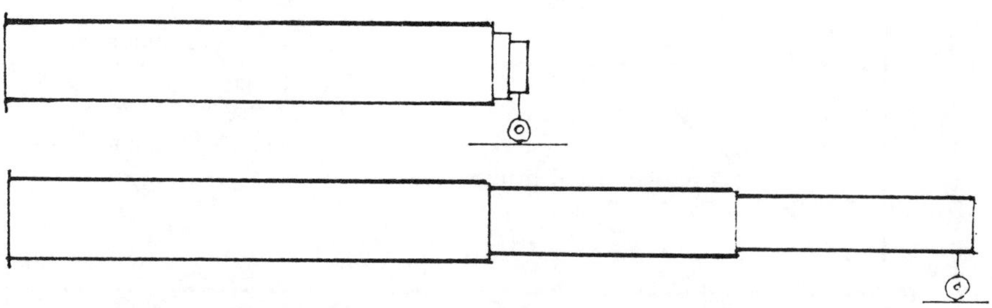

(d) Sliding structure - telescopic walkway

Fig 1 PROPOSED CLASSIFICATION

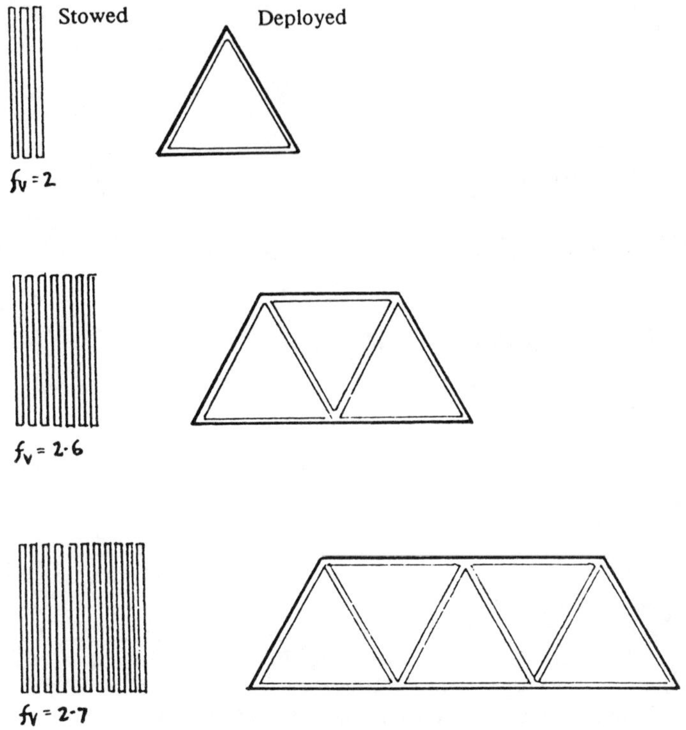

Fig 2 VOLUMETRIC ADVANTAGE, f_v, FOR SIMPLE TRUSSES WITH EFFICIENT STRUTS

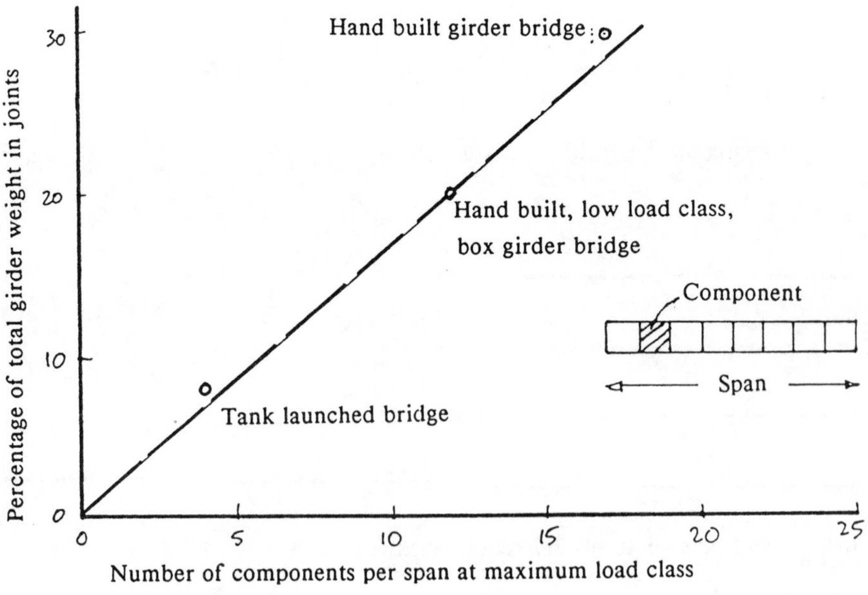

Fig 3 RELATIONSHIP BETWEEN GIRDER WEIGHT AND JOINT WEIGHT

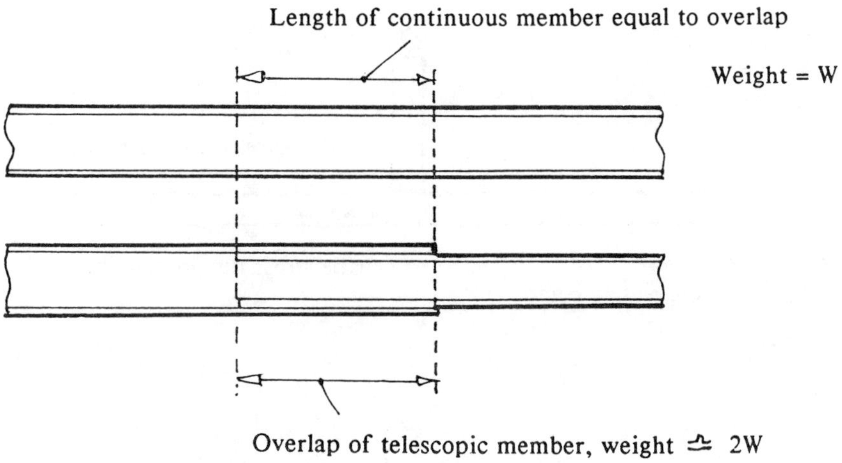

(a) Value of f_j for simple telescopic tube ≏ 0.5

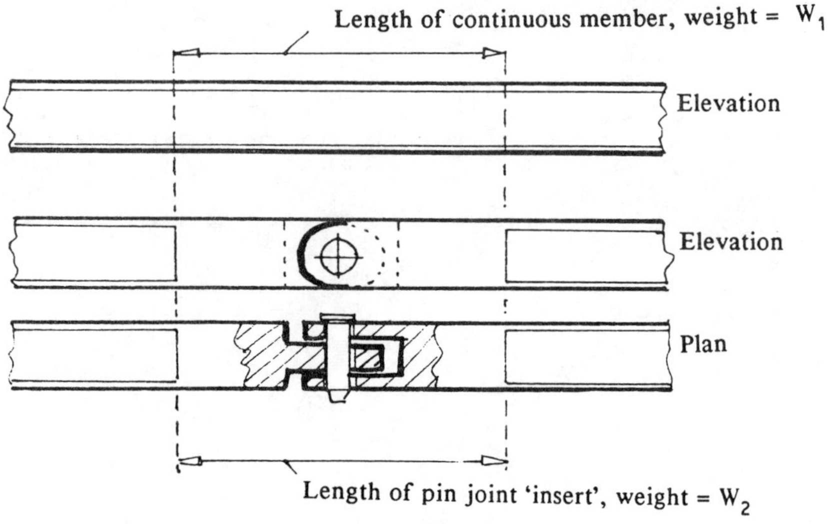

(b) Value of f_j for pin joint = W_1/W_2

Fig 4 CALCULATION OF JOINT WEIGHT FACTOR, f_j.

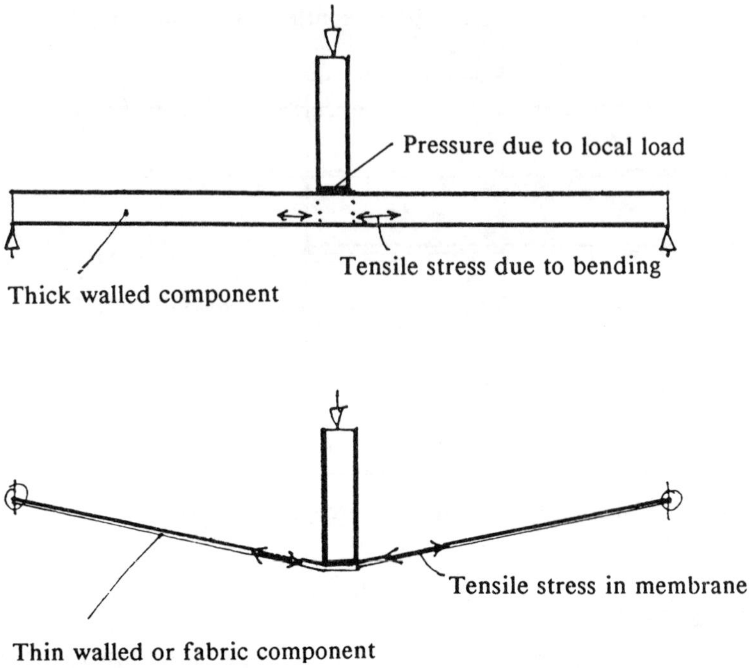

Fig 5 PRESSURE/STRESS RATIO FOR LOCAL DAMAGE, f_s .

SECTION 1: RAPIDLY ASSEMBLED MILITARY STRUCTURES

Connections for Rapidly Assembled Military Bridges, Components and their Performance

D. Webber

Mobile Systems Division 6, Royal Armament Research and Development Establishment, U.K.

ABSTRACT

Lugs are the major means of making field connections. The evolution of UK static design methods is described followed by considerations, other than geometry, that may affect fatigue life. In particular, techniques that may improve fatigue are described together with factors in manufacture and use that may reduce life.

INTRODUCTION

A significant design feature of successful readily transported and easily and rapidly assembled structures, such as military bridges, is the connection between components. These have taken several forms according to the particular application and new ideas continue to be proposed. However, the type of connection most commonly used in the past and still being used in current developments is the pin joint or lug. Joints, whilst they are of paramount importance to the integrity of a structure, since failure would more often than not cause the structure to become a mechanism, add appreciably to the overall weight. There is, therefore, considerable interest in the design of safe and efficient joints.

In the past interest centred on the static performance of the joint, but with the use of stronger materials fatigue has had to be considered as well since fatigue performance does not increase in proportion to static strength. A considerable amount of data exists for the type of lug joint used in the aerospace industry [1], but a feature of these connections is that they are permanent or semi-permanent, so neat fit, or even interference fit pins are used to make the connection and the question is, can this data be used in situations where clearance fits are the norm? The aim of this paper is to address this question, particularly in relation to weldable aluminium zinc magnesium alloys from which the majority of current bridges are constructed.

STATIC DESIGN OF LUGS

Past Practice

The design of lugs at RARDE comprised calculations for tension, bursting, shear, bearing and pin bending. The tension calculation contained, in effect, a fixed efficiency factor of 0.91, the bursting calculation had a variable factor based on head height (head height = distance from pin centre to end) to pin diameter ratio. The shear calculation included a fixed factor and was based on material shear stress, whilst bearing included a fixed efficiency factor of 0.866 for interchangeable parts or unity for permanent connections and used the material bearing strength. Nett cross sectional areas were used except for the shear calculation which used the bursting area. The bursting efficiency factor was constant at 0.45 for diameter to height ratios greater than 1.2. Pins were designed using a plastic, rather than an elastic modulus and a triangular bearing distribution was assumed in order to cut down bending moments. This method appeared to work in practice, but did not take account of the latest research and used assumed material properties such as shear which could be changed for design purposes in an arbitrary manner. A revised method was therefore introduced.

There were no overall geometric limitations except when the same rules were applied to bolts and rivets near plate edges and plug end tongues.

Current Practice

The revised design method follows aircraft design practice. [2,3] Tension is treated as before with a fixed efficiency factor, thus:

Load Capacity = Tension Efficiency Factor x Material Strength x Nett Cross Section

This was considered justified within a limited range of geometries normally used in military bridging, but there are constraints on material form and grain direction. Proposals are made for use outside the geometries and constraints given. A major change was made from the earlier method by considering shear, bursting and bearing together in a single calculation. This calculation comprises using the pin bearing area, the material tensile properties and an efficiency factor which varies with head height to diameter ratio. There are additional constraints for demountable joints. Thus:

Load Capacity = Bearing Efficiency Factor x Material Strength x Bearing Area

The draft for the British Standard on Structural Use of Aluminium embodies similar considerations except for a fixed bearing efficiency factor.

Pin design follows previous practice in bending, but a value of 0.6 tension is taken for shear strength.

FATIGUE DESIGN OF LUGS

Background

The overall situation is summarised in Fig 1. This shows Heywoods [1] collected aircraft lug data, where it was noted that fretting of the pin in the bore at the location of the highest stress concentration appeared to play a significant part in initiating failure. Bridge connections have appreciable clearance to aid pin insertion and it was hoped that they would have significantly improved performance due to fretting not occurring at the point of maximum stress concentration although the stress concentration increases. A series of tests was carried out by Smith [4] on various candidate military bridging alloys in the 1960s at 2/3 and full scale of then current designs which gave rise to the scatter band shown hatched vertically in Fig 1. This band was slightly narrower than the data collected by Heywood but, more significantly, of shallower slope. Therefore, whilst lives were similar to Heywood's at low stress ranges, they were significantly shorter at the higher stresses which is the region of particular interest to RARDE.

In order to investigate possible causes of the differences a preliminary series of tests was carried out on thick plate by Smith [5] to establish a standard geometry. As in the static tests the great importance of head height to pin diameter ratio was confirmed, at least at higher stress ranges. Figure 2 summarises part of this investigation at 155 N/mm^2 stress range where pin diameter was fixed at 25 mm and head height varied. It is seen that with plate a ratio of at least 1.28 is required for optimum life which is more than that required for static design. It is also seen that the mode of failure is also affected by the geometry and therefore the optimum geometry can vary with product form as shown in the next section. At other pin diameters there was no obvious reduction in life until the head height to pin diameter ratio was less than 1.1 or until the lug width to pin diameter ratio was over 2.5 when the elastic stress concentration factor increases rapidly. As a result of these initial tests it was decided that other factors which may influence the life of a lug of a standard geometry should be investigated.

Test Programme

A proprietary alloy, Hiduminium 48 (7039), was selected for this investigation as a logical progression in strength from the weldable aluminium, zinc magnesium alloy 7019, produced to military specification DGFVE 232B then, and still in current use. A rectangular section extrusion was used for ease of specimen manufacture and uniformity of properties yet representing the directional nature of forgings that are generally used for the manufacture of lugs. Typical properties are:

Alloy	0.2% Proof Stress N/mm^2	Tensile Strength N/mm^2	Percentage Elongation
7019	359	415	12
7039	503	544	11

The overall specimen dimensions shown in Fig 3 are taken from the UK Medium Girder Bridge (MGB) and enabled MGB pins to be used with a typical clearance of 1.8% of pin diameter. Provision was made for boring a third hole which appeared to have the following advantages over conventional single or double hole specimens (a) the test could be continued to obtain two results from a single specimen, (b) testing time is reduced, (c) scatter of results under identical conditions can be compared, and (d) unfailed ends can be examined for cracks near the end of their expected life.

The majority of specimens was made at one location, but some were made at another location. The specimens for this investigation were tested under constant amplitude loading at three locations on servo hydraulic testing machines at a frequency of about 2 Hz. A low constant minimum stress was used that was sufficient to maintain tension on the specimen.

The "standard" specimen constant amplitude results at various stress ranges have been used to establish a statistically verified stress life curve against which the effect of various variables on the results can be compared. However, the majority of results have been compared at one stress range (155 N/mm^2) where a minimum life of about 2 x 10^4 cycles was obtained.

Stress Life (SN) Curve

The mean SN curve has been derived on the basis of linear regression by combining the original standard data expressed as $\log_{10}$ stress range and $\log_{10}$ cycles to failure from all sources. Fig 4 stresses are based on the minimum nominal cross section of the lug on a transverse diameter of the bore. No correction has been made for actual cross section which is a factor that will contribute to scatter.

The mean equation is

$$N = \frac{7.739 \times 10^{12}}{\Delta\sigma^{3.757}} \qquad \text{..........(1)}$$

where $\Delta\sigma$ is the stress range in N/mm². The above equation is really only valid for stress ranges between 273 N/mm² and 125 N/mm² where all tests were run to specimen failure and are the only ones included in the analysis. A correlation coefficient of 0.964 was calculated which indicates a very good fit and the standard deviation about the regression line was 0.112.

Also shown in Figure 4 are the lower 95% exceedance limits calculated at 5 stress ranges together with a line drawn by eye through these points. This gives the true lower limit σN curve. It is seen that scatter increases with decrease in stress level (where larger sample sizes are really required) and the minimum curve should be steeper than the mean curve.

Further analyses are shown in Fig 5 where the mean line through first and second hole failures have been considered separately. A steeper curve is obtained for first failures. This curve may be more relevant to a structure where failure depends on the first failure of a number of nominally identical details rather than the mean of all details. Also included in Fig 5 are a number of prediction limits for further results, the most relevant of which is the lower 98 percent limit for which the equation is

$$N = \frac{4.4045 \times 10^{12}}{\Delta\sigma^{3.757}} \qquad \text{......... (2)}$$

Such a curve, being parallel to the mean curve, probably predicts shorter lives than will actually be obtained at high stress ranges and longer lives at lower stress ranges until endurance limit effects predominate.

Standard Data

In order to make comparisons a bench mark has to be established for the standard specimens. Rather than rely solely on the σN data it was decided to analyse the data obtained at a stress range of particular interest (155 N/mm²) in greater detail. The data was a combination of results from RARDE, City University, London, and the Royal Military College of Science (RMCS). The specimens tested at

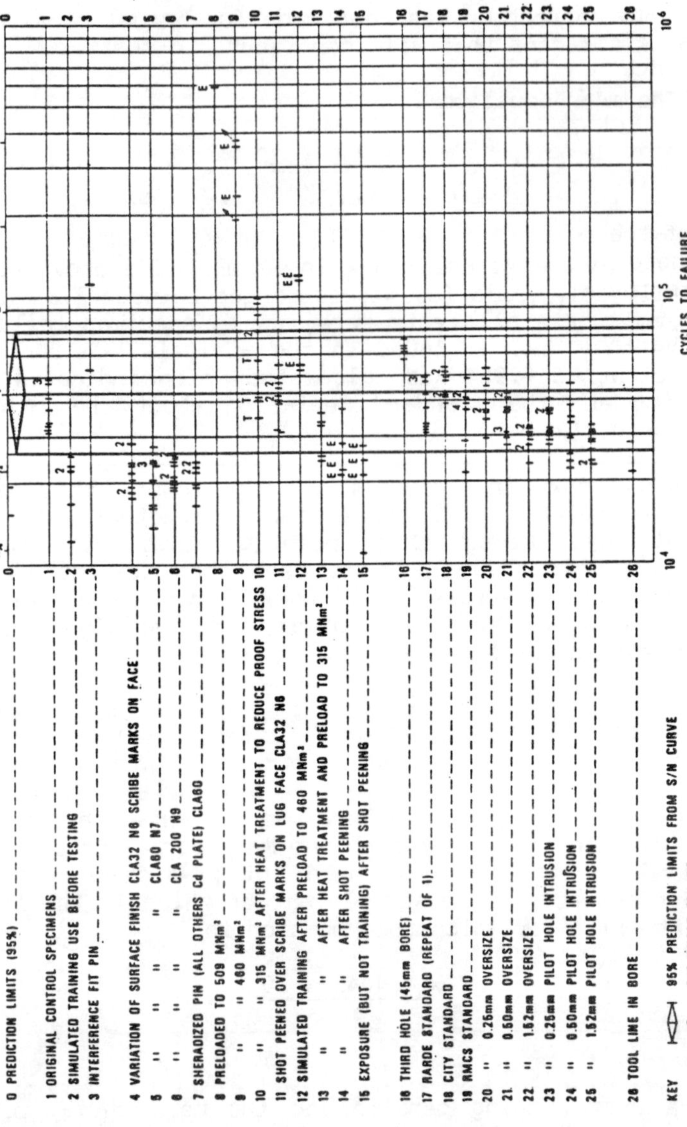

Table 1 Comparison of fatigue lives at a stress range of 155 N/mm² for various lug treatments and conditions

RARDE and City University were identical in all respects, whilst those tested at RMCS were at the upper limit of bore tolerance and are believed to have a finer quality of surface finish produced by the manufacturer of the Medium Girder Bridge (MGB) (but this cannot now be confirmed). Statistical tests showed that all but one result could be considered as coming from a common population. The 95% limit on this data is plotted as Item 0 in Table 1. Nearly all of the following results are compared against this data, or where a factor is varied within a group the samples within the group are compared.

Bore Variables

Bore Sizes Oversizes of 0.25, 0.5 and 1.5 mm, Table 1 Items 20-22 showed progressively shorter lives, however all but one of the 1.5 mm oversize results fell within the scatter band of the control specimens and one standard specimen from the same source had the lowest result. There were statistically significant differences between the sizes, with 0.25 oversize being the best. This indicates that a clearance of 2.4% of pin or bore diameter is slightly more beneficial than the 1.8% specified for MGB, but clearances can be as high as 5.4% without any great loss in life.

Pilot Hole Intrusion Pilot holes are necessary in production and it is important that as little material as possible is removed from forgings after final heat treatment. Consequently, traces of pilot holes sometimes remain after final machining. The results in Table 1 Items 23-25 show that the effect, even when the intrusion is on a transverse bore diameter, is very similar to the same size oversize. Hole intrusions of the size investigated do not have measurable influence on fatigue life although they may reflect on the general quality of manufacture.

Bore Surface Finish It was noted during the preceding investigation that the surface finish of the bores of the standard specimens produced by RARDE(Cc) was not as good as that produced in the MGB bores in production. As a result some 0.25 mm oversize bore specimens with various surface finishes were produced by the manufacturer of the MGB and tested at RMCS with a roughness of about N7 and N9. There was a higher mean and less scatter in the test results Table 1 Item 5 and 6 on the N9 results compared with the N7 results but both sets of results fell at or below the lower limit of the 'standard' specimens.

Another set of specimens was therefore machined by the manufacturer of the MGB, with an N6 finish and tested at RARDE(Cc). These also gave test results, Table 1 Item 4, similar to those described above. However, a feature that was noted from these specimens was that they almost entirely failed from the edge of the bore along transverse scribe lines used for marking out the hole centres. The scribe marks had certainly affected the point of crack initiation and probably the life. A re-examination of the failures of the other surface finish specimens tested at RMCS was not possible due to the time lapse after the possible cause was identified.

There is no clear indication within the results that surface finish alone has any effect. However, such a result would not be altogether surprising since CLA is measured along the length of the bore parallel to the axis across the boring marks and along the line of any fatigue cracks. Circumferential measurements are difficult, and so rarely taken, but they are more significant as will be seen from the following paragraph, Tool Lines. Marking out is also seen to be important and the effects may have completely overridden the effect being investigated but only in the case of Item 6 was a reason for the low life identified.

Tool Lines It was noted on some specimens that failure occurred along a tool line, caused by extraction of the stationary boring tool, when the line was close to a transverse diameter. The results from such specimens were below the mean regression line and sometimes below the lower prediction limit, Fig 4, indicating, as may be expected, an adverse effect. The only results at 155 N/mm^2 stress range are at Item 26 Table 1. The straight nature of the fracture along the tool line was very distinctive.

Third Bores These are failures of the third hole which have occurred before failure of the second hole, indicating either large scatter in life between first and second hole failures, a short life to the third hole or a very effective life improvement method applied initially since these were not applied to the third hole. The third hole minimum head distance is slightly lower than for the standard specimen, but the corners are of course "over square".

When the holes were made by boring to the same size as the other two holes all results in Table 1 Item 16 were higher than the standard mean result indicating improvement but still within the overall scatter. Some specimens had previously been proof loaded, but the stress on the area of the third hole would have been low and prior to the hole being drilled was free from stress concentrations.

Summary A clear and consistent overall picture does not emerge because of the manner in which various changes were made. Nevertheless, it appears that a pin clearance of about 2.4% of diameter is an optimum but variations between 1.9 and 5.4% do not have a great effect on life. Pilot hole intrusions can be treated as similar to a change in effective diameter. Marks in the bore and on the lug face transverse to the direction of stress have a very detrimental effect on life, but have not been quantified. There appears to be no effect due to surface finish as measured along the bore but differences may be undetected because of marking out or machining effect.

Pin Surface Protection

Pin protection was changed from cadmium plating to Sheradising. Both types of finish were used in test and comparable lives obtained. Lug specimens obtained from a commercial manufacturing source, used to investigate surface finish with an N7 finish were used. Samples of the black products of fretting found in the bore of the lug were analysed and found to be largely the pin coating medium. The method of protection has been shown elsewhere [1] to have an effect on the life of the pin, but this is not established in the present application, in fact pins were re-used and no pin failures were encountered.

Preloading

Preloading to the proof test load has previously been shown to be effective in improving fatigue life and had been adopted since the 1970s for all bridge components carrying tension loads. When Heywoods finding [1] that multiple applications of load were more beneficial than a single application was confirmed, ten applications were chosen as the arbitrary compromise between practical expediency and a desire not to fatigue the component.

The magnitude of the preload has a significant effect, as may be expected for a method of life improvement that relies on the introduction of favourable residual stresses. As a general rule the higher the better just short of unacceptable distortion. Some distortion must occur, but this can be so local and so small as to be acceptable. This should be considered in relation to production tolerances and repair tolerances.

The highest preload given in this investigation was of the same magnitude as the 'proof' condition of the lug, (which was taken as 0.2% permanent set on a gauge length representing head to head of inter-engaging lugs). This resulted in a 0.22% reduction in transverse bore diameter and a 0.4% increase in longitudinal bore diameter. In absolute terms that is a reduction of 0.13mm and an increase of 0.30mm which means that all of the manufacturing tolerance is absorbed. This preload resulted in an increase in life of over 10 times, Table 1 Item 8, on average life and a change in the mode of fracture to an end burst. A higher head height to diameter ratio may therefore have resulted in an even greater improvement.

Four specimens were preloaded to a level which corresponds to the structural proof level but based on actual rather than specification material mechanical properties. These lasted at least 5 times as long as the average untreated lug, Table 1 Item 9, and again were end burst failures.

Because of the very high proof stress of the material related to the stress range that resulted in lives of about 30 000 cycles, some specimens were over-aged to reduce their mechanical properties to that of 7019 material. They were preloaded to a stress of 315 N/mm^2 which corresponds to the proof test condition as far as mechanical properties are concerned. The amount of yielding was similar to previous tests but, of course, the value of residual stress was lower. The results, Table 1 Item 10, only show a small but useful improvement over untreated specimens, but the lower results are from specimens with tool line marks which reduced the potential life.

It is unfortunate that no specimens were tested that had been given only the heat treatment in order to establish the effect of strength on life.

Shot Peening

Shot peening is another method of introducing compressive residual stress. Unlike proof loading, which automatically treats the right areas to a largely unknown extent, shot peening can be closely controlled but the right areas for treatment require careful selection. In the present case the bores and surrounding faces were selected and were completely peened to an intensity of 0.12 - 0.16 in Almen A. This was selected as a compromise between a fairly severe intensity for use on aluminium alloy and an intensity which would not cause unacceptable feathering at the edge of the bore.

The results, Table 1 Item 11, fall within the scatter band of the control specimens, but this is not the direct comparison that should be made since these specimens were bored at the same time as the N6 32 CLA finish specimens that had marking out lines on the face. If the comparison is made with these results, which is more realistic, there is a doubling of life.

Becaue of the circumstances it is not clear whether the improvement can be attributed to the residual stress induced or the obliteration of the scribe marks by surface working or both.

It was noted that it was much easier to insert the pin in the peened specimen. This may have useful practical applications. The result is not unexpected when it is considered that one is, in effect, in normal circumstances trying to engage two fine screw threads by sliding.

Environment and Wear

The results discussed so far have all related to tests where the pin is inserted in the bore of the specimen and the test run to failure. Fatigue tests on full size structures are of a similar nature except that the pin might be withdrawn occasionally for purposes of inspection. This is not representative of the use that bridges receive in peacetime, which mostly involves many builds, but little trafficking, particularly at the rated capacity, or the envisaged use of bridges during war where they will be moved from site to site. The results therefore only relate to bridges that are erected and remain on one site.

For this part of the investigation an array of specimens was bolted together to represent an MGB jaw array. The specimens were then exposed on the ground in the bridging field for about a year before testing. During this time a pin was inserted and removed at least once and perhaps twice a working day. There was quite a dramatic reduction in the life of the specimens that had been preloaded to 460 N/mm^2, Table 1, Item 12, the average now being only about twice that of the control specimens with both bore and end failures occurring.

The heat treated specimens preloaded to 315 N/mm^2,Table 1 Item 13, had average lives of less than the control specimens and a reduction in life of about 30% compared with those not subjected to the effects of environment and wear.

The lives of the shot peened specimens were even less than that of the heat treated specimens described above,Table 11 Item 14, but the initial lives of the shot peened specimen were also less. It was interesting that in this group of specimens three end burst failures occurred. This is because the ends had not been peened and no failures occurred at this location for specimens that had not been exposed to the environment or wear. Three other end failures occurred from the ends of these specimens that were bolted together and exposed to the environment but were not subject to wear. End failures are obviously not influenced by wear and must be due to environmental degradation.

Cold Hole Expansion

The effects of cold hole expansion, a proprietary process (Fatigue Technology Inc.) have been evaluated subsequent to the main evaluation as an alternative to preloading or peening to increase life or as a means of minimising the effects of bore damage. The bore is expanded by drawing a mandrel through a split sleeve inserted in the bore. Retained expansions of about 1% were used. Sizes were selected so that the finished hole was the standard size. Tool line damage was applied after expansion. A small amount of thickness belling occurred which would need to be considered in practice.

All tests were conducted at a reference stress range of 155 N/mm² and the results are shown in Fig 6. The standard and tool line damaged results are similar to those previously obtained, whilst all the cold expanded results showed an improvement over the standard specimens. There was more scatter to the cold hole expanded results so the minimum results were not much improved. However, much more significant was that tool line damaged specimens had the same life as standard specimens, that is the damage was rendered innocuous.

The process had a significant effect on the mode of failure. Whilst one end failure occurred in untreated specimens, all but one cold hole expanded specimens failed due top end bursts. Therefore, if larger head heights had been used significantly longer lives may have been obtained.

Comparison with Other Data

The major comparison must be with Smith's suggested design curves [4], which were based on available data at the time of publication, and is made in Fig 1. The width of Smith's scatter band is about 15 to 1 compared with less than 3 to 1 at 155 N/mm² for the present results, because it covers a range of geometries, sizes and material forms. A most noticeable feature is that the slope of the curve for the present results is much steeper and at stresses above 100 N/mm² the lives are much higher. The results of other work by Smith [5] where geometry was investigated generally fall at the lower end of the present scatter band except when the geometry became extreme.

Another comparison must be made with Heywood's curves [1] for lugs made from aircraft alloys of both the Al-Cu and Al Zn Mg types. These covered pin diameters from 5 mm to 70 mm and, of course, the clearances were much smaller than with the present work. The slope of the mean curve was even steeper than that of the present work and the results were as good or better at high stress ranges but the average was similar at lower stress ranges. The minimum results were lower.

The relatively good results of the present specimens may be due to clearance and the very directional nature of the extrusion compared with plate and block forgings that have often been used previously.

Crack Detection

In one series of tests at RMCS, where inspections were carried out every 1000 cycles, cracks were found by means of dye penetrants between 70 and 87% of the life to failure at 155 N/mm² stress range. It is unfortunate that the size of the cracks was not recorded.

No inspections were carried out at City University, but at RARDE(Cc) visual inspections were carried out after the first failure. At 155 N/mm² this showed cracks may remain undetected at 95% of the life and the earliest that detection could be expected was 85% of the life, but 90% of specimens were cracked by 90% of their life. The shortest cracks detected were about 5 mm long.

It was possible to establish a relationship; crack length in mm = 1.42 Percentage of life expended - 111, with a correlation coefficient of 0.677 which is relatively high. This relationship is for a typical case. If safe predictions are to be made a value of beween 125 and 130 should be used instead of 111.

These findings have important implications. Damage tolerant equipment should receive at least 20 inspections during the expected service life if use is consistent throughout the life to ensure safety from unexpected fatigue failure of lugs. This number could be decreased by up to a half by the use of suitable NDI equipment that was more sensitive than visual inspection, or by the accurate monitoring of use.

Residual Strength

Residual strength tests were carried out on specimens which had not failed in the fatigue test because the test machine deflection cut out limits were set very fine. The residual strength of these is given in Table 2, where results are given for both the original nett section and the residual nett section which was assessed by the use of dye penetrant.

Table 2 - Residual Strengths in N/mm²

On Original Area	On Uncracked Area	Uncracked Area / Original Area
184	320	.61
172	271	.64
240	285	.84
170	256	.67
172	271	.64

Parent Metal tensile strength 544 N/mm²

The maximum fatigue stress of these specimens was 170 N/mm^2 so it is seen that most lugs were very close to failure and the lugs had lost approximately one third of the cross section by cracking. An even more significant finding was that the strength of a lug containing a fatigue crack, when the loss of area due to the crack was taken into account was approximately ½ that of an uncracked lug. This indicates the importance of using fracture mechanics in assessing the residual strength of cracked bodies.

Fretting Angle

The fretting that occurred during fatigue enabled observations to be made of the importance of fretting in the fatigue process, where it is normally considered to be very deleterious, and of bearing area.

In the present case, because of the relatively large pin clearance that is used compared with other published work on lugs, the point of fretting is moved away from the point of maximum stress concentration. Although the point of maximum stress concentration moves towards the lug head and increases with increase in clearance the rate of change of the fretting area is greater and the two do not coincide. This might well account for the difference in slope of the stress life curves of the present lug experiments carried out in support of bridging compared with the published data that is concerned with aircraft structures where the lugs have low clearances.

A relationship Fretting Angle = 0.1962 Stress Range + 40.256 (Angle in degrees, Stress range in N/mm^2) was established between fretting angle and load, over the range of fatigue stress used with the excellent correlation coefficient of 0.9196. This showed that at the basic stress that was used for comparison only 57% of the projected area was effective. The effective areas ranged from 46% to 71%. If the relationship is extrapolated to the ultimate strength of the lug the bearing area becomes about 95% of the nominal.

The high correlation means that fretting angle may be a way of assessing the load to which a bridge component has been subjected. Fretting is a process that gradually increases and decreases around the circumference and is not constant through the thickness. The changes through the thickness give some indication of the amount of pin bending that takes place and results in a larger angle of fretting at the faces. Despite the high stresses at the faces nearly all fatigue cracks initiated midway through the depth of the bore, except when there were surface marking out scribe lines.

Variables

The position of the specimen in the extrusion or the melt from which it came has not been identified as a variable. This is because the experiments were not designed to check this possibility as it would not be of practical value.

Present testing procedures or life requirements for military bridges do not make any specific provision for allowing for manufacturing marks, wear or for corrosion. Some allowance for the effects of wear can be made by purposely causing tool marks on a transverse bore since it is not practical to cause damage due to realistic use in a full size fatigue test. The possibility of environmental effects can be checked on a laboratory specimen scale. Before improvement methods are used, the effect of changes in failure mode should be considered.

CONCLUSIONS

The following conclusions are drawn for the fatigue of lugs of fixed geometry made from extruded sections of a fixed chemical composition.

1. Fatigue life can be improved by a significant amount due to induced favourable residual stresses by :

    a. High preloads repeated 10 times
    b. Shot peening
    c. Cold hole expansion

2. Fatigue life may be reduced by a significant amount by:

    a. Wear due to pin insertion and removal prior to fatiguing.
    b. Tool withdrawal marks
    c. Scribed marking out lines on the surface
    d. Environment.

3. The effects of 1 and 2 above may work concurrently.

4. Fatigue life is not affected in practical terms by:

    a. Clearances and pilot hole intrusions up to 3% of pin diameter (only slight reduction occurs up to 5%),

    and may not be affected by:

    b. Roughness of surface finish of bore measured on a straight line parallel to the axis of the bore.

    c. The plating of pins with cadmium or Sheradising.

5. When considering any change consideration should be given also to the possible effect of changing the mode of failure.

RECOMMENDATIONS

Account should be taken of environmental effects and axial marking of bores due to pin insertion and withdrawal. Tool withdrawal marks and setting out marks must also be allowed for.

Preloading or shot peening should not be relied upon as a method of life improvement without calibration and full account being taken of any change in dimensions.

Initial design of lugs should be based on the Smith curve [4] noting that substantial improvements may be possible by careful selection and control of material strength and form, specimen geometry and life improvement methods.

Further investigation is necessary in order to make a quantitative assessment of the effects of material strength and form on fatigue life which will, in turn, be affected by geometry.

REFERENCES

1. Heywood, R.B., Designing Against Fatigue, Chapter 9, Loaded Lugs, Chapman and Hall, London, 1962.

2. Cozzone, F.P., Melcon, M.A. and Hoblitt, F.M., Analysis of Lugs and Shear Pins Made of Aluminium or Steel Alloys, Product Engineering, pp. 113-117, May 1950.

3. Melcon, M.A. and Hoblitt, F.M., Developments in the Analysis of Lugs and Shear Pins, Product Engineering, pp. 160-170, June 1953.

4. Smith, B., Design Data for Lugs under Fatigue Loading, Unpublished MOD(PE) Report.

5. Smith, B., The Effect of Geometry on the Fatigue Life of Pin Joint Connections (Lugs) Fabricated from Aluminium Alloy DGVE 232 Thick Plate Material, Unpublished MOD(PE) Report.

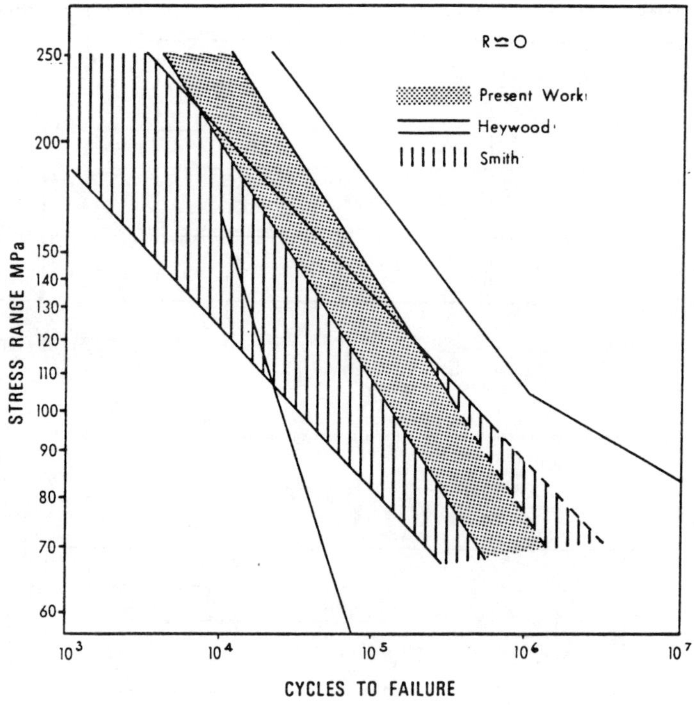

Figure 1 Comparison of Fatigue Life Data for Lugs

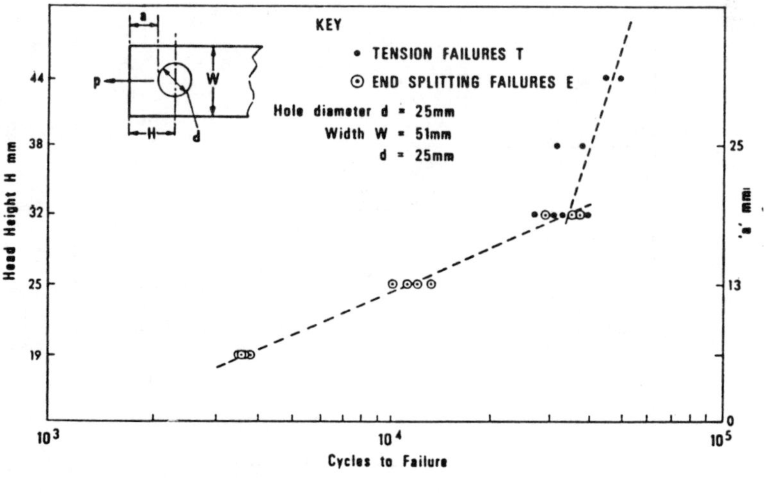

Figure 2 Effect of Head Height on Lug Fatigue Life

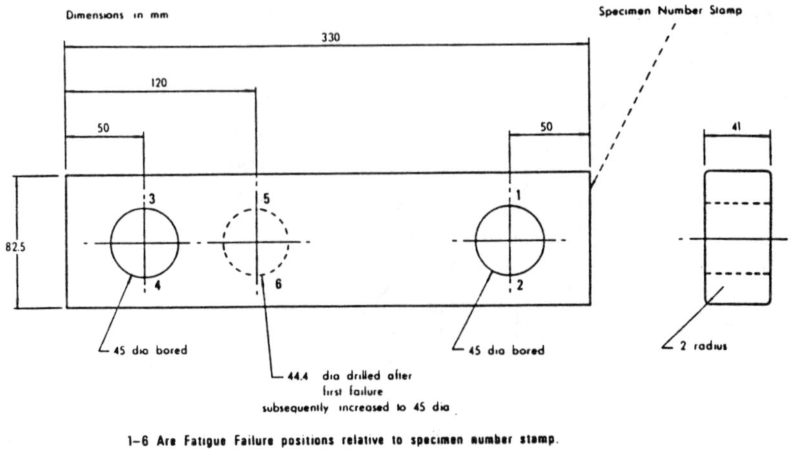

Figure 3 Double ended lug specimen details

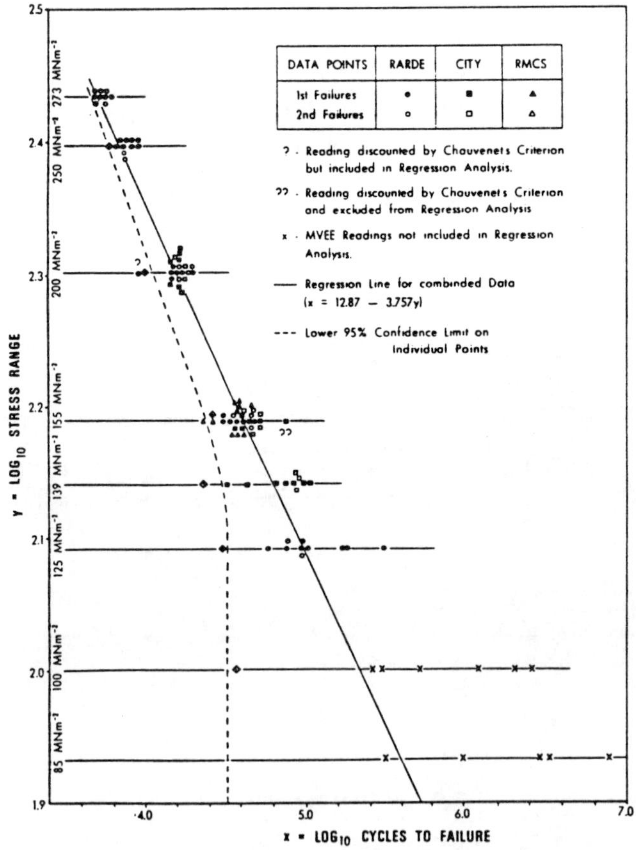

Figure 4 Constant amplitude fatigue life results for "standard" specimens

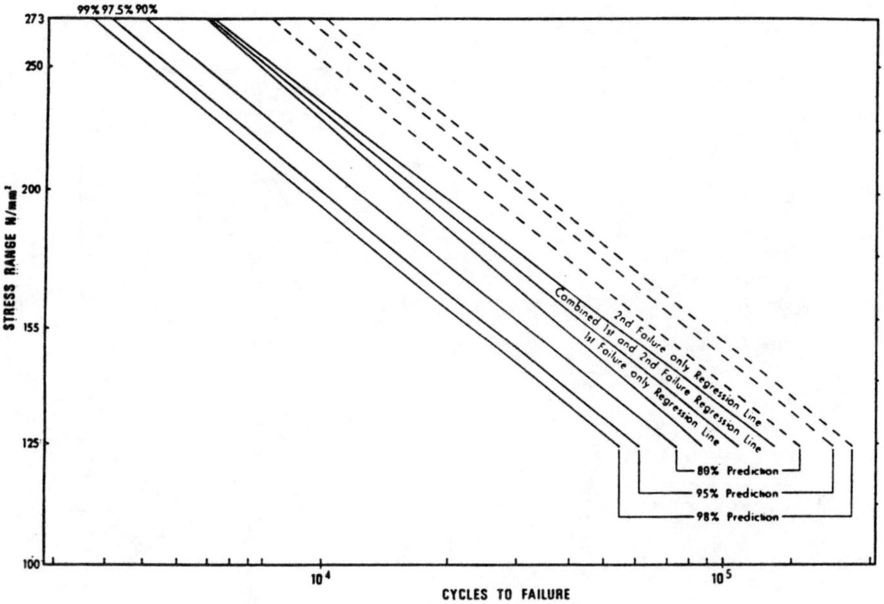

Figure 5 Statistical analysis

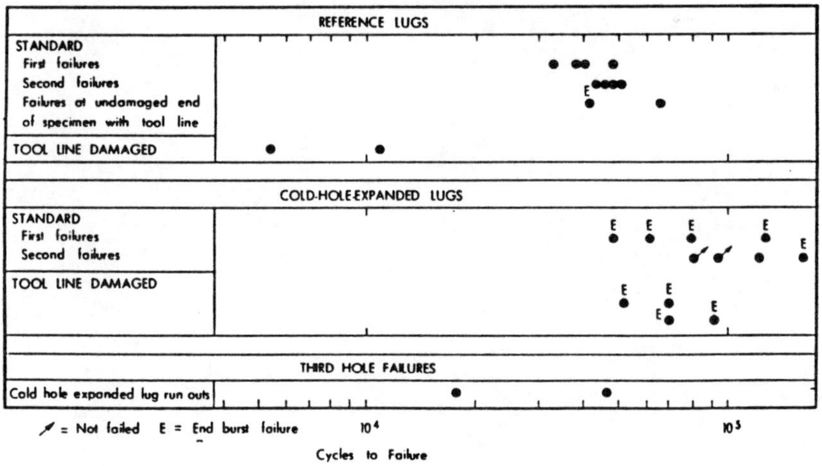

Figure 6 Comparison of cold hole expansion results with "standard" lug results at 155/Nmm^2 stress range

The Launching Rail in Advanced Composites for BR 90

M.B. Snell

Royal Armament Research and Development Establishment, Christchurch, Dorset, U.K.

ABSTRACT

A launching rail is used to launch the deliberate support bridges in the UK BR90 bridge system. Because of the limitations on strength-to-weight ratio, aluminium alloy rails can only be used to launch bridges up to 44m. For greater bridge lengths, up to the maximum of 56 m, a launching rail in advanced composites (CFRP) is required. Two successful research prototype rail modules have been produced and tested to design ultimate load following a programme of materials and critical element research. The research programme and composite rail modules are described.

1. INTRODUCTION

Military bridges must be both mobile and rapidly assembled to give mobility advantages to land forces. Compared with civilian bridges, military bridges are light with a heavy vehicle load whereas civilian bridges are heavy with a relatively light vehicle load. Generally a military bridge is erected and used for a short time and then taken down for use at another site.

Military bridges usually have to be built from one side of the gap only, and special techniques have been developed to carry this out. Considering Deliberate Support Bridging, including Bailey Bridge, Medium Girder Bridge and the Christchurch Bridge (BR90), two principal methods of launching are used: the launching nose and the launching rail. Bailey Bridge is launched over rollers using a lighter assembly of standard bridge panels as an extension to the bridge which makes contact first with landing rollers at the far bank (Figure 1). The lightweight cantilever extension is known as a launching nose, of length 1 bay (10 ft) greater than half the length of the bridge. A launching nose is also used with the Medium Girder Bridge, Figure 2, but in this case it is purpose designed for minimum weight. The MGB launching nose is in two structural sections; the launching nose light at the extreme cantilever position with the stronger launching nose heavy connected to the bridge. Because of its relatively low weight, the MGB launching nose can be longer to make contact earlier with the far bank.

The drive towards greater speed of assembly led to the development of a new launching method for future bridging in the United Kingdom using a launching rail. The launching rail is a preliminary beam spanning the gap, deployed from a special vehicle, from which the bridge is suspended and launched. After launching the bridge the launching rail is taken away to launch or recover other bridges. A single launching rail can be used with many bridges, reducing the overall cost. This method enables the bridge launching operation to be mechanised giving a large reduction in both time and men required. The improvement in bridge launching times is shown in Table 1. The launching rail is the principle method used to launch Deliberate Support Bridges in the BR90 Bridge System. A schematic view is shown in Figure 3.

There are constraints on the application of the launching rail as there are on other launching methods. The length of rail which can be cantilevered from the launch vehicle is dictated by the countermoment provided by the weight of the vehicle (Figure 3). Structurally, the achievable rail length is a function of rail depth and of material strength-to-weight ratio (specific strength).

Both the weight of the vehicle and the depth of the rail have considerable influence on vehicle mobility, and these parameters are limited by bridging system requirements and road traffic regulations. Within these limits the maximum launchable rail and bridge length becomes simply a function of material specific strength.

TABLE 1. TIME TO BUILD A 30 m MLC 70 BRIDGE

BRIDGE	MEN	TIME	MAN HOURS
BAILEY	90	12 hrs	1080
MEDIUM GIRDER	30	80 min	40
BRIDGING FOR 90's	8	30 min	4

The standard BR90 launching rail is in aluminium alloy to launch bridges up to 44 m in length. A longer and lighter Launching Rail in Advanced Composites (LRAC) is under development to launch bridges up to the maximum length of 56 m. The present paper considers the development and use of LRAC. LRAC is produced in carbon fibre reinforced plastic (CFRP) with a weight saving of about 40% compared with the metal rails. By use of low cost industrial prepreg materials and low cost manufacture LRAC is produced at much lower cost than aerospace composites. Tests of prototype research modules have confirmed the feasibility of LRAC.

The Army battlefield environment is severe and places extreme demands on composites materials and structures not found in aerospace or civil fields. Resistance to impact from dropped heavy equipment must be provided. Strength loss from ballistic impact by rounds and shrapnel must be minimised. Resistance to fire and thermal exposure, nuclear flash, etc is required. Structures and materials must be designed to minimise radar and infra-red signatures. Tests to confirm the environmental resistance of the LRAC components are discussed.

2. LAUNCHING RAILS AND AUTOMOTIVE BRIDGE LAUNCHING EQUIPMENT

2.1 Automotive Bridge Launching Equipment

A special vehicle is used to launch and recover both the launching rail and the deliberate dry support bridge. The vehicle is equipped with Automotive Bridge Launching Equipment (ABLE) and has the following main features to carry out these tasks:

a. a telescopic chassis extending the chassis length to give space for assembly of rails.
b. a crane (CALM) to lift and connect rail modules and bridge panels.
c. a tilting launching frame equipped with rollers to support the rail and a drive mechanism to launch the rail.
d. hydraulic jacks to rotate the launching frame to accommodate high or low far banks and to lower the rail bipod to make contact with the far bank.
e. trolleys running on the rail to launch and recover the bridge suspended beneath.
f. winches to traverse the bridge launching trolleys.
g. outriggers with rollers to support the home bank end of the bridge panels during bridge launch.

A schematic view of an ABLE vehicle is shown in Figure 4.

2.2 Launching Rail General Description

A launching rail is simple in concept, being a long slender beam. However, it must withstand a number of combined load cases in both hogging and sagging and must be robust and durable for the military environment. The loads carried in the cantilever rail launching case are vertical bending (hogging) from the self weight of the rail and its appendages (winch cables, trolleys and bipod), and lateral bending from wind pressure. High local restraint forces are applied to the rail by the rollers in the launching frame. The cantilever force system is shown schematically in Figure 5.

During bridge launch, the rail is simply supported between the launching frame at the home bank and its bipod support at the far bank; in this sagging case the stresses in the rail are reversed compared with the cantilever case. The rail can also be given additional restraint from the ABLE launch frame and supported as a propped cantilever to reduce the sagging bending moment. The forces on a simply supported (sagging) rail, shown in Figure 6, are self weight and bridge weight vertically and wind load on the rail and bridge laterally. Concentrated restraint loads of similar magnitude to the cantilever launch are applied in the propped cantilever case. The roller loads from the bridge trolleys are much lower. An additional longitudinal compressive load from winch cable tension occurs during bridge launch.

3. DESIGN CONSIDERATIONS FOR LRAC

LRAC was designed using conservative estimates of the available vehicle countermoment. As with all new UK bridge components, the design loadings and factors are based on the Trilateral Design and Test Code agreed by UK, USA and FRG (Ref 1). The main design loads (static unfactored) for LRAC are as follows:

bridge weight	5.3 kN/m
bridge length	56 m
wind speed	14m/sec
rail drag coefficient	2.0
bridge drag coefficient	2.0
vehicle countermoment	1600 kNm
compression end load (winch lock-out value)	60 kN

Reaction forces are applied to the rail through pads or rollers in the launch frame at 4 m spacing in both the vertical and horizontal plane. The design consideration is that the reaction forces should not exceed those caused by the standard aluminium alloy rail.
Wind load applies to both the rail and the bridge.

Resistance of the structure to impact from misuse or dropped tools, connecting pins, etc is required.

4. DESIGN CONCEPT

4.1 Body Structure

The objective of the design was to achieve a structure of minimum weight and low cost using established composites manufacturing technology and readily available materials. The overall dimensions were set by the dimensions of the BR90 metal rail as:

height = 725mm
width = 475mm
length = 6000m.

The construction form chosen for LRAC was sandwich panel for both webs and chords. This was preferred to alternative monolithic construction for the following reasons:

a. The compression loading intensity in chords and webs gives skin thicknesses for optimal design which are susceptible to local buckling below design load; this must be prevented. Sandwich panel construction eliminates local buckling and is cheaper and simpler to manufacture than the alternative skin/stringer construction.
b. Field joints are inserted into the chords. This is most easily achieved using sandwich panel design.

A schematic view of the LRAC cross section is shown in Figure 7.

The combined bending in the vertical and horizontal planes requires an optimal distribution of the end-load carrying material at greatest distance from the neutral axes. However, the field joints are positioned at the top and bottom of the chords and this directs the load path for both longitudinal and lateral bending through the chord ends; in effect the chords are the main load carriers for both components of bending.

The most efficient chord configuration is a thick skinned sandwich panel with solid booms concentrated at the edges. Under vertical bending the booms each carry about 22% of the total end load, the remainder carried by the central chord skins. Under lateral bending the additional load is carried mostly by the booms, supplemented by the webs away from the ends.

Sandwich panel webs separate the chords, carry vertical shear from rail and bridge weight and local compression from concentrated roller forces. Compared with the chords the loading intensities are low and thin skins are required for efficiency. Buckling under biaxial compression and shear must be suppressed by use of honeycomb core.

The sandwich panel design gives high rigidity against "lozenging" and the need for diaghrams to maintain the shape is minimal. A central diaghram of sandwich construction and closing diaghrams of stiffened plate construction are used.

Resistance to impact damage is provided in two ways:

a. The residual strength of plain (unprotected panels) is designed to be greater than the strength required at design ultimate load.
b. The webs can be protected with a layer of Kevlar cloth and polyurethane rubber to prevent structural damage.

4.2 Field Joints

Field joints are a necessary encumbrance in bridge design; they are heavy and give considerable structural strength problems but are essential for transportation. Metal lug jaws were chosen for LRAC, attached by bolting into the chord ends. High strength steel is used to minimise the lug and connecting pin dimensions. The centre line of the lugs lies on the centre line of the chord panel giving no offset in axial load. The jaw block is a single forged billet with lugs and fingers formed by machining and excess metal removed to reduce weight to a minimum. A bolt group of 44 M12 bolts attaches the jaw to the chord in double shear. The bolts are countersunk to lie flush with the top surface for roller launching.

4.3 Materials

The principal materials used in LRAC are CFRP prepreg for the body structure and high strength steel for the jaws. The prepreg is an "industrial" grade which, although much cheaper than primary aerospace prepregs contains the same high grade fibre (Courtaulds XAS) in an epoxy matrix and has similar mechanical properties.

The jaws are in EN26 steel hardened and tempered. Other materials are epoxy paste and film adhesives, steel bolts, aluminium cleats and honeycomb cores.

5. SUPPORTING TESTS

5.1 Objective

The research programme for LRAC was started without a validated data base of mechanical properties for design . Design data for metals was available from manufacturers specifications, and some manufacturers data on idealised composite material strengths was also available. The areas most lacking were in environmentally degraded material properties, the strength of bolted joints and the residual strength of damaged panels. A supporting programme of materials tests and critical element tests was required to confirm all the main aspects of the design.

5.2 Basic Material Strengths

Testing of standard tension and compression test specimens in XAS/epoxy unidirectional material was conducted at the following temperature and pre-treatments:

a. As moulded, room temperature (23°C) test.
b. As moulded, 71°C test.
c. Conditioned at 50°C and 85% relative humidity, room temperature test.
d. Conditioned at 50°C and 85% relative humidity, test at 71°C.

The specimens were conditioned until the saturation moisture uptake of about 1% was achieved.

The results of strength tests on coupons in conditions a. to d., above, are shown in Figure 8. The RT results agree with manufacturers data. There is a considerable drop in hot/wet strength in compression at the highest temperature (71°C saturated moisture content). However, the degraded strength is well above the design allowable shown. Furthermore, it is considered that the extreme environmental condition is unlikely to be reached because of the thick sections and external protection used and the 50°C condition is more appropriate. The CFRP prepreg system used is considered fully satisfactory.

5.3 Bolted Joint Strengths

Design of bolted joints in composites depends principally on the following strength parameters measured for the multidirectional laminates used:

a. bearing strength
b. filled hole notched strength in tension and compression
c. interaction of notched and bearing strength.

For a given material these basic strength properties are functions of other parameters, mainly:

i. lay-up and stacking sequence
ii. single or double shear fastening
iii. plain or countersunk hole
iv. hole geometry, w/d, e/d, p/d ratios
v. geometry of bolt group and relative stiffness of components.

Tests were carried out with conventional single hole specimens 12 mm thick with 12mm diameter holes to determine bearing and filled hole notched strengths in tension and compression for a range of potential layups and stacking sequences to determine the best layup and its bolted joint properties. The test results were used to define design allowable strengths. It was concluded from the results of strength and failure mode that an acceptable layup in the bolted joint region was:

$$46\%\ 0^{o},\ 36\%\ \pm\ 45^{o},\ 18\%\ 90^{o},$$

with the layers symmetrically distributed through the thickness.
This layup is not optimal although it provides adequate bolted joint performance.

A statistical method based on a characteristic coefficient of variation (CV) in the observed failure mode was used to obtain B allowable values (Ref 2). Only 3 tests results were available for each test and thus the B allowable was the mean strength reduced by 28% for a CV of 7.5%. The B allowable strengths for bearing, notched compression and notched tension are:

bearing strength = 600 MPa
net compression strength = 370 MPa
net tension strength = 428 MPa

5.4 Critical Element Tests

The joint between the chord and the jaw was based on the bolted joint strengths, computer analysis of bolt load distribution and testing of critical elements. The critical element test programme explored the interaction between bolts and the interaction between stress fields. Four bolted joint critical elements were tested, of the types shown in Figures 9 and 10:

a. a static test element representing the chord/jaw at the boom
b. a static test element representing the chord/jaw at the chord centre
c. a fatigue test element of type b.
d. a static test element representing approximately 1/3 of the chord/jaw joint comprising 3 rows of bolts.

The static strengths achieved for elements a. and b. were approximately 40% above design ultimate load in failure modes, shown in Figures 11 and 12, of tension for element a. and bolt shear/bearing for element b. These results confirmed the adequacy of the single row bolted joint strength.

The fatigue test on element c. was continued to failure at 24000 cycles of working load, including full wind load, increased by a 1.15 fatigue load range factor. Failure was by combined shear/bending of the bolts. The results show that fatigue strength is adequate.

Element d. was more complex with 15 bolts in 3 rows, intended to indicate any adverse interaction between bolts in the group. A photograph of the specimen is shown in Figure 10 and the failure mode in Figure 13. The failure load was 15% greater than the design ultimate load. The nature of the failure suggests that the cause was notch tension stress at the first hole in the outer finger. It is considered that the stress concentrations here interacted with the stress concentrations in the stepped boom and at the adjacent hole in the middle finger. The joint design was modified by removing the first bolt from the second row to eliminate the adverse interaction.

5.5 Impact Effects On Web Panels

The web panels are relatively thin (3mm skins) and susceptible to accidental damage and attack by arms fire. A programme of residual strength testing was carried out to determine their post-impact and post-ballistic attack strengths. The results are given in Figure 14. Applied impact energy of 75J represents a medium energy impact, relevant to a connecting pin of weight 2.4 kg dropped through 3.125 m. The impact was applied to the panels by a flat ended indentor of the diameter of the launching rail connecting pin (28mm); slight dishing over an area of diameter 80 mm was caused with no delamination or disbonding. The ballistic impacts were by 7.62 mm standard NATO ball round. Complete penetration resulted with damage over 30 mm diameter for both faces. The residual strength results show that ballistics give the worst strength loss with a reduction of greater than 50%. However residual strength in the worst case still exceeds the design allowable strain.

6. ANALYSIS PROGRAMME

6.1 Engineers Analysis

The analysis programme in support of the rail body design followed a standard pattern based on engineers theory of bending and shear (ETOB). The chord/jaw analysis was based on measured single hole strengths supported by an elastic bolt group analysis computer program to determine the load sharing between bolts in a row. Stresses in the skins calculated by ETOB and FE analysis were further analysed using laminated plate theory and layer failure criteria (Hill-Tsai) to give strength reserve factors. A field strain limit of 4000 ue (microstrain) at Design Ultimate Load was adopted to give high resistance to delamination and fatigue. Strains greater than 4000 ue at DUL were predicted at the corners of the rail under the maximum combined load. This was permitted because of the thick sections at the corners giving high impact resistance.

6.2 Finite Element Analysis

Finite element analysis has not in the past been widely used for stress analysis of military bridges, design being based on traditional engineers theory, testing and experience to produce safe structures. However, finite element analysis is being used increasingly for all types of military bridges to improve and optimise designs. Furthermore, composite materials introduce many new properties and failure modes and place a high requirement on accurate analysis to assist behaviour prediction.

FE analysis, using the NISA II code on a desktop PS/2 computer, covered initial design, dynamic analysis, roller loading and test/analysis comparison of module DUL tests. Initial analysis work included comparisons between NISA II and hand calculations for stiffness and stress levels. A good comparison was obtained which gave confidence in the use of the code for check stressing of the 4 m and 6 m modules. Full advantage was taken of symmetry to reduce the model size. Two models representing vertical and horizontal halves of the structure were used to model vertical and horizontal loading, respectively, superposing the results for combined loading.

Examples of the display output for vertical and horizontal loading for the 4 m module DUL test are shown in Figure 15. Agreement between FE analysis and test results for flexural rigidity was good, measured stiffness being about 5% greater than predicted. Global strains calculated from loading intensity give a similar measure of agreement compared with measured strain gauge data within the limitations of fairly coarse modelling. A later version of the code giving direct output of strain has been used to give a detailed strain comparison.

7. MANUFACTURE AND COSTS

7.1 Manufacture

Full details on the manufacturing method for LRAC cannot be given in the present paper but a general description follows. Manufacture of the composite parts was by hand layup of carbon fibre - epoxy prepreg sheet 0.25 mm thick. Simple aluminium alloy tooling was used for the chord outer skin and CFRP sandwich tooling for the flat laminates, web skins, chord inner skin and booms. Autoclave curing gave the required optimum material quality of 60% fibre volume fraction with less than 2% voids. The modules comprise two main subcomponents, chords and webs, which are bonded in the autoclave from their component laminates. Assembly bonding of the complete structure was done in an oven with the chord jaws jigged to ensure interchangeability. Structural epoxy adhesive with 120°C cure was used to give high bond strength with good elevated temperature performance. Stages in manufacture are shown in Figure 16.

The jaws were machined from annealed forged bar followed by final hardening and tempering. Drilling of the chord skins for the field joint bolt group was accomplished by using the pre-drilled jaw blocks as drill jigs for the CFRP skins.

Two research prototype modules were manufactured. The first, shown in Figure 17, was 4 m long with jaws at one end only. The second was 6 m long with jaws at both ends. The 6 m module had thinner web skins and lower density honeycomb to reduce weight in areas of overdesign. The 6 m module achieved a weight of 441 kg, marginally above the target of 433 kg. A photograph of the 6 m module is shown in Figure 18.

7.2 Costs

The driver for development of LRAC is weight and the programme has achieved the required weight savings compared with aluminium rails. Nevertheless, minimum cost is important and controls materials selection and manufacture. The greatest material cost is in the CFRP prepreg. Standard strength "industrial" prepreg is used, with virtually the same properties as aerospace prepreg but costing about half the price. Materials and manufacture of the prototype steel jaws was found to be relatively expensive, costing about 15% of the total module cost. Jaw costs are expected to fall considerably with production quantities.

Manufacturing by small companies in the industrial sector gives economies in overheads. Combined with lower material costs, this gives manufacturing costs about one third of typical aerospace costs. CFRP modules, greatly superior in performance, are projected to cost about two times that of aluminium alloy modules.

8. TESTING OF 4 m AND 6 m MODULES

8.1 4 Metre Module Test

The objective of the 4 metre module test was to load the module in biaxial cantilever bending to Design Ultimate Load Levels appropriate to the full span cantilever launch at maximum launch wind speed. It is not possible to achieve an exact representation of the launching frame loads with only a 4 m length of rail. Nevertheless, by mounting the rail jaws to a rigid support and applying cantilever forces at the free end, the situation represents quite closely the forces in the last module as its jaws pass over the front roller. This is a conservative representation since any failures are expected to occur at the jaws or the chord/jaw connection which see 7.5% more load in this test compared with the full rail. The general test arrangement and an overall view of the test are shown in Figure 19.

The module was instrumented with strain gauge rosettes, displacement transducers, and a clinometer for vertical rotation. Acoustic emission transducers, 2 on each corner outer flange, were placed near the jaw to record and locate noise emissions following fracture events.

Load was applied in stages to working, proof and design ultimate loads, first in vertical bending and then adding lateral bending to achieve the full combined load cases.

Deflections in both planes were linear throughout. Corrected for rotation the tip vertical deflection at DUL was 42 mm compared with the FE prediction of 44 mm. Measured strains were also linear. The highest longitudinal strain at combined DUL was -3704 ue, measured at the top chord adjacent to the 1st bolt of finger 1. Strains in the centre of webs were low compared with the chords, the maximum principal strains recorded being -1649 ue and maximum shear strain being -2587 ue.

8.2 6 m Module Test

The 6 m module was tested in four load cases as follows:

1. Bi-axial cantilever bending with the same loads as the 4 m module test to represent maximum jaw stresses under combined bending.

2. Bi-axial bending by two simulated roller forces representing the maximum forces applied by the launch frame during cantilever rail launch. The module is supported encastre by its jaws to achieve the correct bending moment under the roller force system.

3. cantilever bending with increased vertical load and reduced wind load representing maximum rail body bending moments during bridge launch without assisting vehicle countermoment.

4. torsion test to determine torsional stiffness.

The force and moment systems applied to the rail module in tests 1 to 4 are given in the table below:

TABLE 2. LOADS APPLIED IN 6 m MODULE TEST

TEST	BM_V	BM_H	SF_V	SF_H	Comments
	kNm	kNm	kN	kN	
1	1939.2	391.7	505	102	Loads at 3.84 m from root
2	2147.2	410.6	617	118	Loads 1.5 & 5 m from root
3	2806	132.2	488	23	Loads at 5.75 m from root

The fourth test was torsion with 101.25 kNm applied by 2 equal horizontal forces of 150kN through the boom centres at the tip.

Full details of the strain distributions or comparison with FE results are not given in this paper, only the main conclusions from the tests.

Test 1 confirmed the strength of the 6 m module incorporating the design changes introduced to save weight. Compared with the 4 m module test the strain and deflection behaviour was similar with no significant differences measured. No failures or inelastic behaviour occurred.

Test 2 was a much more complex test since it involved the application of opposing forces of equal magnitude whose point of application was free to float according to the deflection of the beam. Under the increased bending moment in both planes, the chord strains increased and were greatest at the boom under combined compression (4474 ue). No failures or noise indicating damage occurred. The test confirmed the rail body strength under the maximum launching frame forces.

Test 3 simulated the maximum bridge launching forces with the rail in simple support between the ABLE vehicle and the far bank. Compared with the cantilever cases in Tests 1 and 2, the rail body supports a considerably increased vertical bending moment from bridge mass with a low lateral bending moment from the reduced wind force. The effect on the rail is to increase the strains across the chords while reducing the strain component at the corner booms arising from lateral bending. The maximum longitudinal chord strain was 3945 ue, slightly greater than the strain in Test 1 but less than the highest strain in Test 2. Shear strains in the webs were low, about 2600 ue. The rail body flexural rigidity was calculated from the cantilever deflections in Test 3 as 288 MN/m^2; this agrees closely with the result for the 4 m module. No failures or inelastic deformations occurred at DUL.

Test 4 was a check of the torsional rigidity of the rail body. The torsional rigidity, GJ, calculated from differential deflections at the tip, was 17.6 MNm^2/radian. This is in reasonable agreement with predictions although the high stiffness of the module makes measurement difficult.

9. DISCUSSION

The programme to develop research prototype modules for LRAC has gone far to demonstrate the feasibility of the system. Static strength has been confirmed by module testing and fatigue endurance demonstrated by calculation and element testing. Environmental properties, impact resistance and residual strength have been covered by testing of materials coupons and panels. Overall, the module developed appears fully adequate for its purpose.

A design philosophy for LRAC was to adopt a conservative design strain limit of 4000ue in order to maximise robustness and residual strength. This is a field strain limit typically used in composite aircraft wings for the upper (compression) wing skin. Experience has shown that increasing design strains much beyond 4000ue for compression structure can give problems with post impact delamination. The design of LRAC based on 4000ue gave the required weight saving compared with the aluminium modules. In test the field strain limit was not exceeded except for local areas of the thick corner sections which are intrinsically damage resistant.

It is likely that the research prototype modules have an excess of strength although neither has yet been tested to failure. Some small further savings in both weight and cost could be achieved by reducing the chord skin thicknesses and allowing a higher design field strain, perhaps 4500ue at DUL. However, the upper limit to design field strain is controlled by compression stress. Unlike aircraft wing design, where higher strain limits can be used for the lower (tension) wing skin, LRAC has virtually identical chord loadings,interchanging between tension and compression with the transition from the hogging cantilever rail launch to the sagging bridge launch case. Compression and resistance to impact controls the design and will limit the allowable strain increase.

Because of the module size it is impossible to test LRAC modules in environmentally degraded (hot/wet) conditions. An assessment of degraded strength depends on material test results. Furthermore, the main load paths are carried in sections 50 mm thick which will never achieve saturated moisture content in the lifetime so the full effect of hot/wet strength reduction will not occur. The approach to environmental degradation is to use a calibrated finite element model with degraded properties to predict changes in performance.

Use of finite element analysis was found successful. The results of the strain survey comparison show good agreement for the major strains. It is the intention to develop a fully calibrated F.E. model for LRAC. This will reduce the amount of full scale testing required with design or material changes and, together with mechanical test results, could become part of the supporting data for acceptance.

The design concept allows changes in the section dimensions to accommodate changes in bridge weight or launching vehicle characteristics and, although designed specifically for BR90, can be considered a generic component with application in other bridge systems. Fibres and resins can be changed to suit requirements; hybrid panels of glass/carbon could be introduced for less demanding loads, with thermoplastic or other matrices dependent on the environment.

The next step in development is to demonstrate successful interfacing with the ABLE vehicle, its launching frame, rail drive and overrun features. This will require more modules and their manufacture will also be a major input to production development and cost reduction.

10. CONCLUSION

The LRAC research programme has been very successful. Prototype modules were developed which met their strength targets and achieved the required weight saving compared with the aluminium modules. The module weight was 441 kg, about 40% less than the aluminium alloy rail module (heavy type) which it replaces. This will allow the maximum length BR90 bridge to be launched using ABLE.

A pyramidal programme with a base of materials tests, reducing numbers of critical element tests concluding with the full scale static tests formed the basis of development. Finite element analysis using the NISA II code supported the design and testing.

The module developed is based on sandwich panels with thick chord skins and booms of solid CFRP at the corners. Steel jaws are bolted into the chord core space in double shear.

Manufacture was by hand layup of relatively low cost CFRP prepreg with autoclave curing to achieve the high quality required of 60% fibre volume fraction with < 2% voids. The cured components were oven bonded using 120°C epoxy structural adhesive. Manufacture in the industrial composites sector gave costs estimated at about one third of aerospace composite costs.

Although it is unlikely that LRAC can be produced as cheaply as welded aluminium alloy rails, the gains in bridge system performance are great. These gains can be achieved for relatively little extra in total system cost.

The next step is to carry out interface trials with ABLE and the Christchurch bridge. More modules will be manufactured to produce a 60 m launching rail for trials with the 56 m bridge set.

11. ACKNOWLEDGEMENTS

The successful research programme for LRAC was only achieved by the close cooperation between RARDE, Christchurch and the Prime Contractor, Williams Fairey Engineering Ltd. The author wishes to record his thanks for the unstinting efforts of Mr K D Clare and his team at WFE Ltd.

Manufacturing was carried out by Polymeric Composites Ltd, under Mr D Pairaudeau, whose single minded commitment to this major task ensured that quality and strength targets were achieved.

12. REFERENCES

1. Trilateral Design And Test Code For Military Bridging And Gap Crossing Equipment RARDE, Christchurch, BH23 2BB, UK

2. Def Stan 00-970 Vol 1 Chapter 200 Design Requirements For Service Aircraft Issue 1 (12 Dec 1983) UK MOD, AD(R) 1

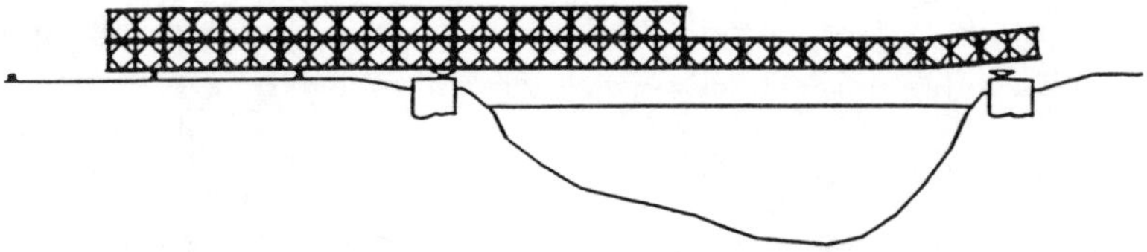

Figure 1. Bailey Bridge Launching Nose

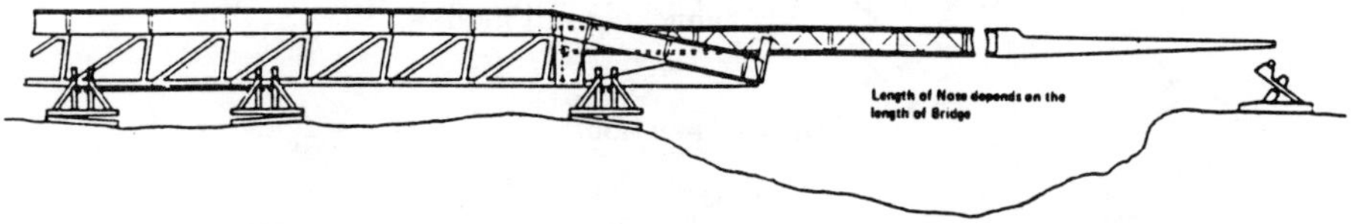

Figure 2. Medium Girder Bridge Launching Nose

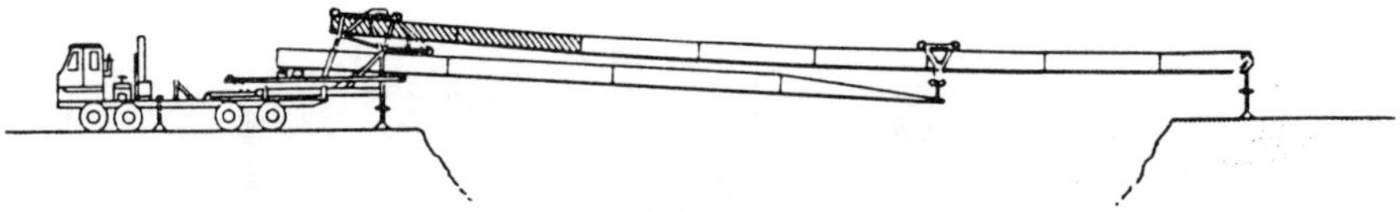

Figure 3. Christchurch Bridge Launching Rail

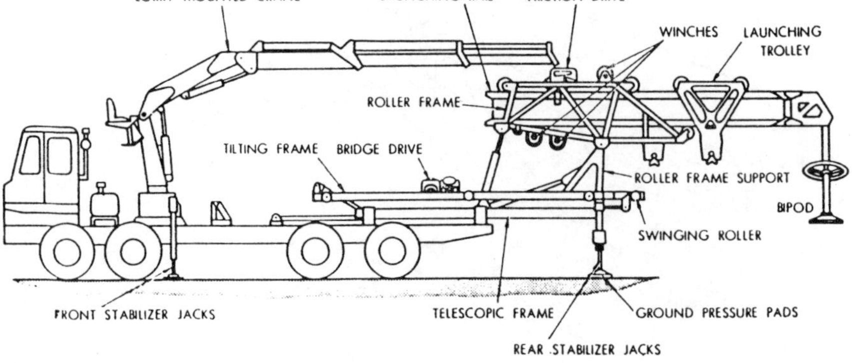

Figure 4. The ABLE Vehicle

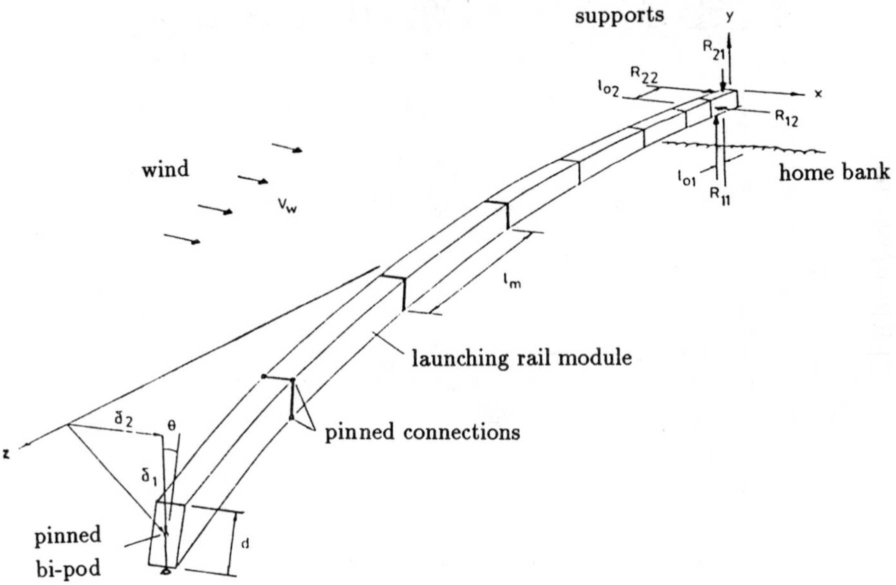

Figure 5. Launching Rail Cantilever Force System

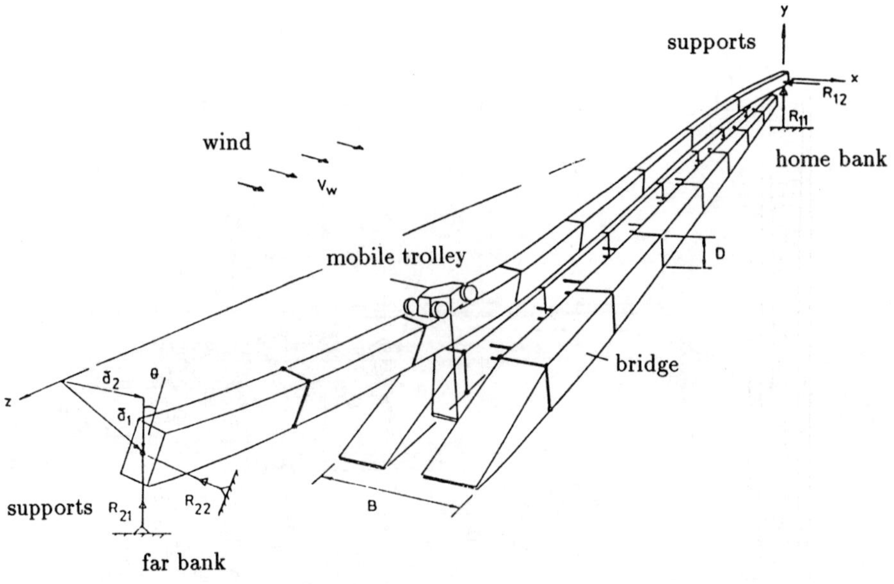

Figure 6. Force System on Launching Rail with Suspended Bridge

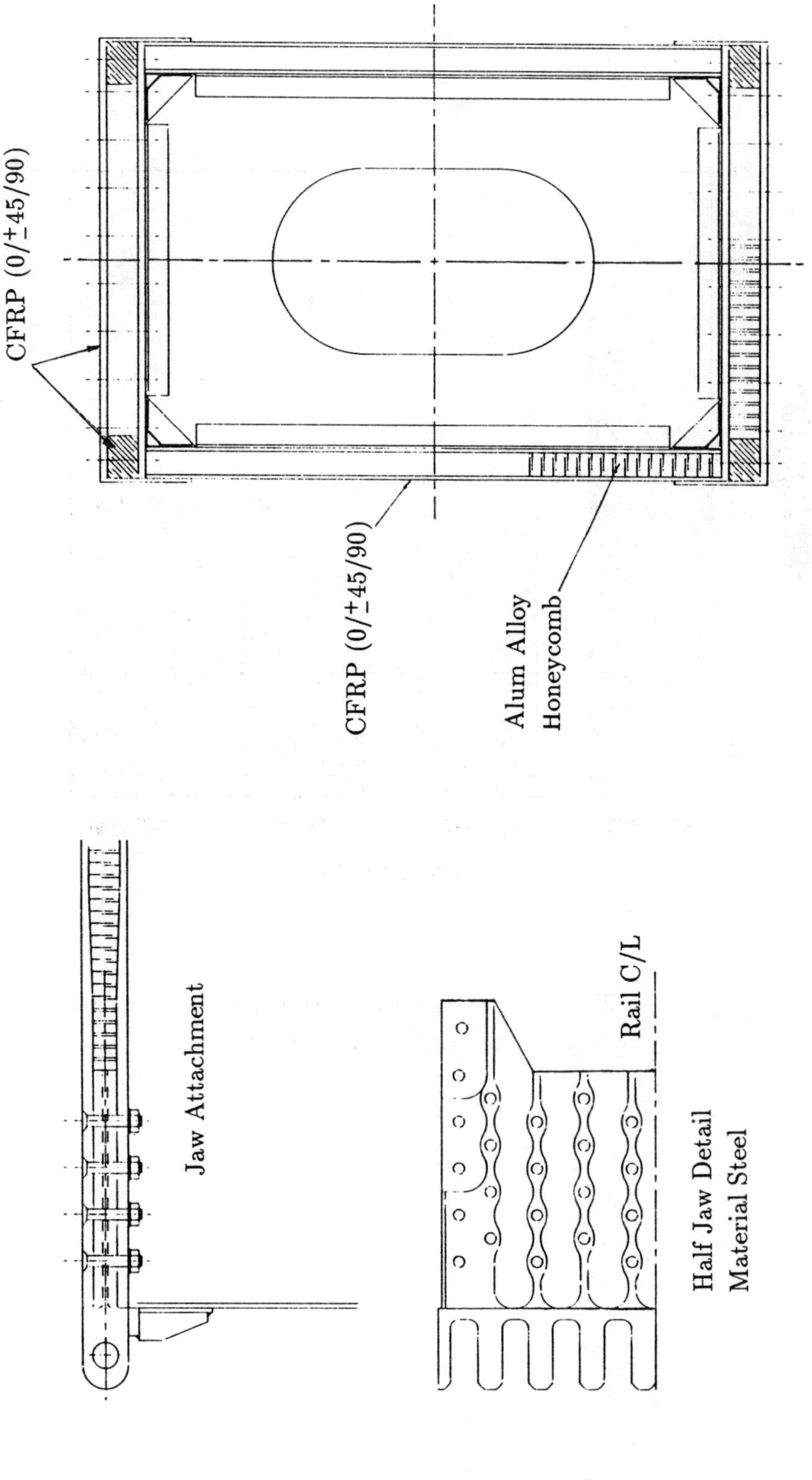

Figure 7. LRAC Structural Details

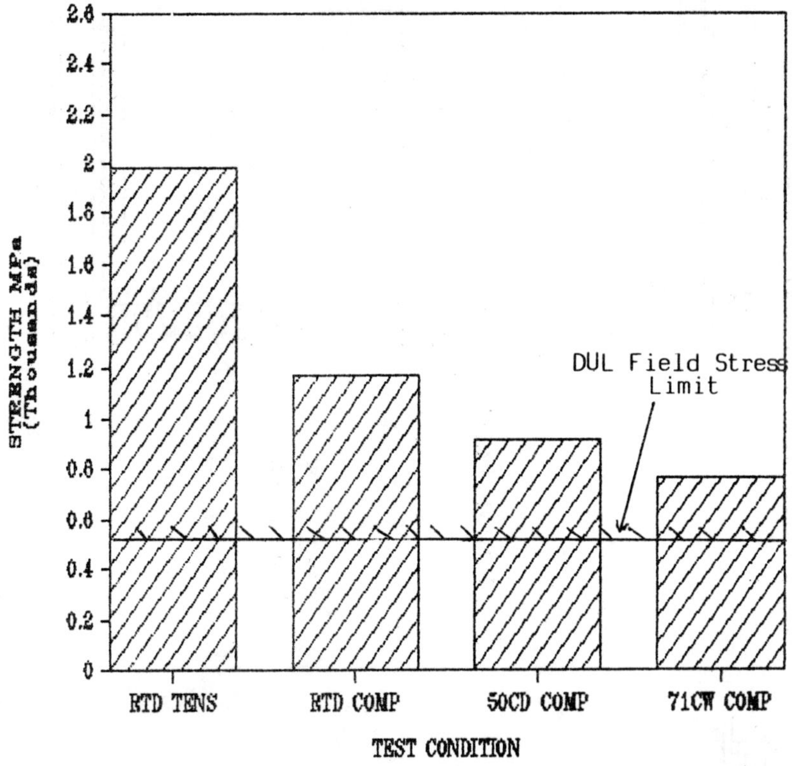

Figure 8. Effect of Environmental Degradation on Strength
Unidirectional Carbon/Epoxy Prepreg.

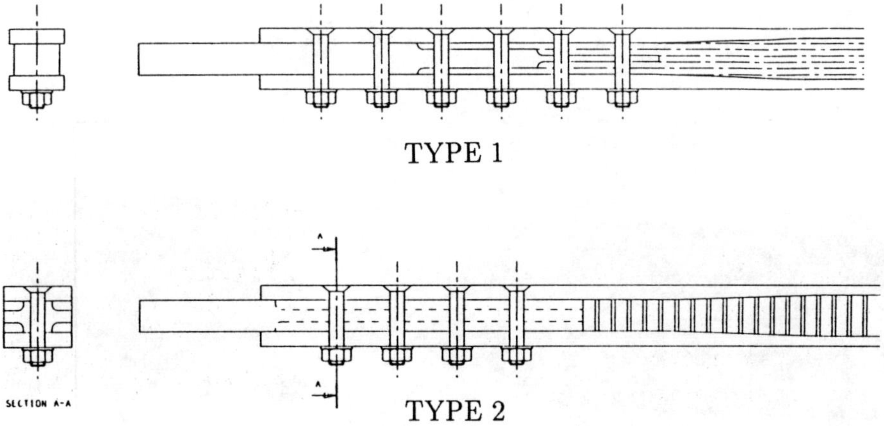

Figure 9. Finger Element Specimens Types 1 and 2

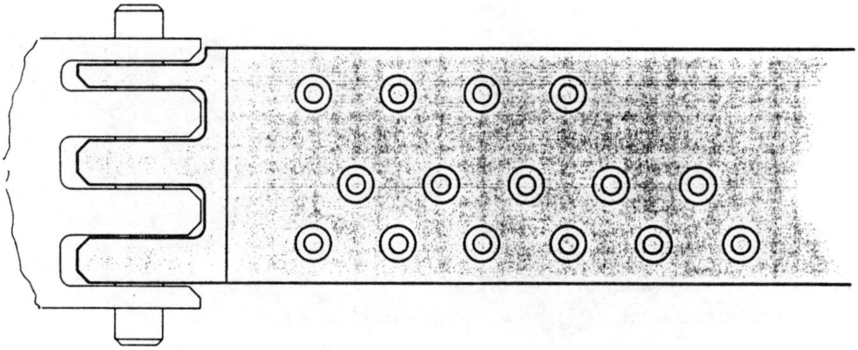

Figure 10. Critical Element for Chord/Jaw Connection

Figure 11. Failure Mode for Finger Element Type 1 - Tensile Fracture

Figure 12. Failure Mode for Finger Element Type 2
Shear Failure of Bolts/Bearing Failure of Composite

Figure 13. Failure Mode for $\frac{1}{3}$ Chord/Jaw Critical Element

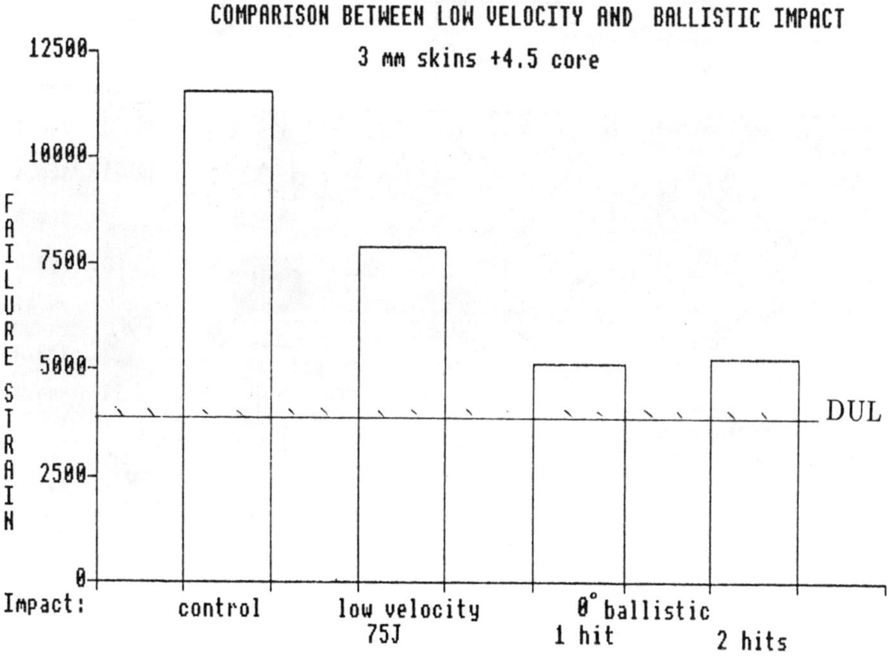

Figure 14. Residual Strength of Web Panel

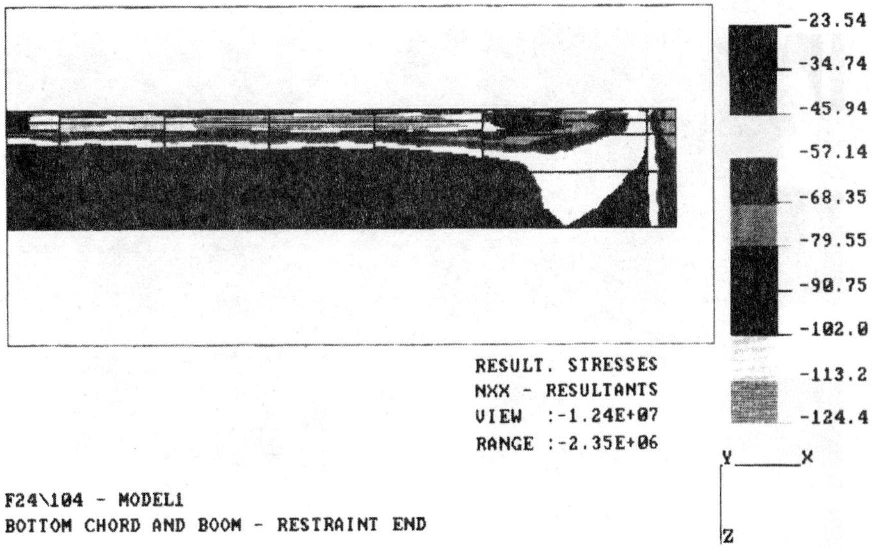

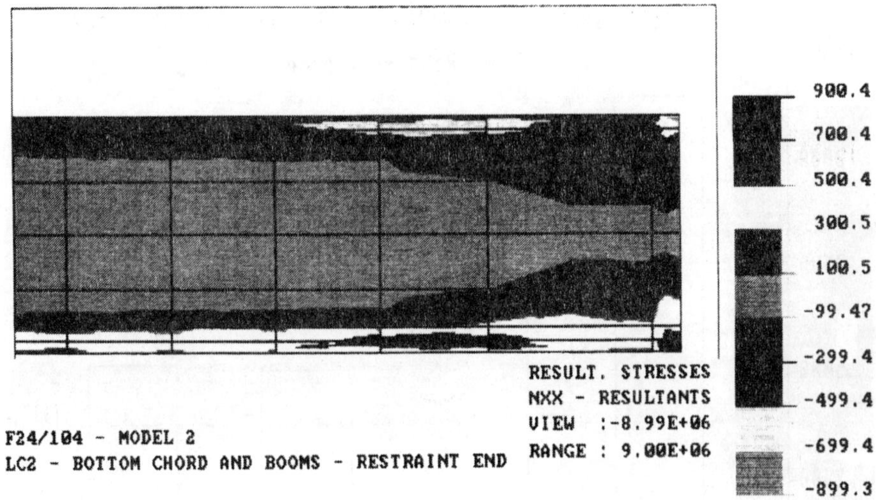

Figure 15. NISA II F.E. Analysis Display Output
Stress Resultants in 4m Chord and Booms at DUL

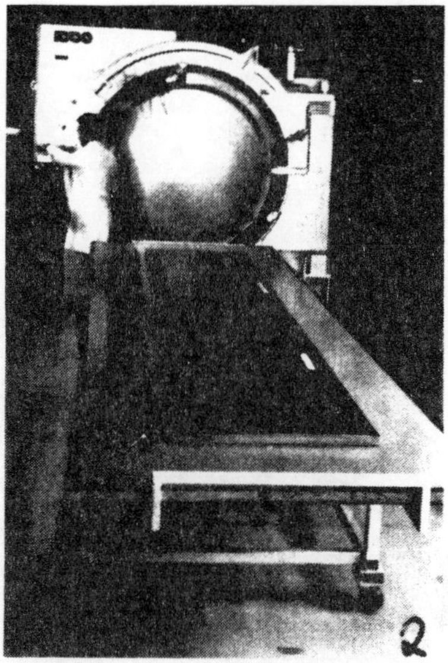

Figure 16. Stages of Manufacture
4m LRAC Module

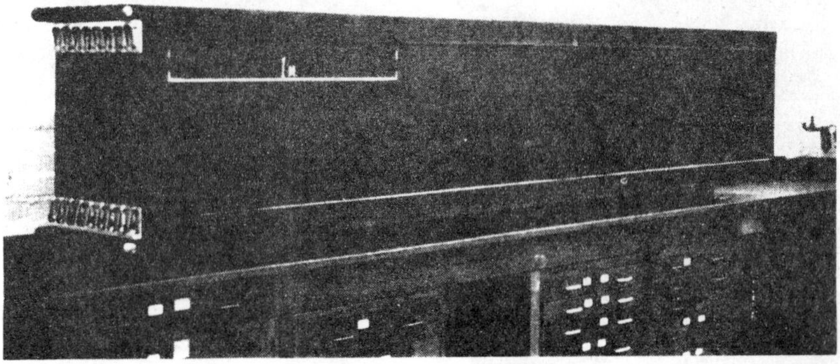

Figure 17. 4m LRAC Module - Jaws At One End Only

Figure 18. 6m LRAC Module - Double Ended With Lifting Beam

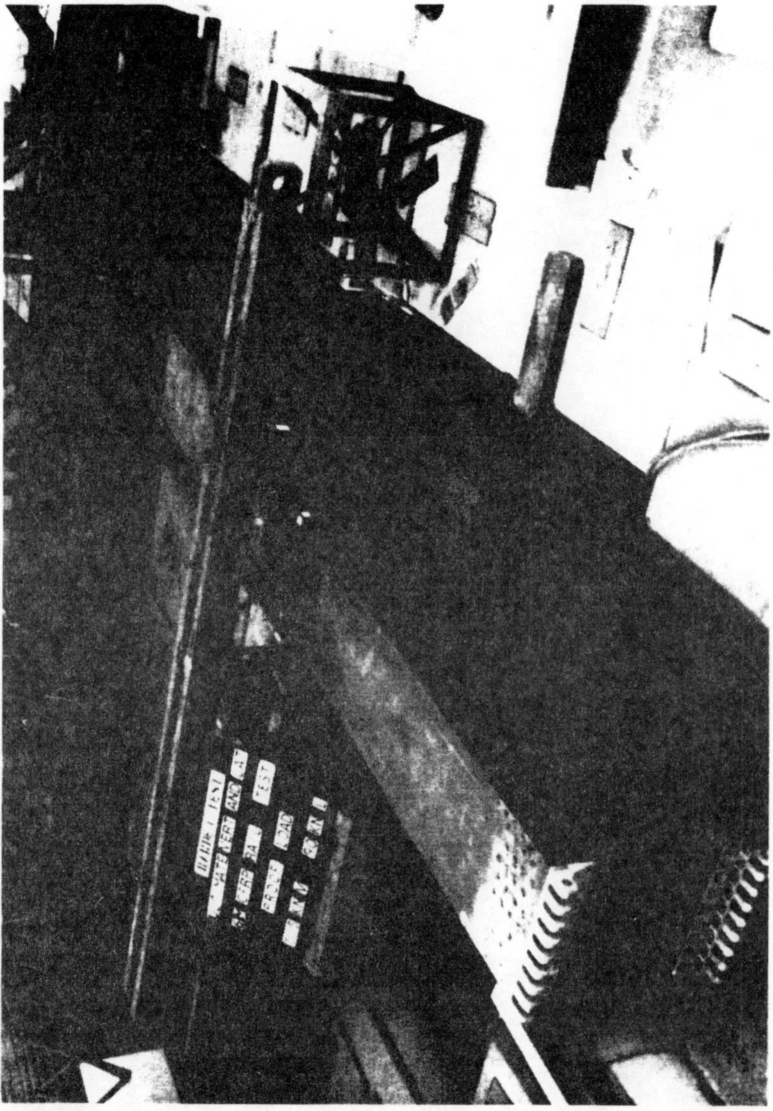

Figure 19. General View of Test Arrangement - 6m LRAC Module

Simulation of the Effects of Possible Local Failures During the Erection of Foldable Bridges

B. Hug, G. Sedlacek, D. Neuenhaus

University of Technology in Aachen, Institute of Steel Construction, D - 5100 Aachen, Germany

ABSTRACT

The requirements for a minimum of transport space, erection personal and assembly time for military bridges have led to systems like the Foldable Dry Bridge that are erected almost automatically.

In order to achieve a maximum reliability in reaching the required performances the different stages of transport, unloading, assembly and launching are simulated in the design phase by using a computer program for multibody dynamics, where the dynamic behaviour in all stages and also the effects of failures of local elements can be studied.

The multibody dynamics program contains a catalogue of particular elements as rigid bodies, joints with nonlinear characteristics, passive and active force elements and sensores, cable elements (taut or slack), to allow the nonlinear element behaviour.

The method is demonstrated with the simulation of a failure of a suspension during the unfolding process of the Foldable Dry Bridge.

INTRODUCTION

This paper presents a recently developed computer method for simulating the nonlinear dynamics and kinematics of articulated multibody systems like cranes or robots or structures that are turned, launched or folded in service or during erection. Furthermore it is used to simulate the dynamic failure of structures in order to study probable failure path's as a consequence of local damages.

The latter problem is of particular interest for the determination of failure causes after the collapse of a structure. Of even more interest are failure path studies a priori, in the project stage, to avoid several failure consequences, that means to design structures with sufficient robustness, so that they will be not damaged by any events to an extent disproportionate to the original cause.

In multibody systems (figure 1) in general all mass-bodies are assumed to be rigid. They are interconnected in arbitrary arrangements with massless joints and force elements that are either passive (springs, dampers) or active (servomotors, hydraulic cylinders).

Joints are purely kinematical elements which allow large relative motions of contiguous bodies with one or more, up to six degrees of freedom. Force elements have in common with joints, that they are mounted between bodies and exert forces and/or torques which act with opposite signs on two bodies.

When the deformation of a body shall be considered, this body has to be modelled as an assembly of several rigid bodies and force elements. Specially developed nonlinear elements allow the perform the simulation of the behaviour within the limits of the service conditions as well as at the ultimate state, i.e. in the failure phase.

THE PERFORMANCE OF THE PROGRAM MEPHISTO

Method applied to the computer program

The theoretical background can be briefly outlined as follows (for details see [1]). For a system of n bodies the principle of virtual power

$$\sum_{i=1}^{n} \int_{m_i} \delta\dot{z}\cdot(\ddot{z}\cdot dm - dF) = 0 \tag{1}$$

in combination with the kinematics relationships (figure 2)

$$z = r_i + \rho$$
$$\delta\dot{z} = \delta\dot{r}_i + \delta\omega_i \times \rho$$
$$\ddot{z} = \ddot{r}_i + \dot{\omega}_i \times \rho + \omega_i \times (\omega_i \times \rho)$$

yields

$$\sum_{i=1}^{n} [\delta\dot{r}_i\cdot(m_i\cdot\ddot{r}_i - F_i) + \delta\omega_i\cdot(J_i\cdot\dot{\omega}_i + \omega_i \times J_i\cdot\omega_i - M_i)] - \delta P = 0 \tag{2}$$

with

F_i, M_i externally applied forces and torques, respectively,

δP total virtual power of all force elements together.

In matrix form the equation (2) reads

$$\delta\dot{r}^T\cdot(m\cdot\ddot{r} - F) + \delta\omega^T\cdot(J\cdot\dot{\omega} + V - M) - \delta P = 0 \tag{3}$$

If q represents a column matrix of conveniently chosen generalized coordinates equal in number to the total number of degrees of freedom of the entire system then the following relationships exist

$$\delta\dot{r} = \alpha\cdot\delta\dot{q}, \quad \delta\omega = \beta\cdot\delta\dot{q}, \quad \ddot{r} = \alpha\cdot\ddot{q} + u,$$
$$\dot{\omega} = \beta\cdot\ddot{q} + v, \quad \delta P = Q^T\cdot\delta\dot{q} \tag{4}$$

where α, β, u and v are matrices defined by the system. When the equations (4) are inserted into equation (3) we obtain

$$\delta\dot{q}^T[\underbrace{(\alpha^T m\,\alpha + \beta^T J\,\beta)}_{A}\ddot{q} + \underbrace{\alpha^T(m\,u - F) + \beta^T(J\,v + V - M) - Q}_{-B}] = 0 \tag{5}$$

Under the assumed conditions the equations of motion read

$$\boldsymbol{A} \cdot \ddot{\boldsymbol{q}} = \boldsymbol{B}$$

This is the wanted differential equation of motion, used in the program.

Application of the theory to failure path checks for steel structure

For the particular use of multibody dynamics for steel structures, the joints and the force elements may have particular features, which are controlled by the behaviour of these joints and force elements as well as in the serviceability state as in the range of the ultimate limit state conditions.

In order to model the formation of plastic zones with local yielding and buckling, a 'plastic hinge' element is introduced, which may act in a plane or in space as given in figure 3 where the force elements for such a plastic hinge may comply with given moment rotation curves. These moment rotation curves may either be obtained by tests or from finite element calculations for monotonic or cyclic behaviour [2]. In figure 5 a comparison between a rotation curve from test and a rotation curve from finite element calculation is indicated.

For the consideration of truss elements in frameworks, in addition to plastic hinges, additional joint elements and force elements have to be applied at the connections. An example for such a joint model is the element 'deformable connection' shown in figure 4. This joint element allows the modelling of the elasto-plastic rotational and translational deformation of any connection. The requirements for such a joint are firstly to represent the nonlinear moment-rotation behaviour of the joint, including the effects of buckling in the post critical range. Secondly it must represent the shear deformation of the connection as well as the deformations due to bar forces in the connected bodies. Finally it must represent the damage under cyclic movements e.g. in form of an evolution curve for the moment rotation hysteresis.

The interactive working process of the input and output

Figure 6 shows the structure of the program package MEPHISTO. MEPHISTO stands for "Multibody systems with Elastic-Plastic HInges and changeable STructure Organisation".

The procedure for the analysis of multibody systems in using the program package MEPHISTO is demonstrated in figure 7. For a better distinction, the working parts for the user are separated optically from the working parts for the program.

First of all the user must decide the physical model for the system to be analysed. In the case of the simulation of the unfolding and turning of the Foldable Dry Bridge modul, it is necessary to consider the crane cantilever, the load frame and the bridge segment together, figure 9. The truck transporting the bridge elements and the vehicle with the crane, would be relevant for the simulation of the unloading of bridge parts and do not need to be taken into consideration when the unfolding and turning process is simulated.

For creating the mechanical model, the structure has to be divided into bodies and connections, which yield essential contributions to the dynamic behaviour with respect to the aim of the investigation. The governing law of motions for the internal force elements and the restrictions to joints needs careful consideration.

The input data for the description of the multi body system can be provided by filling out several menus the preprocessor places at his disposal. The data consist of information about the masses and measurements of bodies, physical values of the connection elements, nodal coordinates and loading. The assignment of the data to the program variables is not carried out by a fixed input process but by keywords. For this reason the data blocks can be freely styled in the input file. Together with any comments to the input values, the input file can be built up into suitable forms for the documentation.

Following the input data parameters, the main program sets up the differential equations of motion for the structural model. The user is asked by dialogue to give instructions for the simulation. These may be: choice of the integration scheme for the equations, the duration of the simulation and the determination for the output files to create. It is possible to realize the simulation in several stages. After reaching the declared number of time steps, the simulation will be interrupted. Now it is possible to intervene in the simulation process interactively, e.g. by changing external forces or declaration of local failure conditions as cable failure etc. This increases the flexibility, because no new input file need to be created.

The outputs can be represented in form of listings, 2-D-plots or in form of 3-D-animations.

SIMULATION OF THE DYNAMIC BEHAVIOUR OF THE FOLDABLE DRY BRIDGE

System description

A foldable bridge module of the Foldable Dry Bridge as illustrated in figure 8 shall be lifted by a crane from the platform of a truck in the following steps:

- The bridge segment suspended by four cables is lifted upward off the platform of a truck and simultaneously the sideparts of the bridge segment unfold automatically due to gravity.
- In lifting, the unfolding segment is turned by the crane to be positioned on an erection rail for further assembly. Before the bridge element is put on the rail the unfolding process is finished.

The computer simulation of the lifting and unfolding process shall verify that it works without collision, unexpected impact or vibration problems or infringing any serviceability or ultimate limit state conditions.

Figure 9 gives a survey of the crane and bridge elements, with all rotational and translational joints, as well as the load frame with cables connected to the bridge segment. The figure indicates the number of the rigid bodies (symbol○) and joints (symbol Δ).In the bridge module, passive dashpots are installed to control the unfolding process between the side parts ⑭, ⑱ and the central deck-plate ⑫, which are not indicated. The cables are considered as force elements which allow the modelling of a cable failure as well.

Simulation for normal service conditions

An example for the simulation of the normal service (without accidental conditions) is given in figure 10.

From the lifting, unfolding and turning process the cable force in cable No 2 is plotted versus the process time the program allows indication in a similar way the time history for any wanted value, e.g. action effects like moments, deformations etc. at any given location.

Simulation of accidental situation

As an accidental situation the failure of cable No 4, two seconds after the start of the normal process is assumed.

Figure 10 shows the time history for the cable force in cable No 2 for this case. The peak value immediately after the cable failure represents the impact that has to be taken into account in the design of the cables and their connections.

The simulation also reveals that the unfolding process due to the cable fracture cannot be completed but stops before the end.

Of course the time history of any other action effect produced by this simulation can be plotted as for normal service conditions.

CONCLUSIONS

By the simulation of the dynamic service conditions of military bridges in the erection and serviceability states by using multi body dynamics programs as MEPHISTO the behaviour of such structures can be predicted in the design phase.

From these predictions the feasibility of the erection procedure as well as impact factors for the design can be derived.

The simulation of accidental situation allows to check the consequences of assumed failures including dynamic effects. These studies allow for measures to be taken to reduce the vulnerability of the structures and to improve the robustness.

LIST OF REFERENCES

1. Wittenburg, J. Dynamics of Systems, Teubner, Stuttgart, 1977.

2. Sedlacek, G., Lopetegui, J., Spangemacher, R. and Stutzki, C. Nichtlineare Berechnungen von Stabwerken aus Stahl. Nichtlineare Berechnungen im konstruktiven Ingenieurbau, Springer-Verlag, Berlin, Heidelberg, New York and Tokyo, 1989.

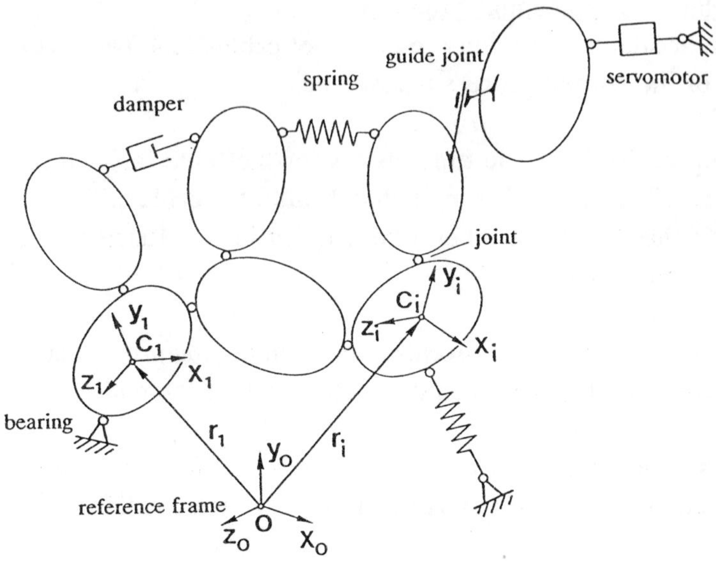

Figure 1. General multibody system.

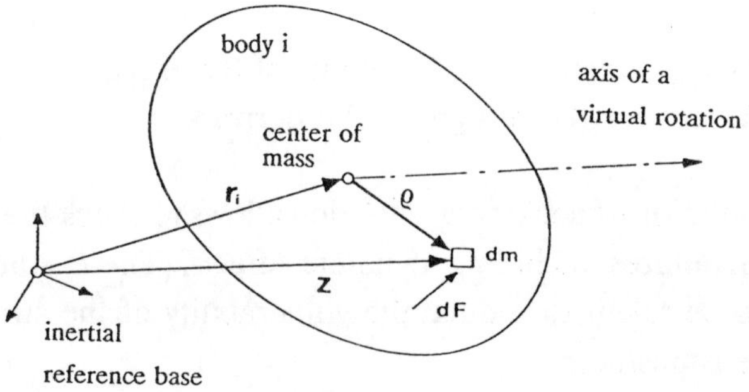

Figure 2. Vectors locating a mass particle on body i.

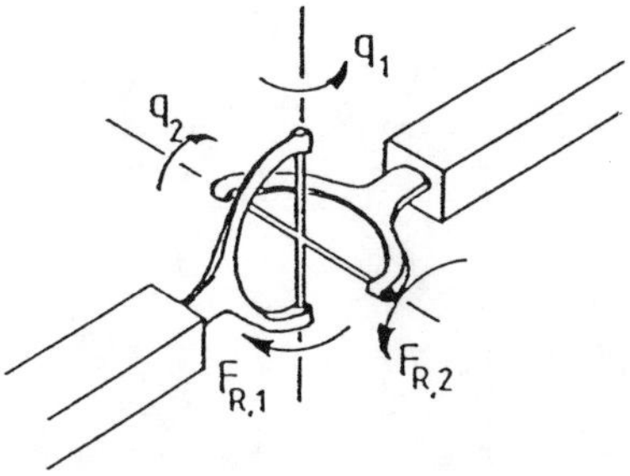

Figure 3. The 'plastic hinge' element.

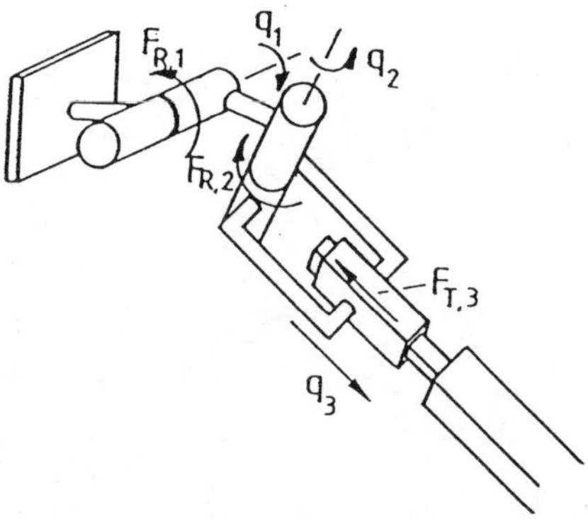

Figure 4. The 'deformable connection' element.

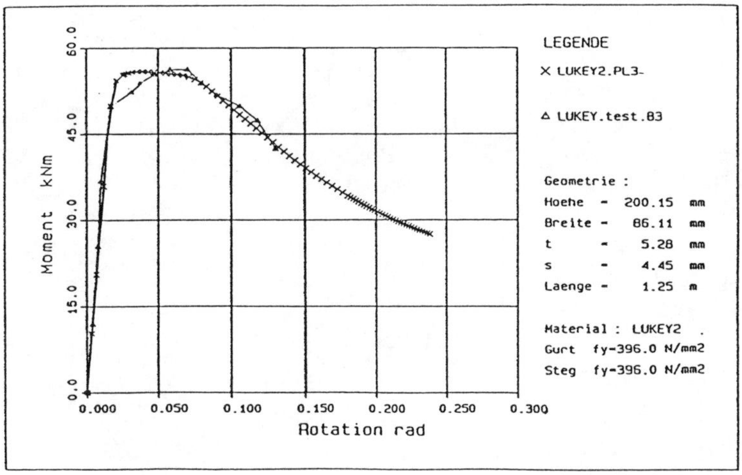

Figure 5. Moment-rotation-curves, comparison between test results and calculation results.

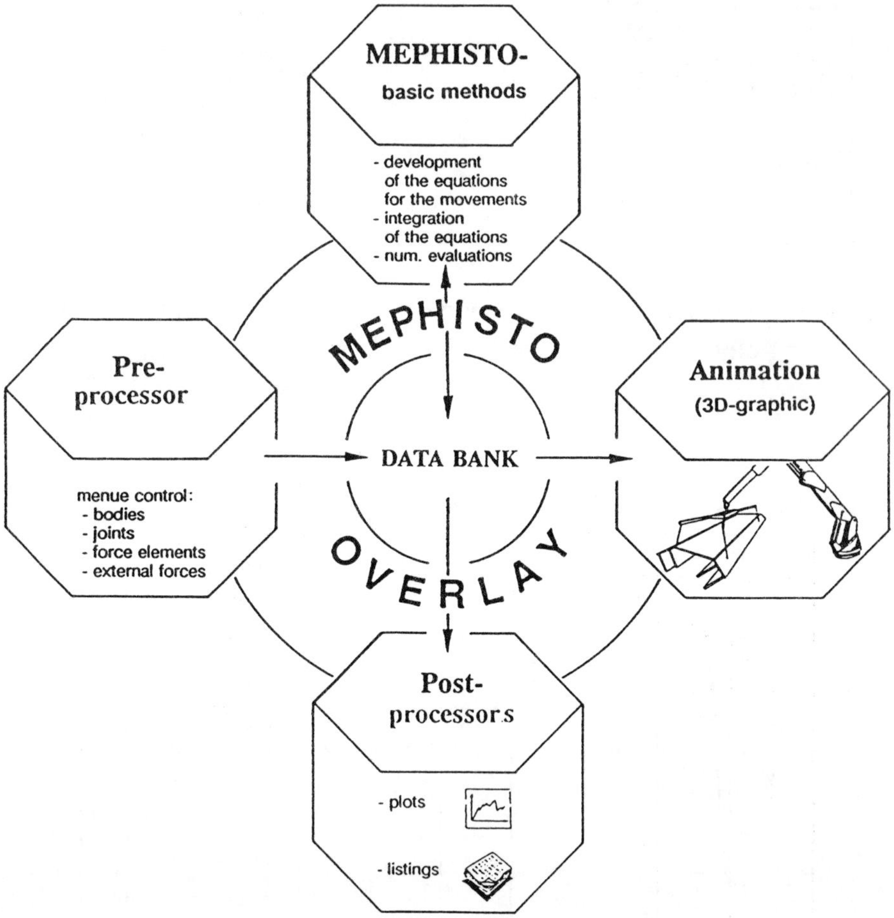

Figure 6. Architecture of MEPHISTO. Relationships of components.

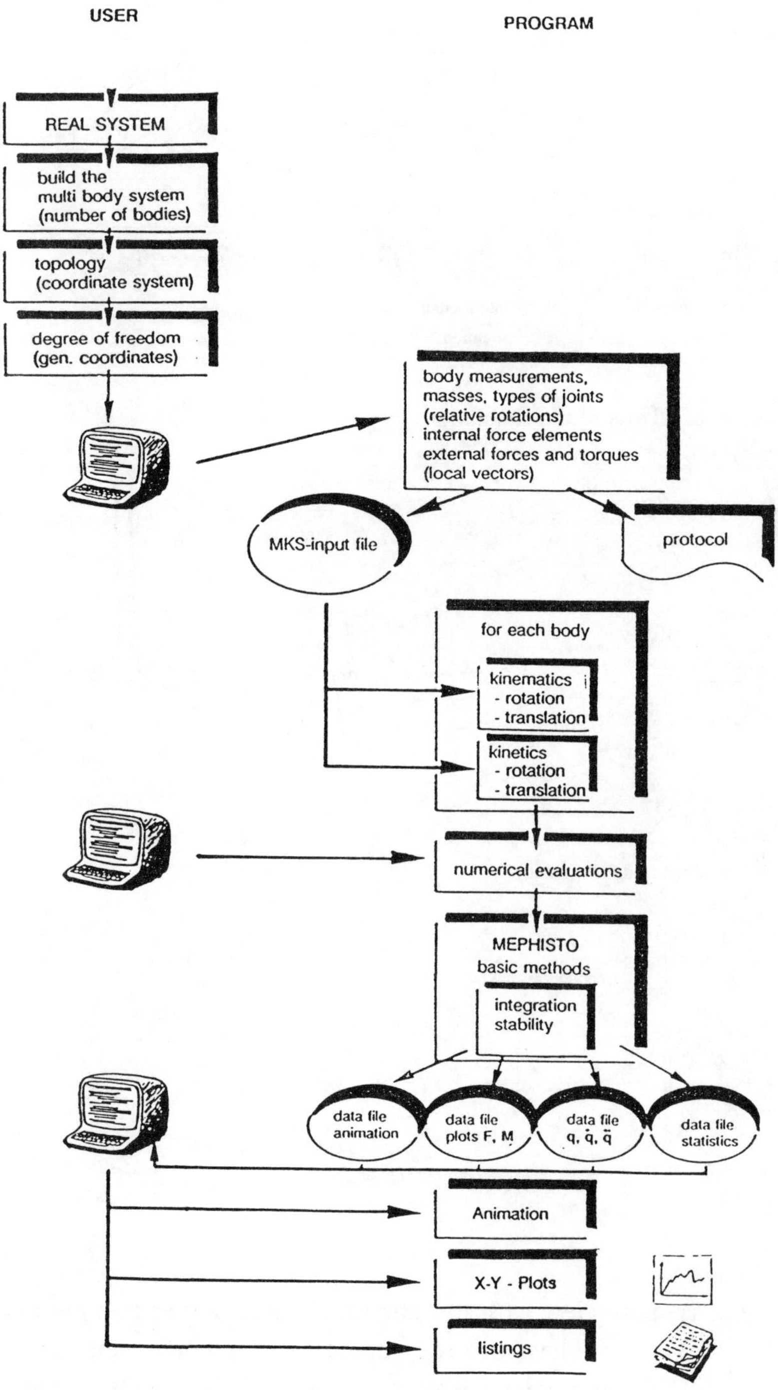

Figure 7. Procedure for the analysis of multibody systems in using the program package MEPHISTO.

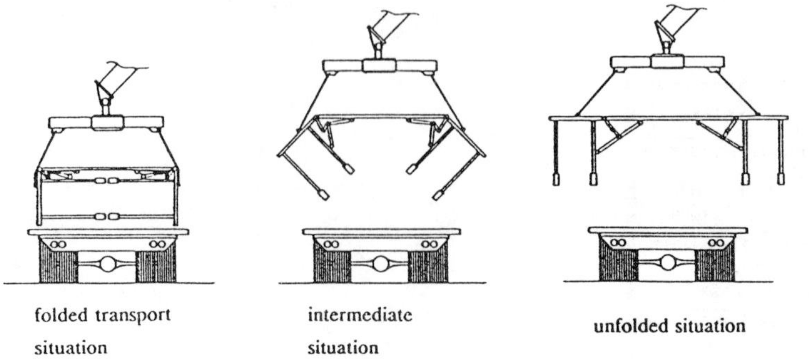

Figure 8. Parts of the simulation.

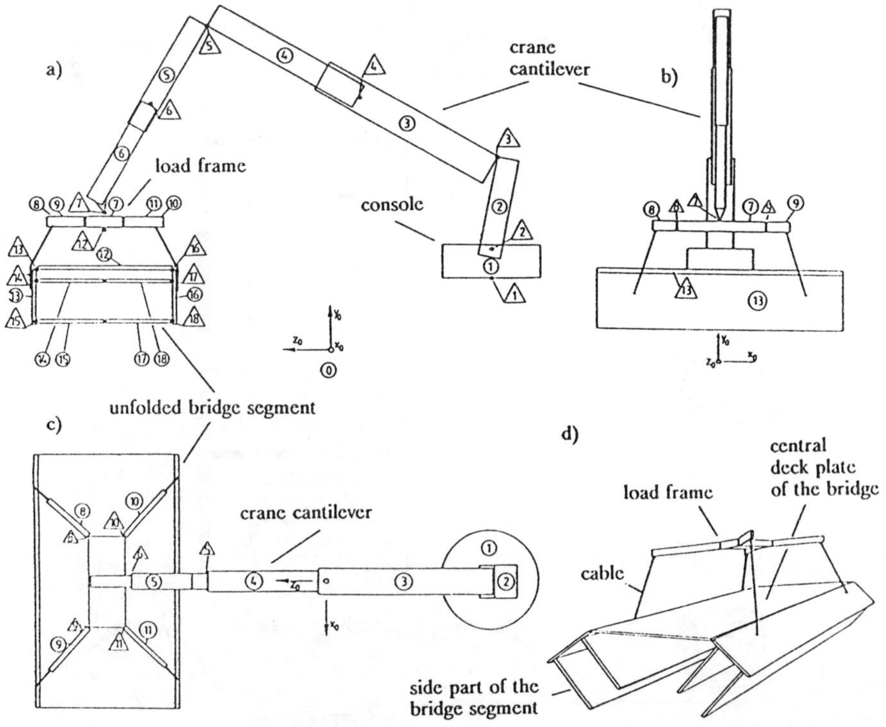

Figure 9. a) Side-view with the numbers of the bodies and joints.
b) Front-view with the numbers of the bodies and joints.
c) Ground-view with the numbers of the bodies and joints.
d) Perspective view of a foldable bridge segment when being lifted up and in unfolding.

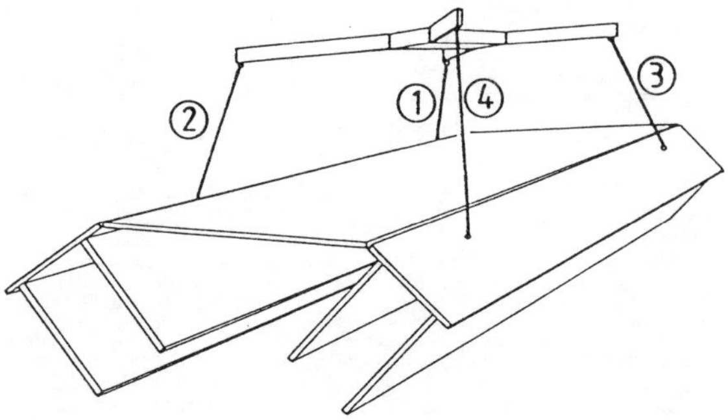

without cable failure

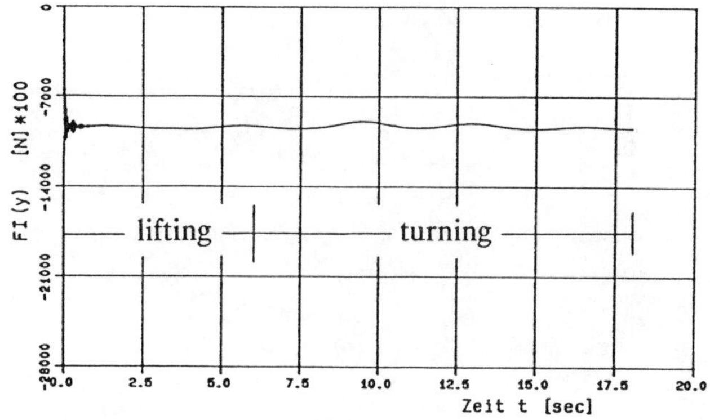

with cable failure

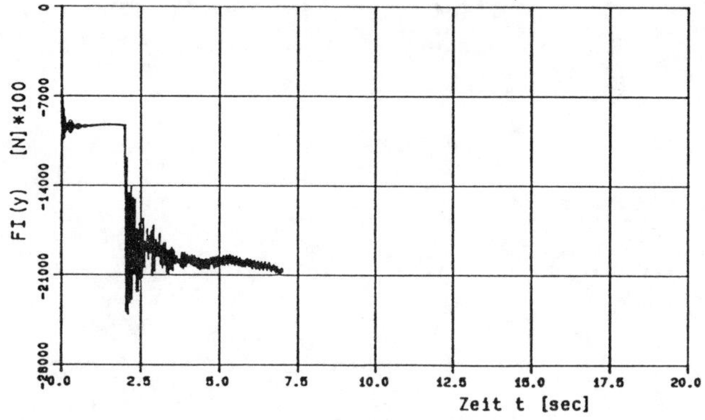

Figure 10. Example for the simulation.

The Design and Manufacture of an Armoured Bridgelayer System

M.A. Smith, D. Turnbull

Vickers Defence Systems, Newcastle Upon Tyne

INTRODUCTION

Since the earliest conflicts, mobility on the battlefield has been crucial to the success of any military operation. Military planners and engineers have therefore devoted a great deal of time and effort to both creating and overcoming obstacles.

This paper will consider one particular obstacle - the simple gap - and one way of overcoming it -the armoured bridgelayer.

THE HISTORY OF ARMOURED BRIDGELAYERS

The humble gap was used to its most devastating effect in the First World War. Lines of trenches and barbed wire protected by fire from concealed machine gun emplacements made successful offensives almost impossible. It was this stalemate which led to the development of the tank, an armoured assault vehicle which could negotiate mud, barbed wire and trenches. The first ever tank action took place on the Somme on 15 September 1916.

These tanks brought their own limitations with them. The upward sweep of the tracks at the front approximated to the curve of a wheel which could climb a 5 ft obstacle or cross an 8 ft gap. To this day the simplest way to deny a tank force mobility is to place wide gaps in its path.

Because tanks invariably lead an attack it was soon realised that they needed to carry their own gap crossing equipment.

Thus, in 1919 the first AVLB (Armoured Vehicle Launched Bridge) went into operation carrying a 30 ft Canal Lock Bridge. (Figure 1).

Figure 1.AVLB circa 1919

This concept was extended and developed over the years, primarily by the Royal Armaments Research and Development Establishment (RARDE) at Christchurch. The first scissors type bridge was launched at Christchurch from a Light Tank MkV hull in 1938 (Figure 2).

Figure 2.Scissors Bridge on Light Tank 1938

This vehicle was the forerunner of the Chieftain AVLB in service with the British Army today. (Figure 3).

Figure 3. Chieftain AVLB

PRINCIPLES OF OPERATION

The major components of the launching system of an AVLB are: (Figure 4)

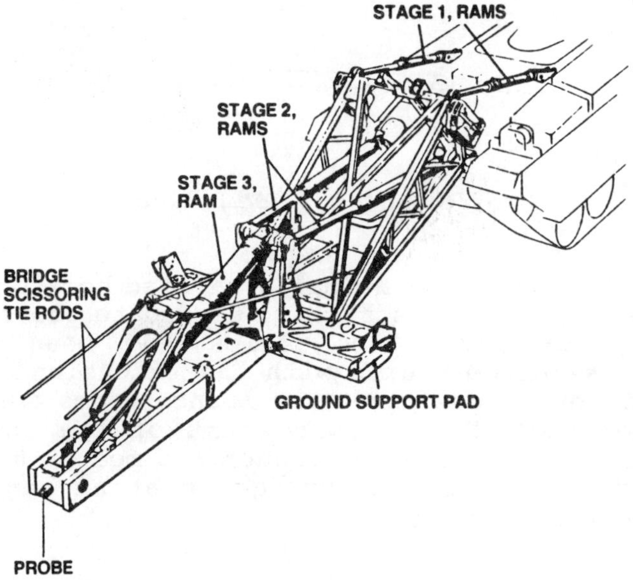

Figure 4. AVLB Launching Mechanism

These are all connected by a system of pins and hydraulic cylinders. The hydraulic system is driven through a power take off by the tank main engine and the bridge is launched in three distinct stages: (Figure 5)

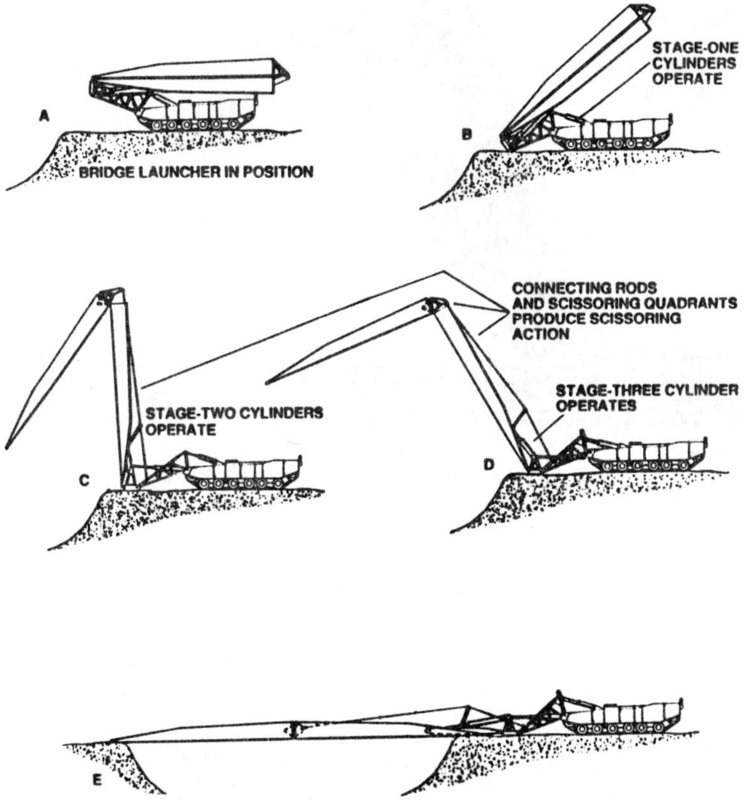

Figure 5. AVLB Launch Sequence

The bridge is normally carried horizontally above the tank where it is firmly secured to prevent movement during transport. Once clamps have been removed the launch sequence starts with the extension of the Stage 1 cylinders.These pivot the launch frame and the folded bridge about the forward part of the chassis. This continues until the the launch pad meets the ground. At this point the folded bridge is at an angle of about 30° to the horizontal.

The Stage 2 cylinders are then extended tilting the launch pad onto the ground. This operation stops when the pad tower is vertical. At this point the bridge is partly scissored by the rods connecting the central quadrant to the launch frame. The scissoring action allows the bridge to be launched on sites where longitudinal slopes or differing bank heights would defeat horizontally launched bridges.

When the Stage 2 cylinders reach the end of their strokes the Stage 3 cylinder is extended causing the bridge to scissor further until the far end of the bridge touches down on the opposite bank. At this point the rods will slacken and, as the stage 3 ram continues to extend, the centre hinge in the bridge will close.

The rods are now free and are detached from the launching structure by a remotely controlled actuator. By reversing the tank the launch mechanism is detached from the bridge. The tank is now able to retract the launch mechanism and either cross the bridge or return to a rear area to collect another bridge. The whole launching operation has been carried out without the crew leaving the safety of their armoured vehicle.

During the recovery of the bridge the launching sequence is reversed and the bridge is folded back onto its horizontal position on top of the tank. The bridge is symmetrical and can of course be recovered from both ends.

The latest Vickers bridges can be launched in under 3 minutes and recovered in under 5 minutes. The vehicle requires a crew of 3 -commander, driver and radio operator.

BRIDGES

There are currently three tank bridges of interest to Vickers. Two are in service with the British Army, one with the Nigerian Army and one about to enter service with the Korean Army. They are respectively the No.8, No.9 and VDS 22m bridges.

Table 1. Bridge Specifications

BRIDGE DESIGNATION	No.8	No.9	VDS 22m
MILITARY LOAD CLASS (MLC)	60	60	66
BRIDGE TYPE	SCISSORS	FIXED	SCISSORS
OVERALL LENGTH	24.4m	13.4m	22.0m
EFFECTIVE SPAN	22.9m	11.9m	20.5m
ROADWAY WIDTH	4.01m	4.01m	4.01m
MATERIALS	MARAGING	BS4360	XABO960
BRIDGE WEIGHT (kg)	12700	8840	13100

The main elements of the VDS 22m bridge are shown in Figure 6.

THE K1 PROJECT

VDS' most recent experience in the design and manufacture of a bridging system has been gained working on a project for the Hyundai company of the Republic of Korea. This project involved converting the Korean army's most modern main battle tank, the K1, to the bridgelayer role.

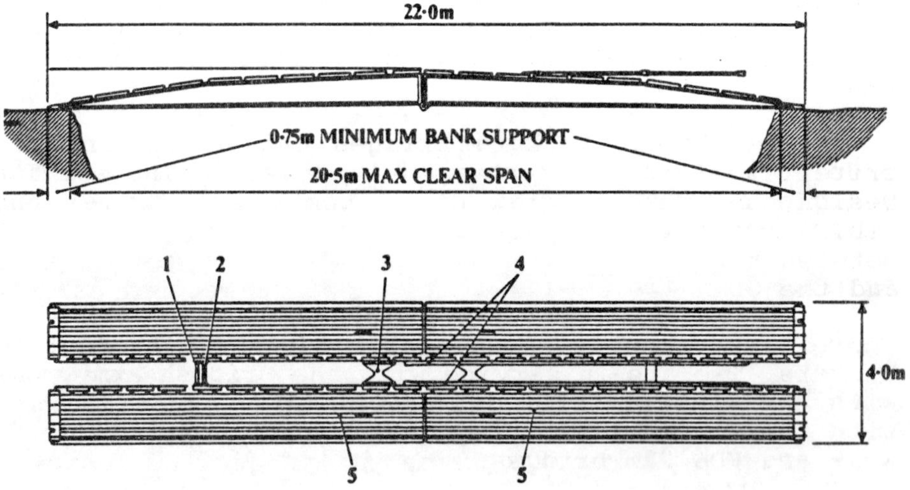

1 Lifting beam
2 Bracing strut
3 Bracing frame
4 Scissoring quadrant and rods
3 Bridge sections

Figure 6. VDS 22m Bridge for K1 Project

This project gave VDS responsibility for the bridge and launching mechanism, the special role of hydraulic and electrical systems and the concept design for the MBT hull conversion.

The K1 bridge system was based on the tried and tested No 8 which has been in service with the British army for many years. However to produce a more commercially competitive product for the K1 project it was necessary to make two very basic changes :

* material
* hydraulic system

The No 8 bridge and launcher were made from Maraging steel. This is a very highly alloyed material which is extremely expensive to buy and to fabricate.

The No 8 hydraulic system used custom built components. The comparatively small production quantities for K1 made this uneconomic. In addition, the system did not take advantage of up-to-date commercial developments.

<u>Structural Design Aims</u>
In the design of the K1 bridge and launcher structures there were two main requirements:

* design for static strength
* design for fatigue life

Design guidelines for the achievement of both criteria are given in the Trilateral Code for the Design and Testing of Military Gap Crossing Equipment (TDTC). This document was the product of collaboration between the military engineers of the UK, West Germany and the USA. The design aims can be quantified as:

1. To survive the application of a proof load of 108 Tonnes with acceptable deflection and permanent set.

2. To survive 8000 crossings of a 60 Tonne K1 vehicle.

3. To survive 1000 launch and recovery cycles

Design for Static Strength

The TDTC recommendations for static design can be very briefly summarised :

For the bridge :

Design Load = $D+A_1+0.8A_2+0.6A_3+0.4A_4$ (1)
where D = the dead load of the bridge
A1-4 = applied loads in order of decreasing severity - vehicle weight, mud, snow and wind.

Loads less than A_4 are considered insignificant and the factors on loads A_1 to A_4 recognise that they are not likely to occur together.

For the launcher :

The rules are not so detailed because of the wide variety of mechanisms in use and the difficulty of producing a comprehensive set of relevant conditions. For K1 the TDTC recommendations were these:

Design load = Deadweight + Windload, taking into account vehicle attitude and wind direction.

To determine the design loads on the launching structures they were considered in 12 positions throughout the launch. The calculations were repeated for three different vehicle attitudes -horizontal, 10 degrees uphill and 10 degrees downhill, all with a 3 degrees transverse slope and wind from a range of angles.

The structural design began with a review of the No 8 analysis; the object being to pick out critical areas and maximum stresses in each of the major structures.

Having identified these maximum stresses, a review of alternative materials was conducted. It was not possible to find a conventional structural steel with equivalent strength to Maraging.

The best alternative was Thyssen XABO 960 with a minimum yield strength of 960 N/mm2, approximately 70% of that of Maraging steel.

Table 2. Comparison of Maraging and XABO960 Steels

		MARAGING	XABO960
0.2% YIELD STRESS	(N/mm)	1313	960
UTS	(N/mm)	1452	1000-1100
NOTCH TOUGHNESS @-40 C (J)		24.4	27
CARBON	(%)	0.030	0.180
SILICON		-	0.500
MANGANESE		-	1.300
PHOSPHORUS		-	0.025
SULPHUR		-	0.020
CHROMIUM		-	0.800
MOLYBDENUM		4.000	-
NICKEL		18.000	2.000
VANADIUM		-	0.100
COBALT		8.000	-
TITANIUM		0.800	0.600

Using the TDTC recommendations the initial analysis showed that with the new material it would not be possible to have the same bridge span as the No 8. The resulting structures would have been too heavy and too large. The decision was taken to reduce the span by 2 metres so that the 60 Tonne K1 vehicle could still be carried. The alternative of reducing the bridge's load capacity was not acceptable.

A revised design for the main bridge girder was produced and a half scale model was fabricated. This was subjected to a simple proof load test and to a fatigue test; the test loads being chosen to reproduce the stresses in the full size component. The test girder was instrumented to record stresses and strains.

The results of the tests were sufficient to confirm the choice of material and to permit the detailed analysis and design to proceed.

Design for Fatigue Life

By limiting the operational life of the bridge and launcher it is possible to work at the low endurance, high stress end of the material's fatigue curve. This allows a highly stressed, low weight structure to be produced. A feature of this type of structure is that it is necessary to make regular checks on the initiation and growth of cracks, just as is done in aircraft structures.

The half scale test girder gave some confidence in the new design and material. However, further action was required to validate the design and reduce technical risk. Advanced Mechanics and Engineering were therefore commissioned to carry out a full fatigue and fracture assessment.

The main objectives of this assessment were:

1. To establish a body of knowledge on the fatigue and fracture mechanics behaviour of XABO 960

2. To assess the structures' abilitity to meet their fatigue life targets

3. To identify any design changes required to meet the fatigue life targets

4. To define acceptable defect levels for production welds

5. To establish the criteria for in-service inspections.

Development of Welding Procedures

The No 8 bridge and launcher was manufactured from Maraging steel. This is supplied in a 'soft' condition and after all welding is completed the whole fabrication is heat treated to achieve a final yield strength of around 1300 N/mm2. The heat treatment is an expensive process and can lead to problems through shrinkage and distortion. The raw material is very expensive because of its high alloy content and limited production. These factors combined to make it necessary to find a way of producing the structures using a less exotic material.

The material chosen to replace it was Thyssen XABO 960. This is a high strength, water quenched and tempered, fine-grained structural steel. In contrast to Maraging steel it has a low alloy content.

Apart from overcoming the difference in strength, the main problem in changing materials was in developing suitable production welding methods.

In XABO 960 the parameter which governs the strength of the weld metal is the time it takes to cool from 800 C to 500 C, referred to as T8/5. The cooling rate is determined by a number of factors:

* The thickness of the plates to be joined
* The geometry of the joint, ie fillet, butt, bead on plate etc
* The form of the weld preparation
* The preheat and interpass temperatures
* The heat input from the welding process

Nomograms are available to assist in estimating T8/5 from a given combination of these parameters. The main question of what T8/5 was required to give the desired weld strength had to be answered experimentally.

The development process started with a series of butt welds made using a tractor welding unit. Each weld was identical except for the heat input. Welding conditions were continuously measured and recorded using a PAMS monitor.

A number of test pieces were made from each of these welds. Standard tensile test pieces were machined from the weld metal alone, transverse tensile test pieces were made containing both parent and weld metal. The former established the strength of the weld metal the latter tested the fusion between the weld and parent material. Hardness surveys of the weld metal and heat affected zones were also carried out as were impact toughness tests.

After conducting a series of such tests it was possible to plot graphs of weld metal strength against T8/5, weld metal strength against hardness and T8/5 against hardness.

Hardness figures were important because few of the production weld configurations could be directly tested for strength. It was therefore necessary to have a means of inferring strength from a more readily measured quantity. It was also much quicker and cheaper to do hardness checks alone.

From these tests it was established that a range of very low T8/5 times would give the required weld strength. The tests also showed that the hard heat affected zones produced by these welding techniques did not result in low toughness.

Having established the target for T8/5 it was possible to estimate the heat input for any weld using the standard nomograms from Euronorm Information Circular No2 [1].

The next stage in the development programme was to attempt to duplicate these results using manual welding techniques since very few of the structures were suited to automated welding. After a few days of concentrated practice the welders were producing welds of consistent heat input. Just as important, they found the welding techniques acceptable and were confident that they could use them in production.

To establish welding procedures for a particular weld involved a combination of calculation, experimentation and testing. First the heat input was estimated from the nomogram, next a set of welding parameters (current, voltage, travel speed) was chosen, then a number of welds were laid down. These were sectioned and examined for fusion, penetration, defect levels and hardness. When a range of suitable welding conditions was confirmed this was logged as the approved procedure for that type of weld.

Qualification procedures for welders on this project were extremely rigorous. They had to demonstate an ability not only to produce low defect level welds but also to lay down welds at very consistent speeds to control heat input.

Controlling the quality of welds during production was all-important. The conditions for each weld were set up and recorded using a PAMS monitor. On the critical welds the entire welding process was monitored and recorded.

This was followed by extensive radiographic and visual examination to check that there were no defects exceeding the acceptance criteria determined through the fatigue and fracture assessment.

Hydraulic System

As previously explained, the bridge is carried on the roof of the vehicle and is launched by hydraulic rams acting in sequence.

Hydraulic fluid is supplied to the rams from a variable delivery hydraulic pump. This is constantly driven by the vehicle engine through a drive shaft and a step down gearbox. Fluid is also supplied for other vehicle systems since there is one fully integrated hydraulic system.

The system operates on a load sensing principle at a maximum pressure of 276 BAR [4000 psi]. A load sensing line feeds the actual load pressure to the hydraulic pump controller. This ensures that the pump provides only sufficient pressure to maintain the required flow. This reduces wasted energy and ensures that heat generation is minimised. In the No 8 system the pump ran at a constant pressure regardless of demand.

In case of a main pump or engine failure the system has an auxilliary pump which is driven by an electric motor. The vehicle batteries supply enough energy to achieve one complete launch or recovery. The auxilliary pump cuts in automatically on detection of low pressure, this ensures that essential vehicle system functions, like braking, can always be carried out.

A number of features are built in to ensure safe operation. Depressurising valves relieve pressure in stage two and three rams during recovery. This prevents the bridge being driven onto the mechanical stops and overstressing the launch structure. All rams are fitted with cushioning valves which operate at both ends of their travel. This slows down the ram speed and reduces shock loads. The rams are also fitted with hose burst check valves so that if pressure is suddenly lost the structure will descend slowly without causing damage.

Dual counterbalance valves are fitted for all three stages. Their function is to prevent the load running away in advance of the speed dictated by the pump flow rate. They also act as relief valves to prevent damage as a result of externally applied loads.

Testing

There are three separate phases of testing in the K1 project.

1. Proving tests carried out on the bridge, launcher and hydraulic system before delivery of the hardware to Korea.

2. Development tests carried out in Korea following the integration of the bridging system and the vehicle.

3. User trials carried out by the Korean army.

   At the time of writing the first phase of testing has been completed and the second is about to begin.

   The first phase consisted of a load test and a launch and recovery test.

Load Test Two sections of the bridge were assembled to form a full span single trackway, ie half of a full bridge. This was supported at either end and loaded at the centre using a hydraulic ram. The load was applied via a spreader structure which simulated the footprint of a main battle tank. The bridge half was then loaded at a controlled rate to the full test load of 54 tonnes. This was derived from the design load plus an overload factor. During this test the bridge was required to satisfy a number of criteria; the deflection during the test must not exceed 270mm, when the load was removed the bridge must not have a permanent set of more than 2mm at the centre and lastly there must be no damage to the bridge sections. To satisfy this last criterion extensive ultrasonic examination of the welds was carried out after the tests.

Launch/Recovery Test Because a K1 hull was not available to VDS a reinforced concrete test rig was built to simulate it. The hydraulic power pack was installed in a converted freight container standing alongside. The hydraulic system was exactly the same as that in the vehicle except that it was powered by an electric motor rather than a diesel engine. The piping to the cylinders and valves was kept as close as possible to that on the vehicle so that pressure drops would be comparable. This rig performed a dual function:

It was used to commission the hydraulic system, which is no simple matter when load-sensing and counter-balance valves are in use. The main problem encountered was instability resulting from the interaction of these two elements of the system. By using the test rig it was possible to try out different valve settings until the problem was eliminated. These settings could then be used when commissioning the system on the vehicle.

It provided a suitable platform from which a bridge could be launched and recovered. This demonstrated that the scissoring mechanism operated, that the geometry of the bridge and launcher was correct and that the hydraulic system could launch and recover the bridge within the specified times. It also tested the strength of the launcher structure and the bridge centre hinge which is most heavily loaded during a launch.

These tests demonstrated that the system performed its basic functions and eliminated any problems due purely to the bridge and launcher.

The second phase of testing is intended to prove the performance of the bridging system in combination with the vehicle. Bridges will be launched and recovered on a number of test sites with widely varying bank conditions. A large number of bridge crossings will be carried out using different types of vehicles. Stresses and deflections in the bridge will be recorded and will be used in comparison with the theoretical strength and fatigue life calculations. These tests will be carried out by Hyundai with assistance from VDS.

The last phase of testing is the 'real thing'. The AVLB will be handed over to the Korean army who will carry out their own trials and assessments. These will cover not only the actual operation of the system but also aspects such as ergonomics and ease of maintainability and repair. These trials will demonstrate how the system reacts to use (and abuse) by the military rather than by teams of engineers. On successful completion of these trials the K1 AVLB will be ready to enter service.

References

Euronorm Information Circular No 2. Weldable fine-grained structural steels. Recommendations for processing, in particular for welding.

Acknowledgements

Hyundai Precision and Ind Co Ltd, Seoul, Korea.

The Tank Museum, Royal Armoured Corps Centre, Bovington Camp.

Fatigue and Fracture Assessment of a Tank Bridge

X.N. Niu(*), G.M. Zintilis(*), M. Smith(**)
(*)Advanced Mechanics and Engineering Ltd, Guildford, GU2 5RE
(**)Vickers Defence Systems, Newcastle, NE99 1BX

ABSTRACT

Fatigue characteristics of XABO steel has been examined and experiments have been conducted to determine crack growth constants and critical CTOD. A fracture mechanics procedure has been developed to assess the fatigue crack growth in the pin-loaded lug components designed for K1-AVLB. It has been demonstrated that using the procedure developed the fatigue damage caused by bridge launches and subsequent tank crossings can be assessed, and therefore the safety of the bridge system can be maintained.

INTRODUCTION

The 'raison d'être' of military bridges is mobility and rapid deployment which enables the rapid crossing of obstacles and gaps by tanks. As a result of these requirements tank bridges and their launch frames have to be both light and strong. A typical main battle tank weighs about 60 tonnes.

Strength and lightness is achieved by the use of high strength or low density materials such as steel and aluminium alloys. The 'No. 8 Bridge' deployed by the British Army in the Gulf War is made of maraging steel with a tensile strength of the order of 1400 MPa. This material is both difficult to weld and expensive. The use of less exotic materials may also enable field repairs of battle-induced damage to be made.

The overall considerations in the choice of material and form as well as aspects of design and manufacture are presented in a separate paper of this conference. The purpose of the present paper is to present the work done by AME Ltd on behalf of VDS for the assessment of the fatigue and fracture behaviour of the bridge and its launch frame.

The choice by VDS of Thyssen's new XABO 960 steel led to the need for fatigue and fracture tests in order to measure the material properties and thus enable the use of fracture mechanics methods. The objective of these methods were to assess and, where necessary, modify the design of joints and to make recommendations for post-fabrication and long term inspection, maintenance and repair.

The tests performed and the fracture mechanics methodology developed specifically for this bridge system, relate to high stress, low cycle fatigue. This is a direct consequence of lightness and strength. Time did not permit large number of fatigue and fracture tests to be conducted. For this reason it way necessary to concentrate the effort on the key and most highly stressed components. Also it was felt necessary to apply statistically derived factors of safety on these limited test results so as to allow for uncertainties. These include fillet-welded plate-to-plate and plate-to-forging joints and the forged central pin joint of the bridge. The latter is regarded as the most important component of the system as its failure results in the total collapse of the bridge. For this reason the present paper is primarily devoted to the pinned joint.

The fracture mechanics tools developed for this bridge system allow for inspection, maintenance and repair criteria to be established under combined launching and crossing stress cycles. In addition, fracture mechanics analyses have led to several detailed design modifications which improve the fracture strength and fatigue life of the bridge/launcher system.

FRACTURE MECHANICS APPROACH

For pin-loaded lugs subjected to cyclic stress, fatigue cracks may develop at the edge of the hole where the presence of the high local stress concentration and pin-fretting action is likely to result in crack initiation. Therefore, it is highly desirable that the fatigue and fracture characteristics of the lugs be considered both in design and in service.

This paper is mainly concerned with fatigue and fracture assessment of the lug components designed for the K1-AVLB bridge system. The major objectives are to evaluate the fatigue characteristics of XABO steel and to provide a practical means of assessing the significance of cumulative fatigue damage in the lugs caused by both the bridge launcher and subsequent tank crossing. These have been achieved by carrying out a linear elastic fracture mechanics study which essentially consists of analyses as follows:

a) Derivation of a stress intensity factor solution
b) Application of a crack growth rule to fatigue life estimations.
c) Application of Miner's law to assessment of the cumulative fatigue damage.

In addition, effort has been devoted to determining the material toughness and fatigue crack growth constants. This was achieved by specifying a number of standard and special fatigue and fracture tests. The details of the analyses and the conclusions drawn from the study are now presented.

The problem of cracked pin-loaded lugs has been fairly extensively studied. Several analytical techniques have been proposed and used to determine the desired stress intensity factor, see, for example, [1]. However, the calculated and published stress intensity factor data are invariably concerned with specific lug dimensions, which differ from those of the lug under consideration. Moreover, some techniques used in these analyses seem to be superseded by recent advances in fracture mechanics, which enable the stress intensity factor to be calculated more accurately and efficiently. The approach developed and used in this study is mainly based on the methodology described in [1], but the formulation used in the calculation is significantly different from that given in [1]. The following presents details of the analysis.

The dimensions of the pin-loaded lug under consideration are shown in Figure 1. However, the cracked geometry of the pin-loaded lug is idealised as a finite-width plate containing a central circular hole with cracks emanating symmetrically from the edge of the hole, as illustrated in Figure 2. Figure 3 shows that a pin-loaded system can be resolved into two relatively simple loading configurations. However, even for those two simple configurations, accurate theoretical solutions have not yet been found, thus approximate solutions have been considered here.

For Configuration I (see Figure 3), a weight function method has been used to calculate the desired stress intensity factor $K_I^{(1)}$. In this method, the stress intensity factor due to mode I loading is derived by evaluating the integral:

$$K_I = \int_0^a \sigma(x) \cdot m(x,a)dx \qquad (1)$$

where a is the crack size, m(x,a) is known as the weight function, and $\sigma(x)$ is the stress distribution normal to crack plane and calculated for the potential crack line in the uncracked geometry subject to the load acting on the cracked geometry. The weight function used in this study is developed [2] for symmetrical edge cracks emanating from a circular hole in a finite-width plate, and the details of the weight function are given as follows:

$$m(x,a) = \sqrt{\frac{\pi}{w}}\, F_0\left(\frac{b}{w};\ \frac{x}{b}\right) f\left(\frac{a}{r}\right) \tag{2}$$

where

$$F_0 = \frac{\left\{1+0.297\sqrt{1-\left(\frac{x}{b}\right)^2}\left[1-\cos\left(\frac{\pi b}{2w}\right)\right]\right\}\sqrt{\tan\left(\frac{\pi b}{2w}\right)}}{\sqrt{1-\left[\cos\left(\frac{\pi b}{2w}\right)/\cos\left(\frac{\pi x}{2w}\right)\right]^2}} \tag{3}$$

$$f = 1 + 0.234\, \exp\left(-2.65\frac{a}{r}\right) \tag{4}$$

$$b = r + a \tag{5}$$

where r is the notch radius and w is the lug width. The required stress distribution $\sigma(x)$ is taken from [3] in the form as follows:

$$\sigma(x) = \sigma_0\left[A_1 + A_2\left(\frac{r}{x}\right)^2 + A_3\left(\frac{r}{x}\right)^4\right] \tag{6}$$

where σ_0 is the nominal stress at the gross section of the lug, and the coefficients A_1, A_2 and A_3 are functions of the ratio r/w and their values are given in [3]

For Configuration II (see Figure 3), the stress intensity factor is approximated by:

$$K_I^{(2)} = C_w K' \tag{7}$$

where K' is the stress intensity factor for edge cracks emanating from a pressurized circular hole in an infinite body, and C_w is a generally-accepted finite width correction factor:

$$C_w = \sqrt{\sec\left(\frac{\pi b}{2w}\right)} \tag{8}$$

The solution for K' is taken from [4] and results of K' are reproduced in Figure 4. A sufficient number of data points are taken from the curve that represents K', corresponding to $\lambda = 0$. A cubic spline technique is used to interpolate K' values for varying crack length.

With the solutions for both Configurations I and II, the desired stress intensity factor K_I, are calculated using the superposition algorithm illustrated in Figure 3. The results of the K_I calculation are presented in Figure 5, in terms of what is known as the geometry correction factor Y, which is defined as:

$$Y = \frac{K_I}{\sigma_0\sqrt{\pi b}} \tag{9}$$

It should be noted that the stress distribution $\sigma(x)$ is a function of the ratio of hole radius r to half plate width (half lug width in this study) w. The values of the coefficients A_1, A_2 and A_3 used in the calculation of K_I correspond to r/w equal to 0.4, which differs from that of the lug under consideration, 0.43. However, it is believed that the inaccuracy in the stress distribution caused by the difference in these parameters would not lead to significant error in the calculation of the stress intensity factor.

MATERIAL TESTS

Fatigue Crack Growth

To investigate the characteristics of the fatigue crack growth in XABO steel, fatigue tests have been carried out on both standard compact tensile specimens (CTS) and notched specimens. The main objective of these tests is to determine the material constants C and m of the Paris equation (see below).

Compact Tensile Specimens

Two 10.15mm thick CTS were fabricated and tested under cyclic loadings with a stress ratio of 0.1. One specimen was tested under constant amplitude load and the other under varying amplitude load. In the constant load test, the applied load was chosen to give the stress intensity factor range of approximately 10 $MPa\sqrt{m}$; whereas in the varying load test, the applied load was reduced in steps to be in accordance with, as far as possible, the following relationship:

$$\Delta K = \Delta K_0 10^{c(a-a_0)} \tag{10}$$

where c is a constant equal to -0.06, a_0 is the fracture length, and ΔK_0 is chosen so that a threshold is reached within a suitable range of crack lengths. Additionally, a total of four 12.5mm thick CTS were fabricated using XABO forgings, and also tested under cyclic loadings with a stress ratio of 0.1. Three of the four specimens were tested under constant amplitude load and one under varying amplitude load. The applied loads were chosen in much the same way as those for the 10.5mm thick specimens.

The purpose of conducting the varying load test was to assess the threshold stress intensity factor range and measure the slow crack growth. All the specimens were pre-cracked to produce appropriate initial cracks.

Notched Forging Specimens (NFS)

The geometry of the three specimens tested is shown in Figure 6. The specimens were made of XABO forgings. The fatigue tests and subsequent analysis of test results were in general conducted following the recommendations of "Draft British Standard method for the measurement of threshold stress intensity factors and fatigue crack growth rates in metallic materials". It must be pointed out, however, that the specimen geometry and grips were non-standard for the measurement of crack growth rate.

All three specimens were tested under cyclic tensile load with an approximately zero nominal stress ratio. The applied nominal stress range at the gross section was 147MPa for the first two tests, and was reduced to 66.3MPa from the initial value of 120.8MPa for the third test. Starter cracks were prepared for the specimens at the root of the notches using a spark-eroding technique.The development of fatigue cracks were monitored during the fatigue tests using a potential drop technique. The purpose of testing specimen 3 under smaller and varying cyclic stress ranges was to measure the slower crack growth rate.

Test Data Analysis

To calibrate the desired crack growth constants C and m of the Paris equation (see below), the crack growth rate is plotted in Figure 7 against the stress intensity factor range on a log-log scale. The crack length used to calculate ΔK and crack growth rate is the mean in the interval concerned and the geometry correction factor required was calculated using eqs.(1-6) for the notched forging specimens, and the formulation used to determine the stress intensity factor range for the compact tensile specimens is taken from ASTM 647 "Standard test method for measurement of fatigue crack growth rates".

Figure 7 indicates a fairly evident linear correlation, as anticipated, between the measured crack growth rate and stress intensity factor range. No distinct and consistent indications, however, can be observed from Figure 7 to suggest that the fatigue crack growth is affected by the difference in the material yield stresses of the XABO plates and forgings. Figure 7 also shows that the threshold stress intensity factor range, denoted by $(\Delta K)_{th}$, is about 7.3MPa$\sqrt{m}$ for the XABO steel.

Certain degree of scatter can be observed from Figure 7, but seems to be acceptable as far as fatigue test data are concerned. The crack growth data from individual test series appear to fall into the same scatter band, suggesting that it may not be necessary to derive C and m values for XABO plates and forgings separately. Moreover, as the crack growth data have been obtained from limited number of fatigue tests, it would not be appropriate to calibrate the constants C and m following a common procedure of fitting a 'best curve' to the data. Instead, the following approach has been adopted in analysing the test data.

For most steels it is often reasonable to assume that the exponent of the Paris law m is 3. The three parallel solid lines shown in Figure 8 are of a slope of 3, and represent the mean, upper and lower bounds respectively. It should be noted that these lines were not drawn based on any statistic analysis. In fact, some engineering judgement was made in deriving the upper-and lower-bounds, while the mean was derived by taking the average of the upper- and lower-bounds on a logarithmic scale. The following lists the values of the parameters C and m, together with the approximate threshold stress intensity factor range:

$$(\Delta K)_{th} = 7.3 \text{ MPa}\sqrt{m}$$

$$m = 3, \quad C = \begin{cases} 1.0\times10^{-8} & \text{Upper Bound} \\ 5.12\times10^{-9} & \text{Mean} \\ 2.73\times10^{-9} & \text{Lower Bound} \end{cases}$$

(units: MPa$\sqrt{m}$, mm/cycle)

Material Toughness and Maximum Allowable Crack

To investigate the material fracture toughness, CTOD tests have been conducted using specimens made of XABO forgings. A total of seven bend test-pieces were tested under bending load following the procedure specified in BS 5762: 1979. The average value of the critical CTOD has been found as follows:

$$\delta_c = 0.415 \text{ mm}$$

The formulation given in PD 6493: 1980 is used to determine allowable equivalent crack size for the applied stress of 533MPa at the notch root:

$$a_m = 37.8 \text{ mm}$$

It should be noted that a_m represents the maximum allowable crack size corresponding to *half length* of an equivalent through-thickness crack in an infinite plate loaded in tension. For cracks other than such a through-thickness crack, the actual maximum allowable crack dimensions (depth or length) need to be determined in accordance with the relation [5] given below:

$$a = \frac{a_m}{Y^2} \tag{11}$$

where a is crack length and Y is the geometry correction factor.

The dimensions of the pin-loaded lug components and notched forging specimens have been shown in Figures 1 and 6, respectively. The required solutions for the Y correction factors are taken from the previous Section above. Using eq.(11), the corresponding allowable crack lengths have been found to be 4.2mm for the pin-loaded lugs, and 7.3mm for the notched forging specimens tested.

FRACTURE MECHANICS PREDICTIONS

Fatigue Crack Growth Law

The fatigue crack growth has been assumed to follow the Paris' law:

$$\frac{da}{dN} = C(\Delta K)^m \tag{12}$$

where da/dN is the crack growth rate, N is the number of load cycles, C and m are experimentally determined material constants. ΔK is the range of stress intensity factor, defined as:

$$\Delta K = \Delta\sigma_0 \, Y(a) \sqrt{\pi a} \tag{13}$$

where $\Delta\sigma_0$ is the nominal stress range. It is easy to show that the remaining life N of a component containing an initial crack of size a_i, can be determined by evaluating the following integral:

$$N = \int_{a_i}^{a_c} \frac{da}{C\left[\Delta\sigma_0 Y(a)\sqrt{\pi a}\right]^m} \tag{14}$$

where the upper limit a_c denotes the critical crack length. The Gaussian numerical integration procedure has been used to evaluate eq.(14).

Notched Forging Specimens

To verify the fracture mechanics modelling procedure, fatigue propagation lives of the notched forging specimens have been calculated using the above formulations. Both the mean and upper bound of C values have been used in the calculations, and the following lists the values used for the variables relevant to the analysis:

$C (\times 10^{-8})$	m	$\Delta\sigma_0$ (MPa)	a_i (mm)	a_c (mm)
1.0	3	147	0.5	7.3
0.512	3	147	0.5	7.3

The calculated crack growth data are compared with those obtained from the relevant tests, as shown in Figure 9. It can be seen that the crack growth in the notched forging has been reasonably predicted when the mean value of the constant C is used. On the other hand, use of upper bound C results in over-estimation of the crack growth rate, thus the fatigue life is, as expected, considerably under estimated.

Pin-Loaded Lugs

The prediction of the remaining life for the lug components is also based on eq.(14). The required stress intensity factors are calculated following the procedure described above. The following values have been used for the variables relevant to the analysis:

$C (\times 10^{-8})$	m	$\Delta\sigma_0$ (MPa)	a_i (mm)	a_c (mm)
1.0 0.512	3 3	143 143	0.5 0.5	4.2 4.2

The remaining fatigue life of the pin-loaded lugs are plotted in Figure 10 with respect to the crack length. The crack length of Figure 10 can also be seen as the maximum tolerable initial crack length for a specific number of bridge launches or tank crossings intended. Note that a complete bridge launch is considered to consist of both deployment and retrieval operations, hence one launch is seen as two equal-amplitude load cycles.

As can be seen from Figure 10, the remaining lives decrease sharply with the increase of fatigue crack length, especially in the case of small cracks. This illustrates how important it is to ensure that no detected crack is greater than that predicted in order that the safety of the bridge and launcher is maintained.

Cumulative Fatigue Damage by both Launches and Crossings

A methodology has been developed of assessing the cumulative fatigue damage due to both bridge launches and tank crossings. Using the principle of Miner's cumulative damage law, it is easy to demonstrate that the following inequality can be used to assess whether the bridge system will survive the numbers of launches and crossings to be carried out:

$$\frac{n_L}{N_L} + \frac{n_C}{N_C} \leq D_S \tag{15}$$

where

N_L is the maximum allowable number of times for which only the bridge deployment and retrieval would be allowed

N_C is the maximum allowable number of times for which only the tank crossings would be allowed

n_L is the number of bridge launches, each launch including two stress cycles

n_C is the number of tank crossings associated with n_L times of bridge launches

D_S is allowable fatigue damage with a recommended value of 0.8.

If no fatigue cracks are detected before operating the bridge system, N_L and N_C are the fatigue endurance of the component to the launches and crossings respectively; n_L and n_C are equal to, respectively, the numbers of the launches and crossings which have been and will be performed. In the case of a fatigue crack of size a being detected before the operations, N_L and N_C must be calculated as the component residual lives due to the launches and crossings respectively; n_L and n_C are the intended numbers of the launches and crossings respectively. The value of D_S varies but must not exceed unity.

For structural components containing cracks, N_L and N_C can be determined from fracture mechanics results like those presented in Figure 10. N_L and N_C can be conservatively approximated by fatigue propagation lives of the component containing a small initial crack, e.g. 0.5mm deep. It is suggested that using eq.(15) and results in Figure 10, cumulative fatigue damage caused by bridge launches and tank crossings can be assessed for the components investigated in this project.

It must be noted that the results in Figure 10 have been derived for the components subjected to a specific stress range, therefore can not be directly used for estimations of N_L and N_C associated with other stress ranges. However, it can be shown that for a given initial crack, the remaining life related to a new stress range $\Delta\sigma'$, denoted as N', can be determined from the following equation:

$$N' = \left(\frac{\Delta\sigma}{\Delta\sigma'}\right)^m N \tag{16}$$

where N is the known remaining life in relation to the stress range $\Delta\sigma$ and m is the exponent of the Paris equation and equal to 3 in this study. N and $\Delta\sigma$ can be obtained from Figure 11.

SUMMARY AND CONCLUSIONS

A series of tests have been conducted to examine fatigue and fracture behaviour of XABO steel. A procedure has been developed to calculate the stress intensity factors for the pin-loaded lug components used in the K1-AVLB bridge system. Using Paris' equation and the material constants determined from specially commissioned tests, critical crack lengths and fatigue propagation lives have been estimated. Finally, using Miner's rule, a simple relation has been proposed and incorporated into the framework of fracture mechanics to account for the cumulative fatigue damage caused by the interaction of bridge launches and tank crossing. Some conclusions have been drawn from this study:

1. Results of the fatigue crack growth tests show that the rate of fatigue crack growth in XABO steel is not distinctly affected by the difference in the yield stress between plate and forging.

2. The critical value of CTOD indicates the XABO steel is of high fracture toughness.

3. The analytical technique developed provides a simple and effective means of calculating the stress intensity factor for a pin-loaded lug configuration.

4. Using the mean values of C and m and stress intensity factor solutions derived, reasonable fatigue life predictions can be obtained for pin-loaded lugs.

5. Because a small increase in crack size can cause a sharp decrease in remaining fatigue life, it is important that NDT crack inspection be carried out regularly during service and no detected crack should be greater than that predicted in order to maintain the safety of the bridge and launcher. In-service inspection is, of course, standard practice for military bridges.

REFERENCES

1. Parker, A.P., "The Mechanics of Fracture and Fatigue: An Introduction", published by E & F.N. Spon, pp.159 (1981).

2. Xiao, S. T. et al., "Stress Intensity Factors in Notched Finite Plates Subjected to Biaxial Loading", Fatigue Fracture Engng. Mater. Struct. Vol. 8, No. 4, pp. 349-372, (1985).

3. Schijve, J., "Stress Gradients around Notches", Report LR-297, Dept. of Aerospace Engineering, Delft University of Technology, (1980).

4. Rooke, D. P. and Cartwright, D. J., "Compendium of Stress Intensity Factors", London, HMSO, (1976).

5. "Fatigue Handbook - Offshore Steel Structures", Edited by A. Almar-Naess, pp.388 (1985).

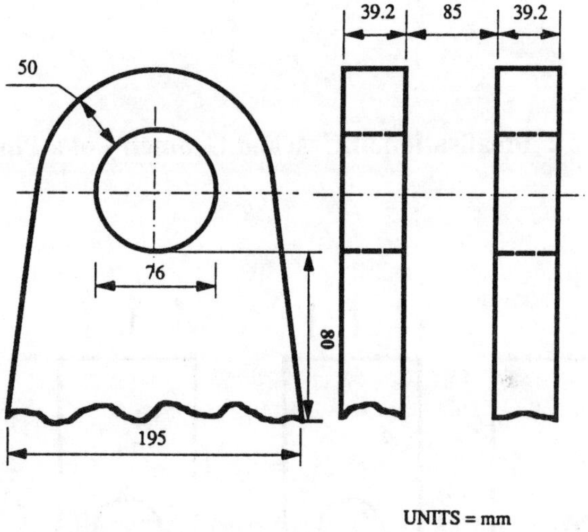

Figure 1 Dimensions of Pinloaded Lug Components

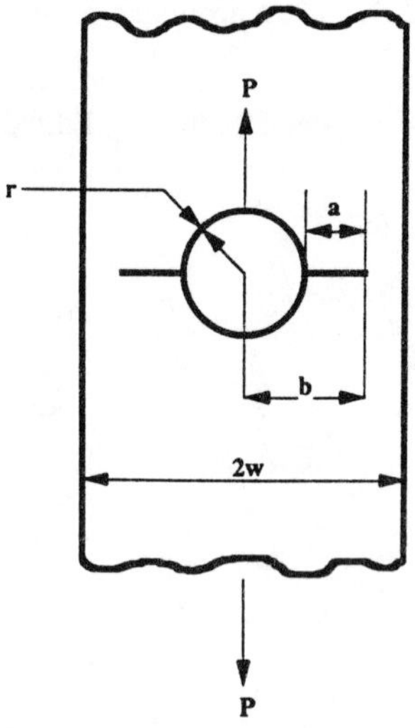

Figure 2 Idealisation of Cracked Geometry of a Pin - Loaded Lug

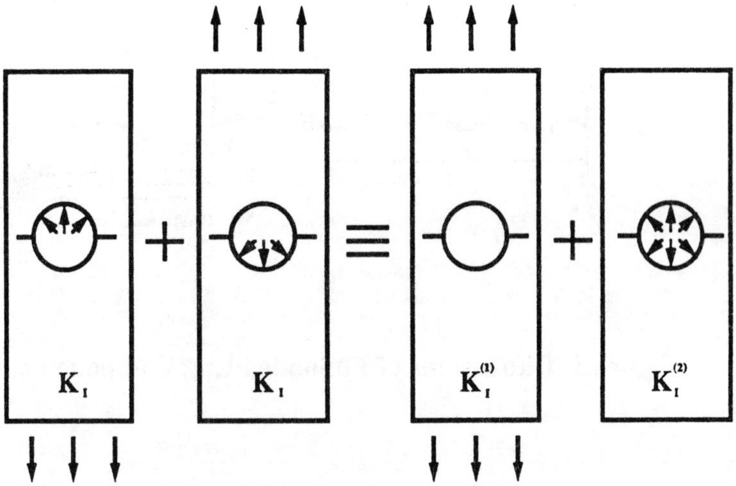

$$K_I = K_I = \frac{1}{2}\left(K_I^{(1)} + K_I^{(2)}\right)$$

Figure 3 Superposition for a pin-loaded hole with radial cracks.

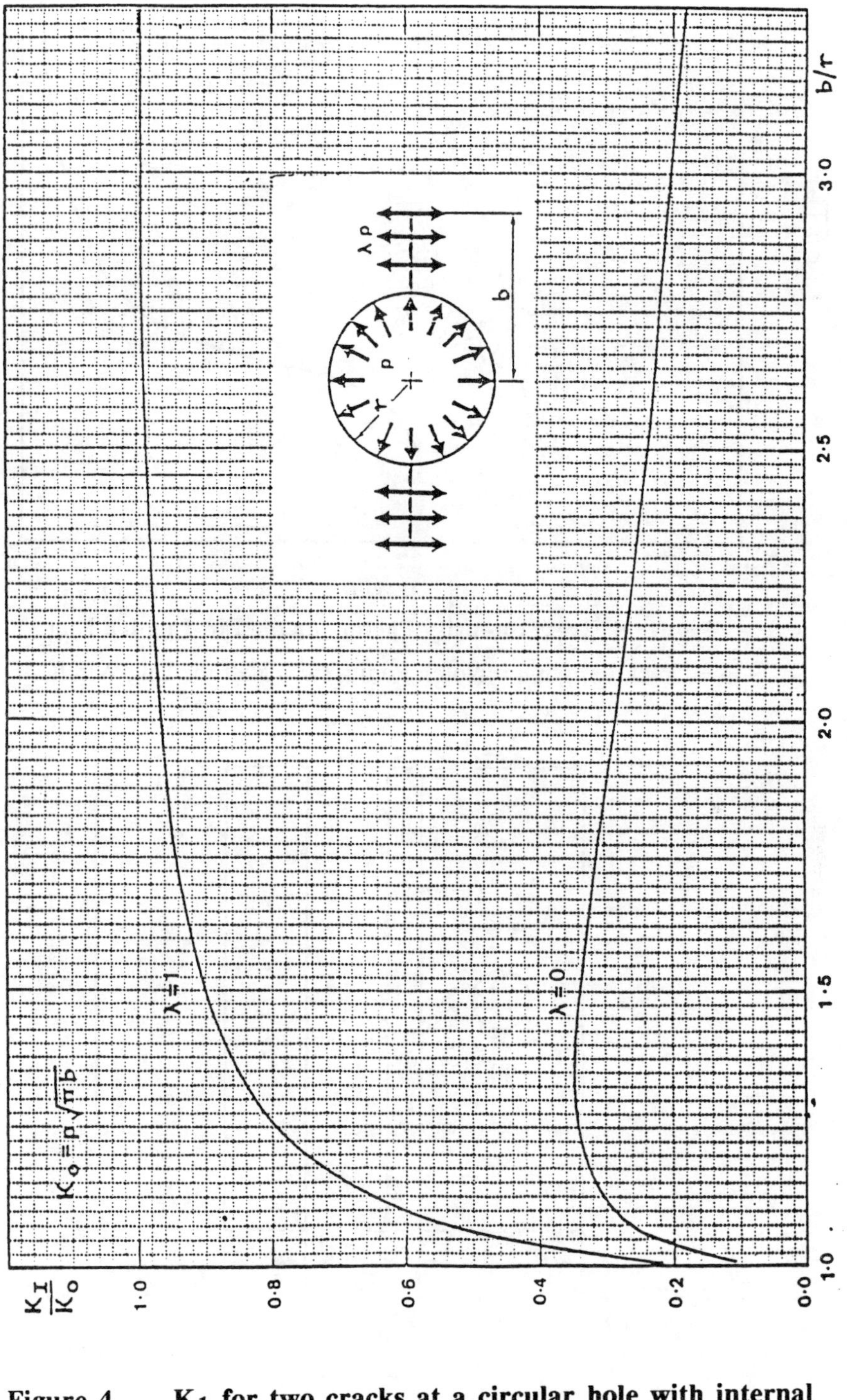

Figure 4 **K_1 for two cracks at a circular hole with internal pressure on the hole and crack**

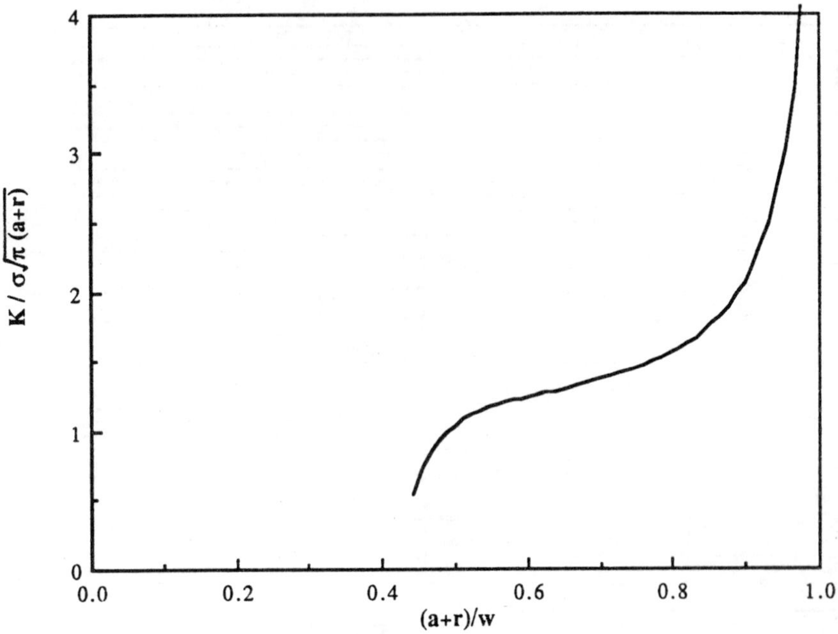

Figure 5 Stress Intensity Factor for the Pin-Loaded Lug

a = Crack Length, r = Notch Radius, w = Half the Lug Width

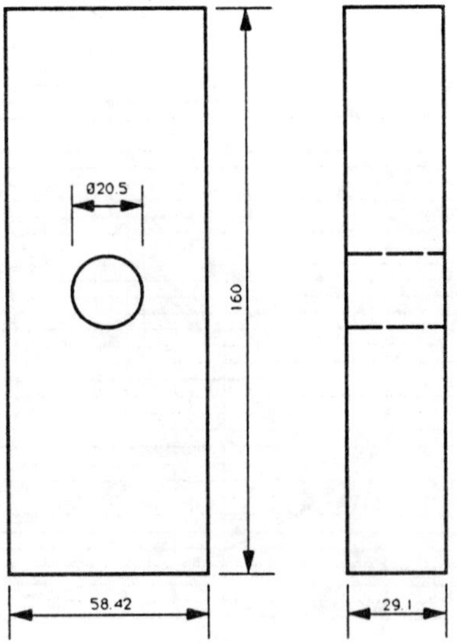

Units = mm

Figure 6 Dimensions of Notched Forging Specimens

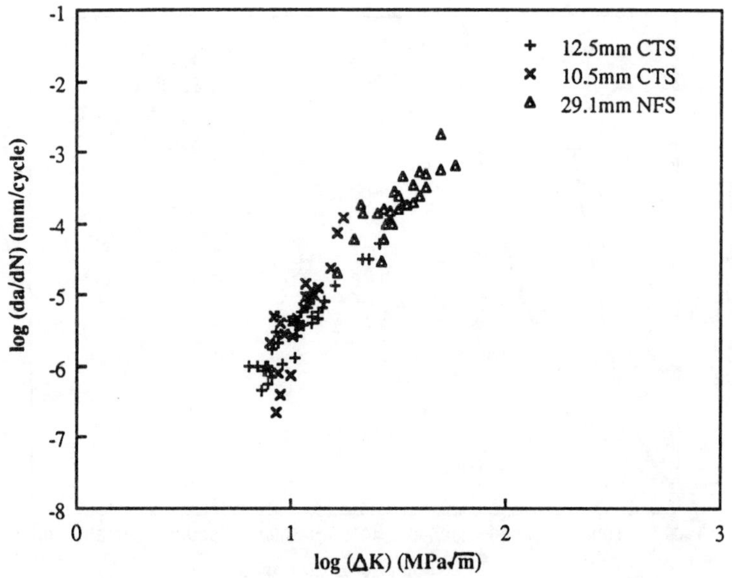

Figure 7 Summary of Crack Growth Rate for XABO Steel

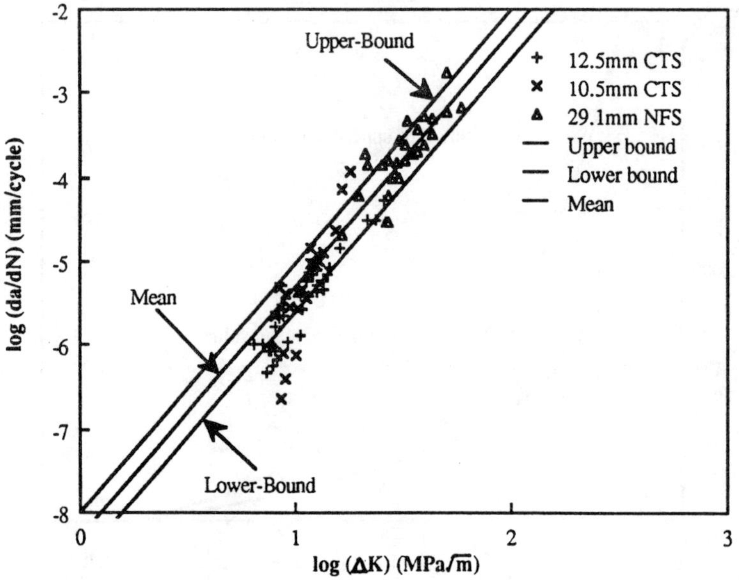

Figure 8 Characterisation of Crack Growth Rate for XABO Steel

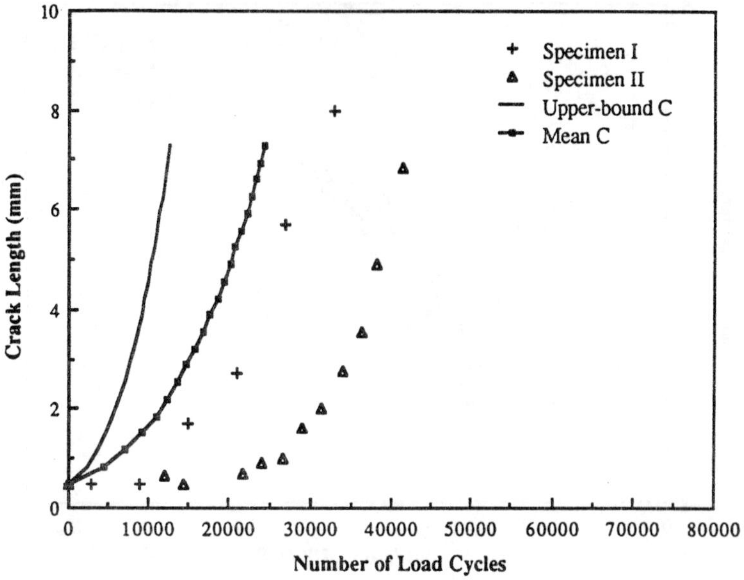

Figure 9 Predicted and Measured Crack Growth in the Notched Specimens

Nominal Stress Range = 147MPa, Stress Concentration Factor = 3.55

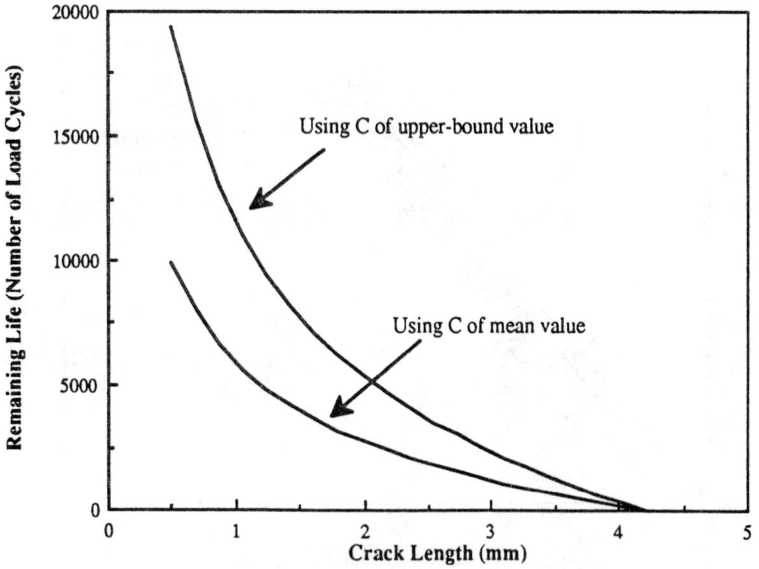

Figure 10 Relation between Crack Growth and Remaining Life of the Pin-Loaded Lug

Initial Crack Length = 0.5mm, Nominal Stress Range = 143MPa

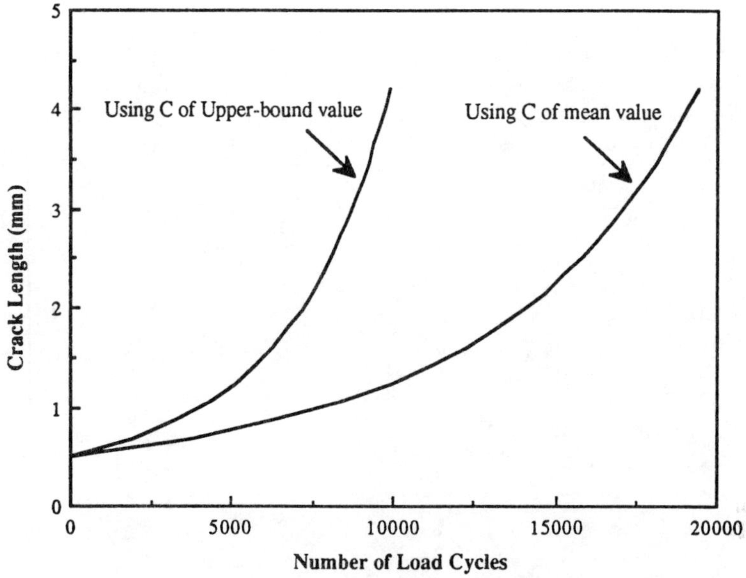

Figure 11 Predicted Crack Growth in the Pin-Loaded Lug

Initial Crack Length = 0.5mm, Nominal Stress Range = 143MPa

New Swedish Bridge System Development

O. Jönsson(*), J. Nilsson(**)

(*)The Swedish Defence Materiel Administration (FMV)

(**) Karlskronavarvet AB

SUMMARY

In the spring of 1991, a complete prototype set of a new 48 m single span bridge will be delivered from Karlskronavarvet to the Swedish Defence Materiel Administration. This is part of the R&D program carried out in close cooperation between the parties with the aim of developing a new cost effective family of bridges for the Swedish Army.

The solution includes the use of

- thin plates and rolled plate profiles of extra high-tensile steel
- large 3D bridge sections with large sectional height and fixed width
- mechanised handling
- standard trucks without cranes

BACKGROUND

River crossing operations are an intergral part of land warfare. Sweden is divided into 118 main river basins. The ten largest rivers in Sweden are situated in the north and cuts the country by running from the mountain ridge border of Norway in the west, in eastern direction to the Bothnian Sea. In the event of military conflict, the communications over the rivers in northern Sweden will be vital for the build up of combat power and equipment from central Sweden to the north of the country. Specialized Line-of-Communication battalions are to be organized in the 1990s. The main task for the battalions is to keep the permanent bridges open using camouflage and repairment measures, and if neccesary, by building heavy wet and dry support bridges.

The heavy wet support bridge component is provided by Ribbon Bridge companies already organized. Bridge equipment suitable for rapid deployment wide gap multi span dry support bridging is presently not operative anywhere in the world. In the present organization for the armoured brigade, the gap crossing equipment for deliberate dry support bridging is the Bailey Bridge, which was introduced in the 1940s. This is a serious insufficency in mobility of the brigades. Fast Bro 2000 is also - among other new bridge systems - an alternative for new dry support bridging for armoured brigades. Bridging systems currently used by the Swedish Army include:

- Dry support bridges	Bailey Bridge	SE Bridge	Krigsbro 4
- Wet gap support bridges	Uniflote	Pontoon 3	Ribbon Bridge

Only the Ribbon Bridge system fulfills existing operational requirements.

REQUIREMENTS

The Fast Bro 2000 program aims at the development of a multispan bridge system on telescopic supports. The bridge is, in the first place, intended for river crossing operations in the northern part of Sweden on bridge sites featuring:

- high water current velocity
- low water level and/or stone blocks
- difficult ice-conditions
- high and/or steep entry - and exit banks

ie, making the building and operation of floating bridges impossible. It should also be possible to construct Fast Bro 2000 as a single span bridge, eg, for overbridging a damaged civilian permanent bridge. Construction time on a prepared site:

- Multi span 100 m in 2 h
- Single span approx 50 m in 1 h

The bridge should allow traffic with all normal civilian and military wheeled and tracked vehicles, load class 60 tonnes. In the development of the single span bridge and multi span bridge, consideration must be taken to utilizing main bridge components common for both systems. Short bridge construction time and a low life cycle cost for the bridge system are of utmost importance.

SCHEDULE

The program started in 1985. Five companies submitted proposals for concept studies. Project definition studies, phase one and two, were undertaken independently by two different companies. In 1986 an UTTEM (Army Staff Requirement) was issued for Fast Bro 2000.

In 1988 Karlskronavarvet was choosen as the prime contractor for the design and development of a prototype of the next generation of fixed bridges for the Swedish Army. Basic tests with the complete 48 m single span bridge are to begin in May 1991. User trials with single span bridge is scheduled 1992/93. A series of single span bridges is likely to be ordered in 1994. Bridge equipment will be entering service in the armoured and mechanized brigades from 1995.

Basic tests with the additional components of the multi span bridge, e g, launching and support equipment, are to begin in May 1995. A series of multi span bridges is likely to be ordered in 1998.

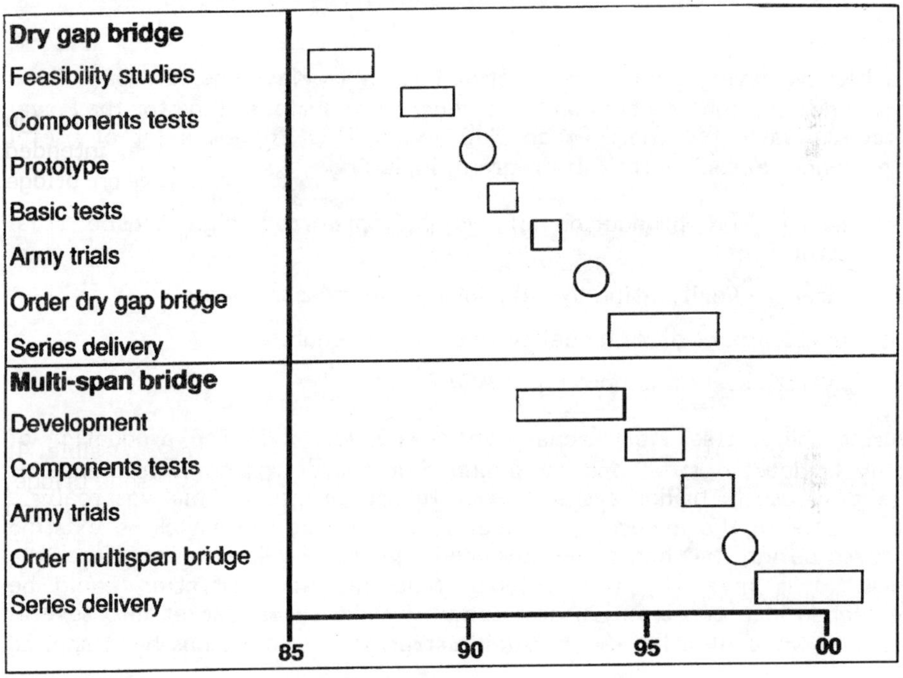

Figure 1. Schedule for the Fast Bro 2000 program

DEVELOPMENT

The time taken to assemble different bridges for a given gap is heavily dependent on the number of bridge components to handle.

The earlier requirement of manual assembly of man-carried smaller elements has been dropped. This facilitates the use of

- large 3D bridge sections with large sectional height and fixed width
- mechanised handling
- small manpower (truck drivers only)
- rational production methods and lower price per running metre

Karlskronavarvet, today a subsidary company to Kockums, has specialized in the design, construction and maintenance of surface ships for the Royal Swedish Navv for more than 300 years. This implies a lot of useful experience that supports this program, including

- design and production of weight optimized high tensile steel structures
- strength verification by calculation and measurements
- development of high quality welding procedures
- QA/QC system according to AQAP 1

During the sixties, Karlskronavarvet was involved in the production of army bridges in steel and aluminium but there was no expericence of design or use of bridge systems prior to this program. This was really a key factor in the necessary innovative process as there was no existing inhouse design that had to be defended against new ideas. Therefore, the basic ideas given by the Defence Materiel Administration could be developed in a constructive atmosphere by Karlskronavarvet into several variants for evaluation by the Administration. A patent has been applied for.

SINGLE SPAN BRIDGE

General

The single span bridge is described below, including overall dimensions (length x width x height) in metres and weight in tonnes for the components.

Bridge section

The basic component is the bridge section, 8.0 x 4.0 x 1.5 m, weighing 5.5 t. It is built up by two upper runway plates, with longitudinal U-shaped stiffeners, and two lower plates with similar stiffeners, all separated by a lattice structure of square-shaped hollow profiles to form a large rigid three-dimensional structure with high section modulus. Most of the plates and profiles are of 5 mm extra high tensile steel. The section is equipped with male/female couplings of two different types. The upper coupling consists of a manually operated horizontal pull/push rod by which a total of 12 attached plungers can be moved into/out of the holes of a series of eye brackets. The lower coupling consists of a total of 8 vertical plungers connected to hydraulic pistons arranged in a hole puncher-like configuration. The pistons are controlled by temporary connection to the hydraulic system of the inner launching unit.

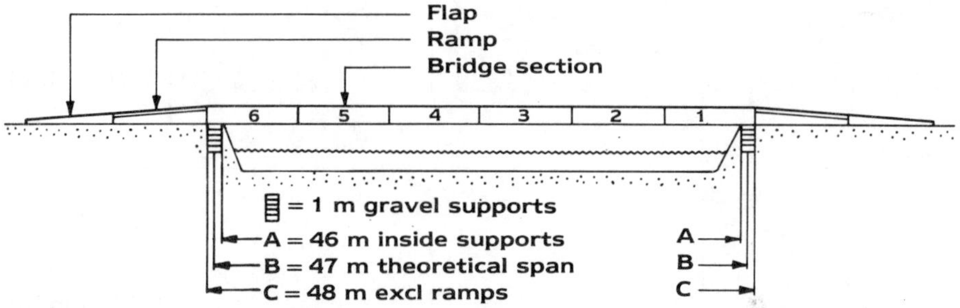

Figure 2. Completed single span bridge

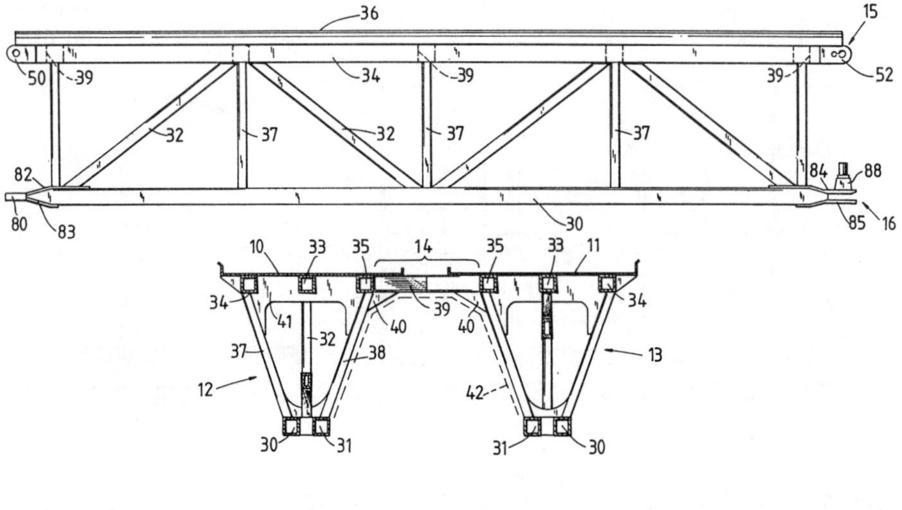

Figure 3. Bridge section. Profile and section

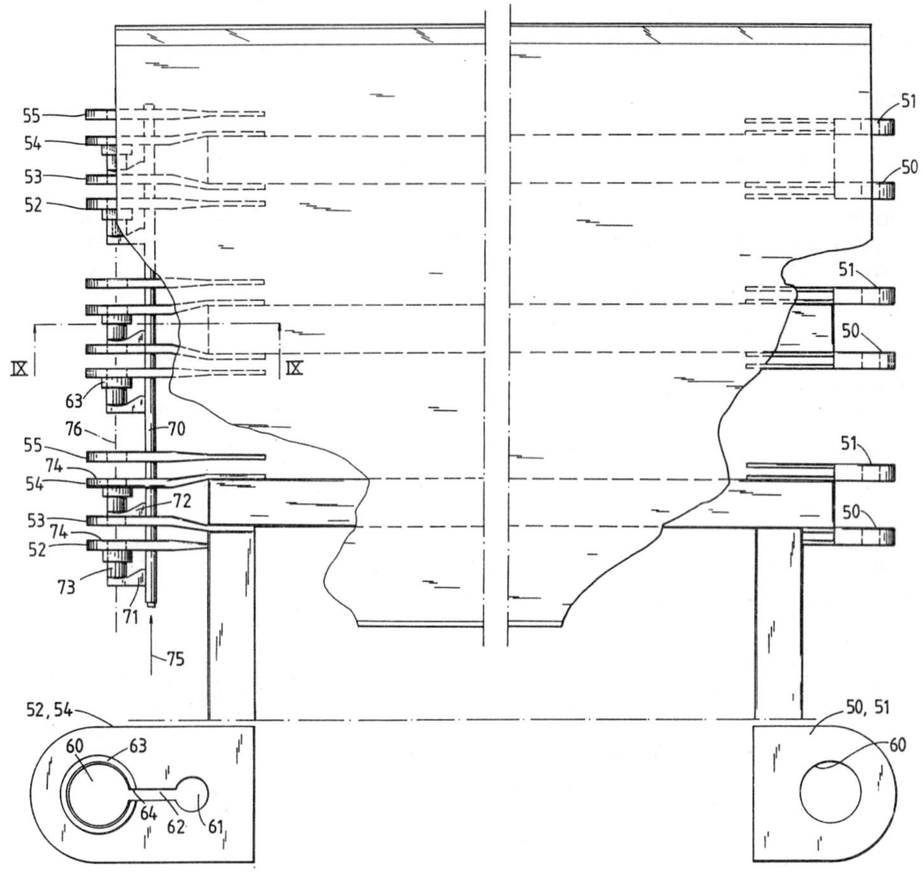

Figure 4. Bridge section. Detail of upper coupling

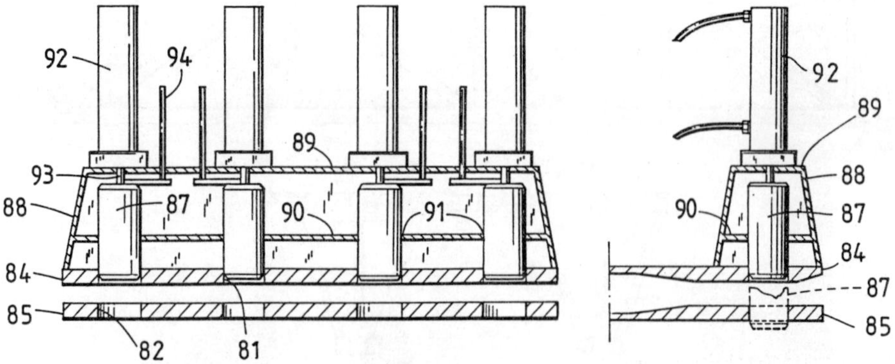

Figure 5. Bridge section. Detail of lower coupling

Ramp with flaps

The bridge set includes two ramps, each 8.0 x 4.0 m weighing 3.0 t.The ramps consists of two upper runway plates with U-shaped longitudinal stiffeners hinged to the upper male/female bridge section couplings. The ramp slope can be adjusted by 4 hydraulic piston supports but the nominal slope is 1:11 to suit civilian buses etc. The far and near ramps are connected by hydraulic hoses to the end support and the inner launching unit respectively. At the lower end, each ramp is also equipped with two hinged runway flaps, each 7.0 x 1.8 m weighing 1.3 t, with a nominal slope of 1:11.

Launching beam segment

The bridge set includes eight launching beam segments, each 8.3 x 0.8 x 0.4 m, weighing 1.7 t. The segments consist of a rectangular box of extra high tensile steel with male/female couplings, each comprising 4 vertical plungers arranged in an appropriate configuration and connected to a common crankshaft like mechanism at one end.

Figure 6. Launching beam segment

Inner and outer launching units

The launching beam is launched by the inner and outer launching units.The overall size of the inner launching unit is 8.0 x 3.4 x 3.5 m with crane folded and the weight 12 t. It consists of a tank-shape structure in which a 36 kW diesel driven hydraulic package is arranged to manoeuvre four hydraulic piston legs and a pair of push/pull hydraulic pistons for the moving of the launching beam segments. An identical second diesel is used to power a 45 ton hydraulic crane with which the total of 21 t counterweights, outer launching unit, launching beam segments, end support, section feeder and bridge segments are handled.

The outer launching unit, weighing 2.0 t, is a support that is adjustable in height and side angle by hydrualic pistons. It is attached to the inner launching unit by an 8.5 m spacer beam and hydraulic hoses.

End support

The far end of the launching beam is supported by an end support that is adjustable in height and side angle by hydraulic pistons fed from an autonomous 7 kW petrol engine driven hydraulic package. The hydraulic package can also be connected to the far ramp for its lowering and adjustment.

Section feeder

For the transportation of bridge sections, a section feeder is attached on the near end of the completed launching beam. The feeder consists of two sequencially acting hydrualic pistons, each with a flip-over mechanism by which the bridge section is fed out or in while riding on the launching beam. The section feeder is connected to the inner launching unit by hydraulic hoses.

Trucks

In order to keep the price down and flexibility up, all 8 trucks involved are of standard 3-axle type with maximum 13 t payload to give the maximum single front/tandem rear axle load of 10/16 t which is the normal limit on Swedish roads. This also gives the potential to use any other similar truck commonly available.

Construction and launching sequence

The complete construction and launching sequence is described below. Based on tests, see below, the total time has been estimated to maximum 60 minutes when carried out by the truck crew (totally 16 men). This will be verified during field tests in the autumn 1991. Recovery of the bridge including launching beam can be done from either side.

Truck no	Time min	Operation
I	-1	Dieseldriven powerpacks started
	0	Inner launching unit positioned
II	2	3 t counter weights positioned
	3	Outer launching unit positioned temporary
	4	Spacer beam connected
	4	Launching beam segments A-B connected & pushed
	8	End support connected
III	10	Final positioning of outer launching unit 13 t counterweights positioned

Truck no	Time min	Operation
IV	14	4 t counterweights positioned
	15	Launching beam segments C-H connected & pushed
	27	Man transfers to far side
	28	End support started and adjusted
	29	Launching beam segments G-H dismounted
V	29	Section feeder connected
	31	Far ramp + flaps connected & pushed
	35	Bridge section 1 connected & pushed

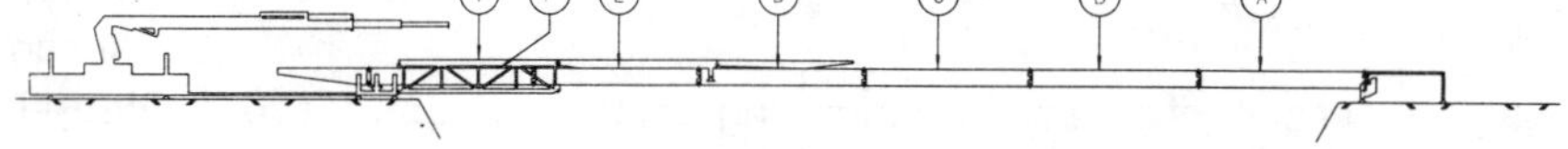

Truck no	Time min	Operation
VI	37	Bridge section 2-3 connected & pushed
VII	41	Bridge section 4-5 conected & pushed
VIII	45	Bridge section 6 connected & pushed

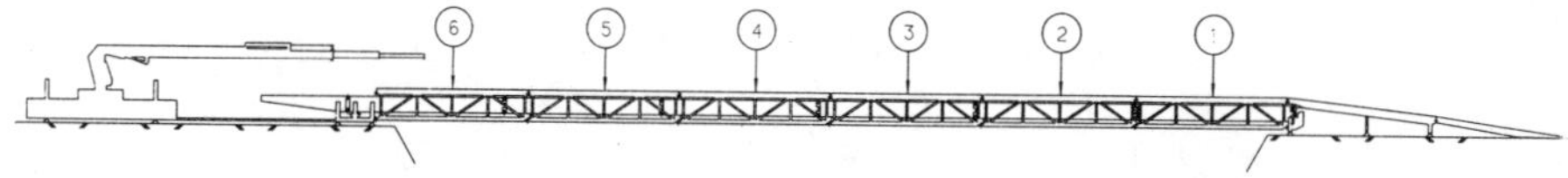

47	Launching beam and end support lowered
48	Section feeder dismounted
49	Near ramp positioned
51	20 t counterweights offloaded
55	Inner launching unit loaded on truck
57	Flaps connected
60	Bridge ready

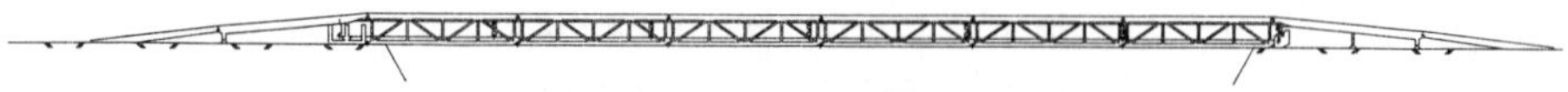

Figure 7. Construction and launching sequence

CALCULATIONS AND SCANTLINGS

As a general guidance for sizing of scantlings, the Swedish Regulations for Steel Structures, BSK, was used. The design principles of BSK are in agreement with ISO 2394-1986 'General principles of reliability for structures'. The verification procedure is carried out with partial coefficients in ultimate and serviceability limit states. This includes dimensioning formulas etc for different structural members. The BSK standard also make it possible to select smaller scantlings by the utilization of extra high tensile steel. In order to do this, however, a comprehensive quality control program was adapted for the production.

In addition to the calculations required by BSK, various finite element computer programs were also used to carry out a wide range of calculations including:

- Actual stress levels, stress distribution and deflection for all major components involved, with scantlings selected according to BSK
- Static wind deflection for the launching beam just before far end support touchdown
- Natural frequencies for the launching beam and completed bridge
- Risk of wind induced oscillations for launching beam and bridge
- Resistance to air explosion for the completed bridge

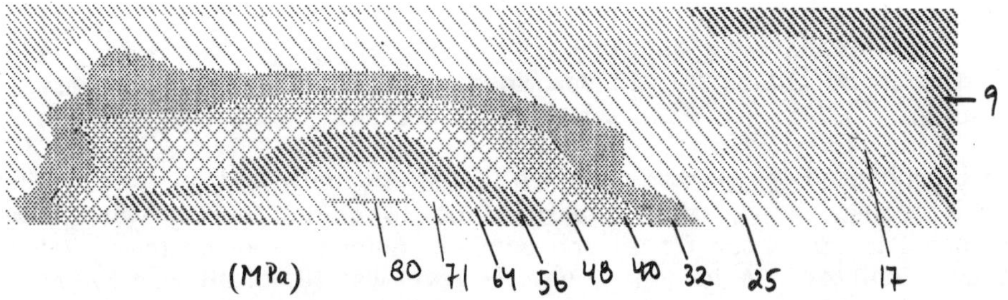

Figure 8. Stress distribution in runway plates under the asymmetric load of a 60 t battle tank

One result from the weight optimization process was that 5 mm plates could be used. Thinner plate was found impractical for welding and quality control. Thicker material is used only where necessary, eg, to reduce bearing stress in couplings and local buckling of launching beam segment flanges. Strength properties for the different high tensile steel material used are given below.

Mill denom	**DOMEX 690 XPD**	**HARDOX 400**	**WELDOX 700 E**	**SS 2135**	**SS 2174**	**SS 2541**
Steel type	Extra high tensile	Extra high tensile	Extra high tensile	High tensile	High tensile	Quenched
Yield stress R_eH N/mm^2	min 690	1000 (min 900)	min 700	min 350	min 320	720--1080
Ultimate stress R_m N/mm^2	min 750	1250	780--930	470--630	470--590	860--1240
Elong A_5 %	min 15	10	min 14	min 20	21	15--21
Hardness H_B	abt 270	370--440				260--380
Utilization	U-shaped stiffeners in bridge sections	Runway plates	Launching beam and couplings	RHS-profiles in bridge sections	Inner launching unit	Plungers in couplings

TESTS

Tests carried out so far are described below together with summary of results.

Two connected bridge sections were tested statically and in load cycling 5000 cycles. Stress levels measured at 16 points, during worst case combination of shear forces and bending moment, showed lower values than calculated. The heighest measured level was 183 of allowed 350 MPa. Deflection was about the same as calculated. Overloading at 120% was carried out during abt 50 cycles without problem.

Figure 9. Load cycling of bridge sections

Handling and connection was tested using two bridge sections. Handling by frontlift trucks and truck mounted cranes presented no problem, eg, due to mass of inertia. Connection was easy to carry out also with simulated heavily soiled couplings.

Two connected launching beam segments were tested statically. Stress levels measured at 15 points, during worst case combination of shear forces and bending moment, showed good correlation with calculated levels. The heighest measured level was 361 of allowed 700 MPa. Deflection was about the same as calculated. There was no risk of local buckling of launching beam flange from the outer launching unit support load. No plastic deformation of coupling plungers or holes was found.

The completed launching beam was tested in static and dynamic conditions. Stress levels measured at 16 critical points showed good correlation with calculated levels. The heighest measured level was 375 of allowed 700 MPa. The deflection, corresponding to 4.11 m vertical translation of the far end, was close to the calculated and the vertical/horizontal natural frequency 0.33/0.15 Hz also correlated well with the calculated value.

The functions and capacity of the inner and outer launching units were tested. Assembly and launching of complete launching beam was achieved in about 20 minutes. Recovery and disassembly went without problem. A gantry crane was used to simulate the hydraulic crane not yet fitted on the inner launching unit.

Figure 10. Testing of launching operation

THE MULTI SPAN BRIDGE

Based on the bridge section used for the single span bridge, a multi span bridge system concept has been developed. It will be used to bridge rivers in the northern part of Sweden. Hydraulic leg pairs will be used to support the bridge in up to 40 m span width for a 60 t battle tank load. The detail design work and prototype production of the new components will start in autumn 1991. A patent has been applied for also for this multi span bridge system.

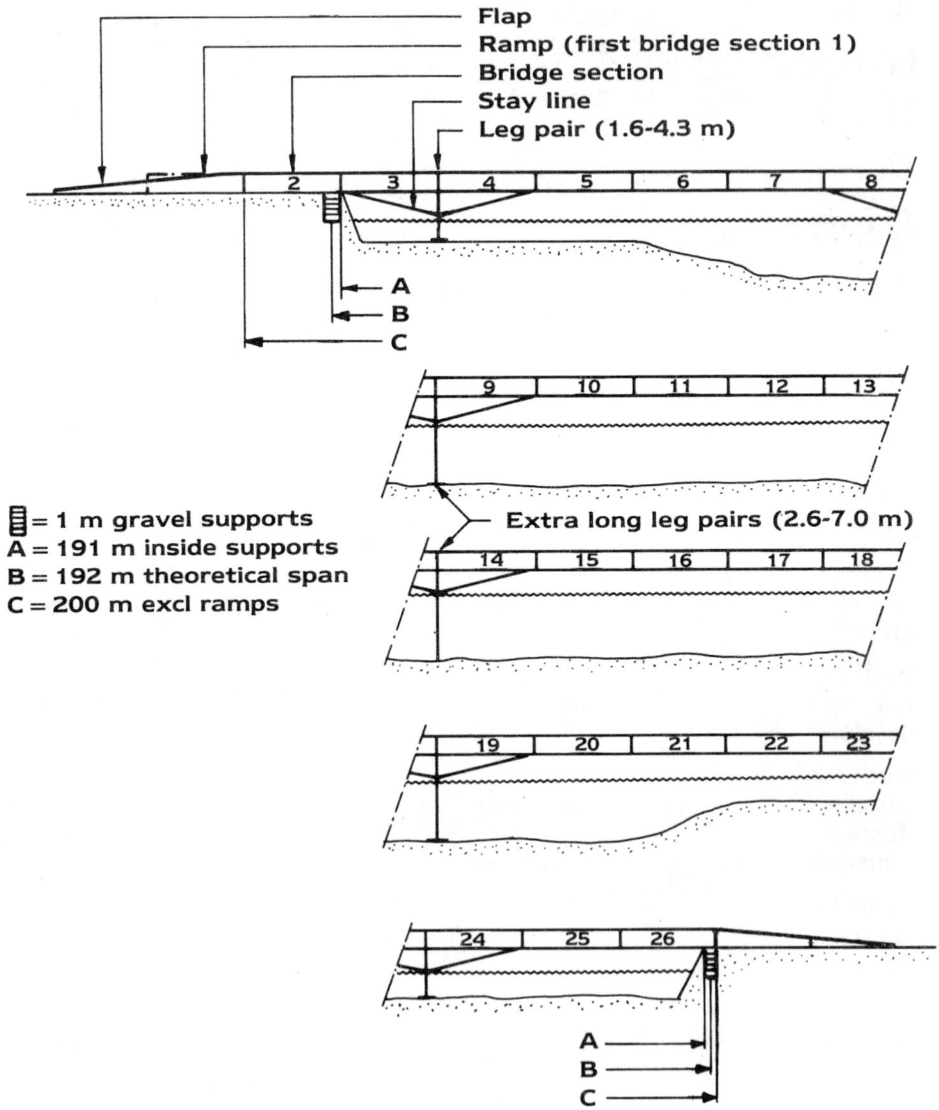

Figure 11. Completed multi span bridge

A Telescopic Mast Family for Multiple Ground Based and Space Application

M. Schmid, F. Guter

Dornier GmbH, P.O. Box 1420, 7990 Friedrichshafen, Germany

INTRODUCTION

The continuous demand for deployable and retractable booms for a variety of applications drove forward the development of a family of masts at DORNIER.

In the frame of the development activities performed during the last years, the original concept idea was refined in order to make the concept suitable for both, ground based and space bound applications as well.

Basically the ground based masts (PATRIOT masts) are related to military radio link applications while the so -called ERM (Extendable and Retractable Mast) is designed to deploy and retract solar generators or antennas in orbit.

Due to the different requirements for ground and space application, the ground based mast is built out of aluminum while the space mast is designed as a very lightweight carbon fibre structure which is optimized for the specific space environment.

In the following, the design approaches which are based on the common mast concept are discussed under consideration of the different environmental and performance requirements.

The developments which were performed in the field of deployable and retractable telescopic masts at DORNIER were based on the idea to act the deployment and retraction movement of the mast by means of a spindle located in the center of the telescopic tube stack.

By developing and refining this original conceptual approach under consideration of the specific space requirements a mast design also satisfying a wide range of space applications could be derived.

FUNCTIONAL PRINCIPLE

Basically the mast structure is defined as a telescopically stowed tube stack consisting of a number of circular tube sections. The circular cross section was mainly chosen in order to obtain maximum and uniform bending stiffness in each bending direction at minimum mass. The mast is driven by a spindle powered by a suitable electrical or hydraulic motor (depending on the application).

Each of the nested tube sections is provided with a threaded nut at its lower end. In the storage mode the nuts are retained by the unthreaded storage section of the spindle with only the topmost nut engaging the thread.

Deployment starts by virtue of the spindle pushing forwards the engaged nut. Rotation of the nut and of the corresponding tube-section respectively is prevented by longitudinal stringers mounted to the next outer tube-section. Shortly before the moving nut leaves the spindle, an end- stop draws out the next outer tube of the storage position and engages its corresponding nut to the spindle thread, whilst the nut of the preceding tube section leaves the spindle. At the same time a special latching mechanism located at the upper end of each tube section locks the extended tube to the next outer one in order to assure a defined deployed position.

This procedure can be repeated until the mast has partially or completely deployed. During retraction the process is reversed. Fig. 1 depicts the functional principle of the mast.

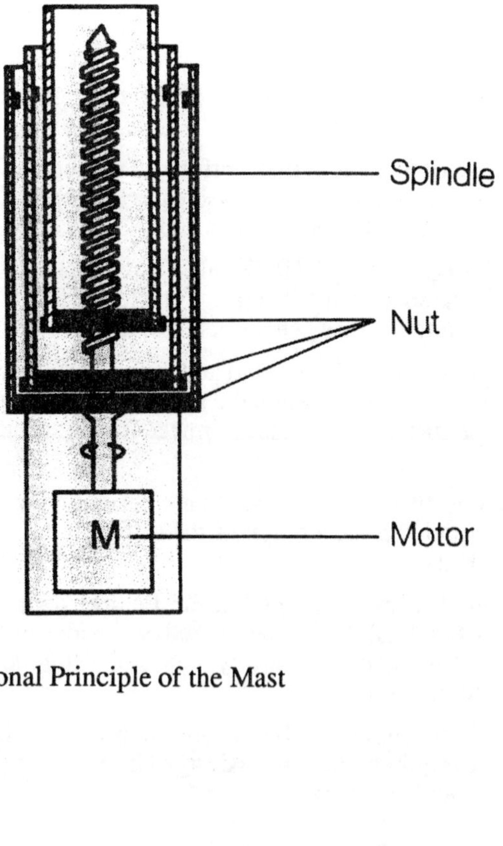

Fig. 1 Functional Principle of the Mast

FIELDS OF APPLICATION AND ACTUAL DEVELOPMENT STATUS

The traditional application of the DORNIER mast concept is its use as radio link mast for ground application.

Due to its good tip pointing accuracy other ground applications like positioning of experiment- or observation platforms with high alignment accuracy requirements are envisaged.

In parallel to the earth bound applications the trend to large and retrievable spacecraft demands a new generation of strong and stiff lightweight masts.

Based on the mast design concept which was also found to be a favorite solution for space application, the so-called ERM (Extendable and Retractable Mast) was developed to deploy and retract large solar arrays, to point unfurlable or fixed antenna dishes or to position a variety of scientific experiments away form the spacecraft vicinity.

In the following table 1 the different fields of application and the actual development status is shown. Fig. 2and 3 depict examples for the most important actual and future applications of the mast concept.

MAST TYPE	APPLICATION	DEVELOPMENT STATUS
Ground Based Mast	Military Radio Link Appl.	Manufactured in Series
Space Mast (ERM)	Solar generator deploymt. Communication antenna pointing e.g. on COLUMBUS	Development incl. detailed design , structural and thermal analysis was performed. An engineering model was built and tested under thermal vac. conditions

Tab. 1 : Development Status of the Mast Family

Fig. 2: Mast for Ground Based Application

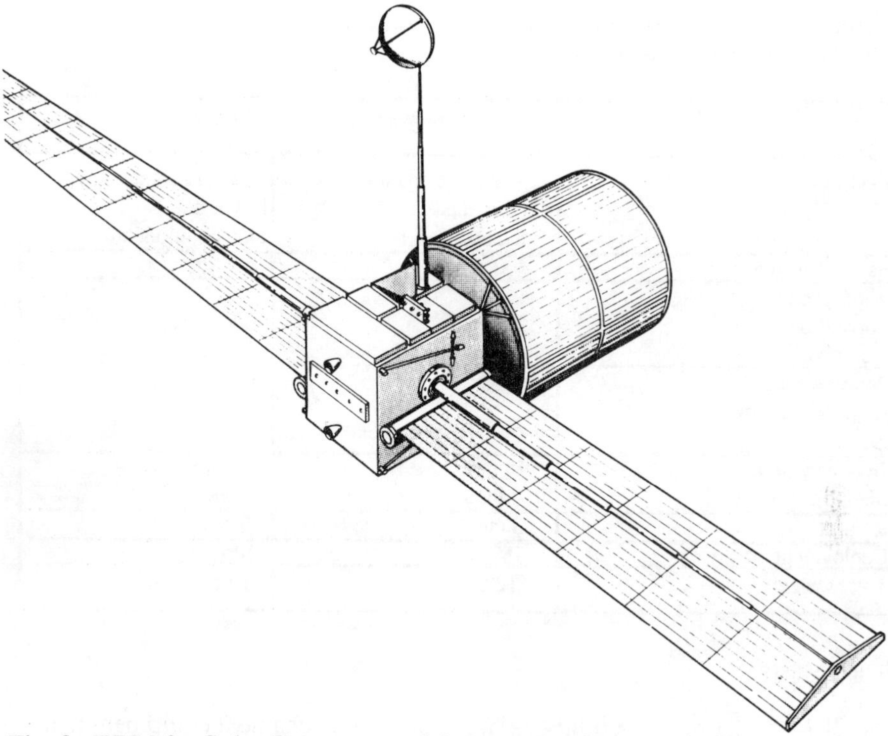

Fig. 3: ERM for Solar Generator and Antenna Deployment in Space

ENVIRONMENTAL AND PERFORMANCE REQUIREMENTS

Basically the environmental and performance requirements of the mast are completely different for the ground based and the space versions.

Whilst for the military ground version requirements for wind-, ice- and shock-loads, dust and humidity have to be taken into account, special environmental conditions for space application are e.g. high vibration loads occuring during launch, acceleration loads during operation due to thruster firing and operation under vacuum conditions over a wide temperature range and at severe radiation influences.

In the following, the most important design requirements for the ground based- an the space mast are discussed.

Ground Based Masts

The general task of a ground based mast is, to elevate a given payload in a certain time with limited manpower and space and to stabilize the payload within certain limits under different environmental conditions.

To meet these tasks the mast, the carrying vehicle and the kind of stabilisation have to be designed as an integrated system. The following examples for two different sets of requirements therefore include the vehicle and the stabilisation method as well.

	27 m mast	34 m mast
Payload	2 parabolic antennas 1, 8 m dia. incl. positioners	4 parabolic antennas 1, 0 m dia. incl. positioners
Payload weight	400 kg	500 kg
Windspeeds - operation - survival	 27 m/sec 33 m/sec	 37 m/sec 37 m/sec
Pointing accuracy at operation wind-speed	+ 0,5°	+ 2,5°
Deployment time Manpower required	45 min 4 men	20 min 2 men
Ice accretion	3 cm	3 cm
Deployment on slope	15°	10°
Total weight	12 t	19 t

Space Masts

The ERM is defined as a lightweight structural / mechanical equipment that can be stored in a minimum volume whilst yielding maximum deployed length, strength and stiffness.
The baseline for the design and analysis of the space masts was a set of requirements covering the application of a solar array mast with a deployed length of 40 m and of an antenna mast with 20 m extended length.
Most of the required performance data like high stiffness and strength, good tip pointing accuracy, minimum thermal deflections and low mass can only be obtained by application of carbon fibre technology in combination with a structurally optimized mast concept.
the following table 2 shows the most important performance requirements valid for the 40 m solar array and the 20 m antenna mast.

	40 m Solar Array Mast	20 m Antenna Mast
Launch Characteristics		
stowed envelope length	3 m	2,25 m
stowed envelope diameter	0.5 m	0.4 m
overall ERM-mass 1)	90 kg	50 kg
number of tube sections	18	13
1st natural frequency, stowed conf.	115 Hz	140 Hz
deployment speed,nominal	20 mm/s	10 mm/s
gear stage concept	spur gear 1 : 10	planetary gear up to 1 : 25
power consumption,nominal	80 W	
peak power consumption	200 W	
drive	brushless DC	
orbiter interface	basic flange and two hardpoints on top	
payload interface	hole pattern	
Mission Characteristics		
deployed length	40 m	20 m
payload capability 2)	40 kg on top plus 10 kg/m line load plus 100 N excentric load	120 kg on top
max. bending moment	2800 Nm	2300 Nm
average bending stiffness	$7 \cdot 10^5$ Nm^2	$4 \cdot 10^5$ Nm^2
1st natural frequency, depl. conf. 3)	0.06 Hz	0.1 Hz
random translational deployment error	< 100 mm	< 30 mm
random angular deployment error	< 2 m rad	< 1.5 m rad
power lines to payload 4)	4 + 1 grounding	
signal lines to payload 4)	15	

1) without thermal hardware and cable follow-up mechanism
2) offset of excentric force 0.5 m to CL / offset of Antenna Mast tip mass 6 m
3) 240 kg on tip and 70 kg distributed / 120 kg on Antenna Mast tip with 6 m offset and 40 kg distributed along Mast.
4) spring driven cable follow-up mechanism with slip rings is provided optionally for both Mast concepts

Tab.2: ERM Requirements and Performance Characteristics

GROUND BASED MASTS

Technical Concept

Mobile mast-structures have to meet two main criterias with partly competing objectives.

These are:

- The structure has to be very stiff to obtain the required vertical and torsional deflection at the mast top under high loads
- Mobility on the other hand requires a structure which is as compact and light as possible for easy transportation

Mast Structure Apart from a few exceptions, the present-day market for mobile antenna masts is dominated by two types:

. the lattice telescope mast and

. the tubular telescope mast

In the past, lattice constructions were mainly used for higher masts, tubular constructions for smaller heights and loads. The reason ist that lattice constructions have a relatively higher structural stiffness than tubular masts in the vertical axis. The drawbacks of large transport dimensions and large surfaces attacked by the wind have to be accepted.

Investigations of structural deformations under wind load with the finite-element method, however, led to the result that the apparent disadvantage of a tubular structure can be transformed into an advantage if one succeeds in controlling the mast excursion by a special arrangement of the guying points and by balancing the tensions so that the mast top remains nearly vertical under all wind loads or rather adopts an S-shaped bending line. The Dornier telescope mast with its tubular structure fulfils this objective by its design where the guys which are also required in lattice masts - are not directly attached to the mast top but to a cross-brace which generates a moment contrary to the wind direction, largely compensating structural bending under transverse loads. This tubular structure thus combines the advantages of small transport volume and low weight with that of a high alignment precision of the antennas.

The other advantages of a tube structure remain unchanged:

. high torsional stiffness

. complete protection of the drive elements

. ideal behaviour under ice formation

favourable production conditions as complete tube sections can be manufactured with the extrusion process.

Fig 4. Deflection of the Guyed Mast Structure Under Transversal Load (fig.turned by 90 deg, tip deflection shown is not to scale)

Drive System The following requirements are made on the drive system of atelescope mast:

. it should be as simple, easy to maintain, and as reliable as possible
. it should be effective in both directions in order not to leave to gravity alone the forces generated while retracting the telescope

. extension and retraction should be continuous as far as possible
. both procedures should be automatic, that is, initiated by push-button action

Mechanical rope drives, mainly used for lattice masts, have proven their value in this respect with the only drawbacks that the ropes must be checked and adjusted in regular intervals, that tensioning "against the ropes" is problematic due to their elasticity, and that temperature and hence length variations have to be taken into consideration.
For tubular masts, usually pneumatic deployment systems are used, which are however, satisfactory for small equipment only and always sensitive to external impact as well as relatively demanding in maintenance.
The mechanical spindle drive used for the DORNIER mast presents a solution which strikes the observer by its simplicity: a lift spindle of the same length as the basic tube drives the first tube segment upwards. Before it leaves the spindle, the second segment is pulled upwards as well and "threaded" into the spindle from below. An automatic latching mechanism between segment 1 and 2 now carries segment 1 further upwards, and this is repeated until all segments are extended. Retraction is implemented in the same way by reversing the sense of spindle roration. This drive fulfils all the requirements quoted above.

Design Description - 34 m Antenna Tower on 7 t Truck

Performance Requirements The development criteria for this system which is briefly described here, are as follows:
* Four parabolic antennae (0.9 m dia.) shall be elavated to a height of 34 m at each place which can be reached by a cross-country truck.
* The mounting time from the transportion position up to the start of radio operation shall not exceed 20 min.
* The available personnel consists of two persons.
* The system shall be operable up to wind velocities of 37 m/sec (83 mph).

Technical Solution The criteria described show that the fulfilment of the performance depends not only on the selected mast, but also on the carrier vehicle, vehicle superstructure, stabilization and mechanization of the deployment process.

The chosen solution is characterized by the following criteria:
* A 7-ton cross-country truck was selected as a carrier vehicle whose weight is used to stabilize the whole system.

*To enlarge the basis and to level the vehicle (terrain gradient up to 10°), hydraulically
operated outriggers with automatic clamping devices are employed.

* All deployment functions (leveling, erection, extension, and bracing of the mast) are carried out hydraulically and controlled simply by pressing buttons.

* The complete payload (in this case four parabolic antennae adjustable in azimuth and elevation) remains connected so as to be ready for operation even in transportion configuration.

* All cables and ropes remain attached and are coiled automatically so as to ensure permanent operability.

* The antennae can be adjusted already during the mounting phase.

* To avoid incorrect operation, all deployment steps are interconnected electronically and are released only if they can be performed without any risk.

* Demounting takes place in the inverted order as in erection.

SPACE MAST

Technical Concept

Baseline for the design was a 40 m solar array mast and a 20 m antenna mast. In order to fulfill the stringent requirements concerning mass, stiffness and pointing accuracy, a design was evolved which is based on the use of thin walled tube sections with circular cross- sections manufactured from Carbon Fibre Reinforced Plastics (CFRP).
Comparision with hexagonal and triangular cross-sections and frameworks have shown that the circular cross section provides a good ratio between stiffness and mass and, in addition, good thermostability behaviour.
The good pointing accuracy at the ERM tip is achieved by use of the accurate and thermostable CFRP structure and by designing for minimum backlash between the tube joints. The design of the tube joints incorporates special guiding elements and these allow adjustment to very small backlash.
Furthermore the given mass budget requires a lightweight design of the electronics and electromechanical components and of the chosen thermal hardware.
A payload interface is foreseen at the mast tip and additional line-load attachment points can be foreseen on top of each tube section if required by the application (e.g solar array).
The ERM design optimization process has shown that the only significant design difference between the solar array and the antenna application is choice of thermal hardware defined according to mission requirements.
The maximum theoretical deployment length which can be achieved by the mast is mainly given by the required stowed envelope (length and diameter) and by the choice of the overlapping length among the tube sections.

To optimize the tube stack w.r.t. maximum stiffness and deployed lenght at minimum mass, a tube stack was chosen which has variable overlapping length, thus providing an almost constant ratio among tube diameter and overlapping length (see fig. 5). The dependency among deployment length and number of tube sections of the chosen design is such that with a 3 m stowed envelope length a theoretical deployed length of up to 70 m can be obtained whilst with a stack length of e.g. 2.25 m a deployed length of approximately 30 m can still be achieved. However to come to an optimal design it is not adequate to go to such high deployment lengths at the stowed envelope discussed. The limitation for the optimal design is given by the chosen overlapping length which becomes longer for increasing tube diameters, that means that the deploying part of each tube section becomes shorter with the chosen number of tube sections.

To minimize mass for a required deployment length the stowed envelope length should be chosen in a way that the dependency between deployment length and necessary number of tube sections to obtain this length is almost linear (see fig. 6).

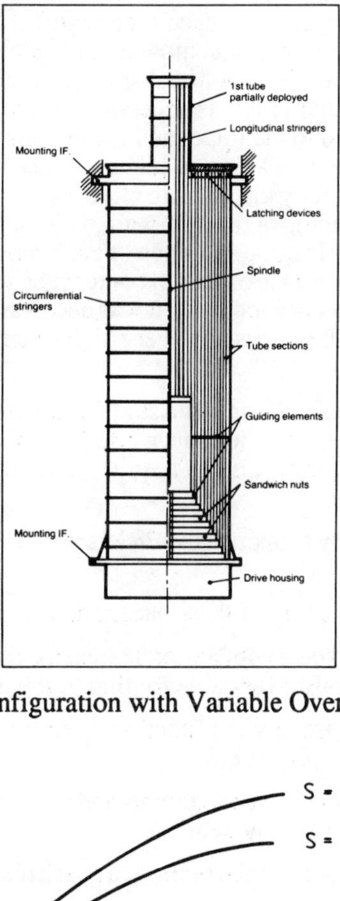

Fig.5 Chosen ERM Configuration with Variable Overlapping Length

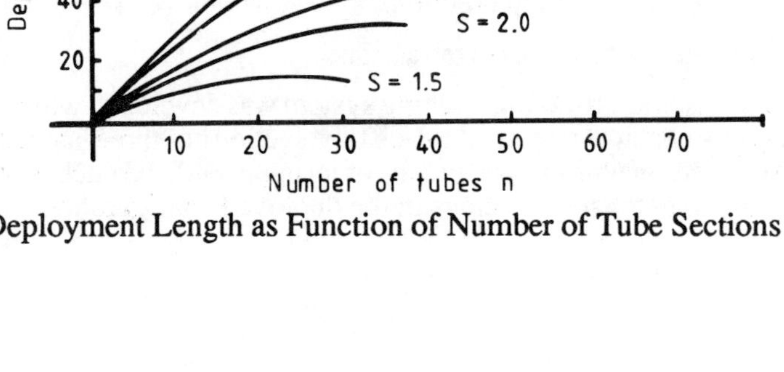

Fig.6 Deployment Length as Function of Number of Tube Sections Required

Design Features

In fig.7 the overall design of an ERM is depicted. The tube structure is manufactured out of 6 layers of high-modulus carbon fibre epoxy resin in a symmetric fibre lay-up. The resulting wall thickness of the CFRP tubes is only about 0.5 mm. The lay-up is optimized to reach high bending stiffness at a near zero thermal expansion coefficient.

A CFRP sandwich-plate carrying a teflon-bronce thread is glued to the bottom of each tube section. At the other end of the tubes a stiffening carbon fibre end cone is mounted. On the inner surface of each tube segment three longitudinal stringers are attached to prevent rotational movement of the next inner tube segment during extension and to provide torsional stiffness in the tube joints. To guide each tube relative its neighbour it was necessary to develop special lightweight guiding elements mounted to the upper and lower ends of the tube overlapping length. The space between two adjacent tubes was chosen to 10 mm, this being the minimal value required by the guiding elements and the inter-tube latching system. To avoid local loading of the thin-walled tube segments the guiding elements attached to the whole circumference of each moving tube are mating directly to the smooth inner surface of the next outer tube section. Due to the big influence of backlash on pointing accuracy it was necessary to shape the CFRP tube sections and their guiding surfaces to very high accuracy (corresponding to a backlash of 0.1 mm).

In order to minimize tip deflections, a mission dependant thermal design has been derived. Depending on the application, white thermal paint will be used while, to fulfill the extreme pointing accuracy requirements, multilayer insulation can also be applied.

The mechanisms necessary to operate the ERM are mainly those in the following assemblies:

- latching mechanism (latching of the adjacent tube sections)

- drive mechanisms (gear stage with motor, launch locking device and spindle
- cable follow -up mechanism for cable routing to the mast tip.

Special problems were posed by the functional requirements for the ERM latching system for the deploying tubes:

- mutual latching of the stowed tube segments and achievement of a defined tube by tube extension during deployment

- sequential latching of adjacent tube sections as each reach their required position

- delatching of the retracting tube sections

- latching of the tube sections after retraction.

To assure these tasks a self-acting latching system was developed which functions by a single mechanical system. As each tube section has three latches mounted equidistantly around the circumference of its upper end, the latches are accesible for ground maintenance routines in the deployed configuration.

The hollow aluminum spindle used to deploy the tube sections is coated with a wear out resistant coating and is reinforced by an inner CFRP-tube to reduce the axial coefficient of thermal expansion and, at the same time, to increase bending stiffness. As a hardware option, the spindle can be driven either by a normal spur gear stage or by a planetary gear, thus providing flexibility w.r.t. application dependant speed, torque and power.

The brushless DC-motor is controlled by a fully redundant ERM deployment electronics unit (DEU). Inductive position sensors provide signals to the DEU so that each required ERM position can be reached and monitored with high accuracy.

To avoid inadvertent deployment of unsupported payloads during launch, a magnetic clutch is incorporated to fix the gear stage in the required position therewith avoiding movement of the tube stack.

Depending on the ERM mission, it may be necessary to provide power and signals to the payload. This task is performed by a cable follow-up mechanism which can be attached to the upper end of the outermost tube. A flat cable is guided along the extending tube sections in a manner rather like a fishing line on the rod. The mechanism is driven by a constant torque spring motor so that the cable is wound onto a drum during retraction process.

Use of an external guidance system for the ERM cable gives advantages of accessibility not only during assembly (and disassembly) but also in the event of failure in orbit. To cover the latter eventuality, an astronaut-override is included in the design.

In order to verify functional performance of the ERM design an engineering model, consisting of three tube sections, was built. This was tested in vacuum over the temperature range +85 to -100 deg C at ESTEC. The overall number of duty cycles (1 cycle = 1 extension + 1 retraction) was more than 600. The two extending CFRP-tube-sections had diameters of 405 and 425 mm respectively and each were of about 1.2 m length. The tubes were shaped according to the chosen ERM-design and equipped with design representative guiding and latching systems. All friction interfaces, including the spindle-nut interface, were design representative. The model was driven by a brushless AC-motor, modified for vacuum application.

CONCLUSIONS

By consequent application of the spindle deployment concept in the telescopic mast design and by including the tubular cross section for the deploying tubes, a mast family could be introduced which satisfies a wide range of actual and potential ground and space demands.

Due to its design flexibility the chosen concept can be adapted to heavy duty- as well as to extreme lightweight applications by means of adequate detailed design according to the specific requirements.

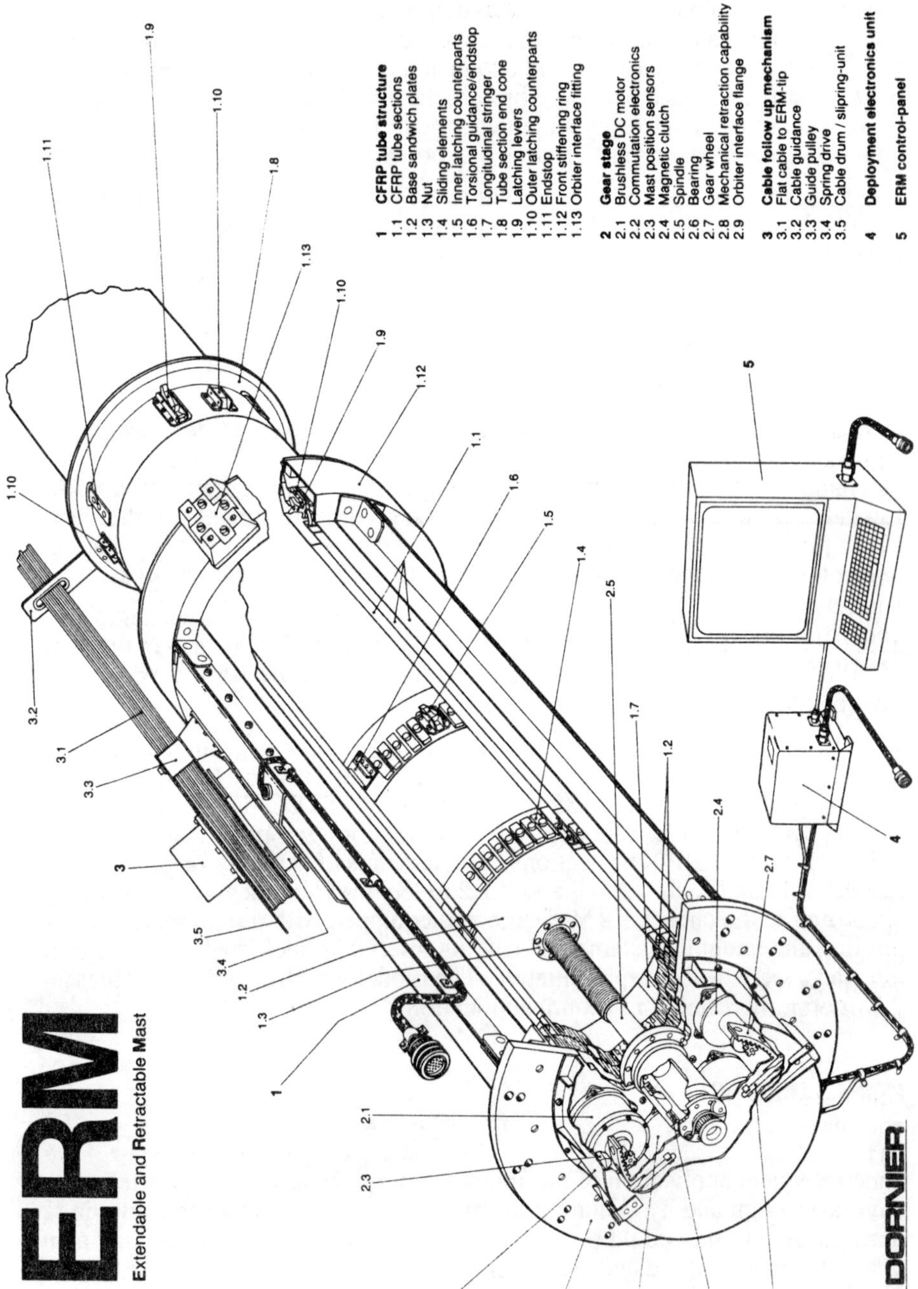

Fig. 7 ERM Design Overview

Rapid Runway and Highway Repair Using Precast Concrete Raft Units

J.W. Bull, J.D. Clarke

Department of Civil Engineering, The University, Newcastle upon Tyne, NE1 7RU, U.K.

ABSTRACT

When runways and other heavy duty pavements have been cratered following an attack, it is necessary to effect repairs within a few hours. During the last five years the Ministry of Defence have developed such a repair method using precast cement concrete pavement units [raft units]. Research into the design parameters and certain constructional requirements have been carried out at The University, Newcastle upon Tyne. The research is developing a design method such that from known loading requirements, subgrade conditions and cratering parameters, the raft unit support requirements can be specified. The design method is applicable to motorway contraflow systems where vehicles must be switched between carriageways and to port areas where container stacking yards need rapid reorganization. A design example is included.

INTRODUCTION TO RAFT UNIT PAVEMENTS

Natural soil is unlikely to have sufficient strength to withstand repeated vehicle loading. Consequently, a pavement layer or layers, traditionally classified as flexible or rigid, is placed between the soil and the vehicle's wheels. The flexible pavement consists of a compacted granular subbase laid onto the soil. On top of the subbase is placed an asphalt concrete surface. In the case of a rigid pavement, a layer of cement concrete is placed over a thin granular subbase. This simplified description of a flexible/rigid pavement indicates different mechanisms of load carrying [1].

This paper considers rigid pavements. In normal rigid concrete pavements, wet concrete is laid on site, allowed to dry and gain strength before use by vehicles. In the case of raft units, a sand bedding layer is placed on top of the granular layer and the raft units, approximately 2m x 2m placed on the sand layer. The raft unit pavement is ready for immediate vehicular use.

ADVANTAGES OF RAFT UNIT PAVEMENTS

The use of precast concrete, for structural and non-structural purposes is a well established technique. Many products such as beams, columns, railway sleepers, bridge beams and paving slabs have been produced in precast concrete over the past ninety years. Although more expensive than cast in-situ products, precast products are manufactured in quality controlled environments, where strength and quality can be monitored and recorded. Raft units are manufactured in accurately built moulds. The reinforcing bars are placed with precision, the concrete poured into the mould and after twenty four hours, the raft unit removed for storage. After 28 days, the raft units can be used on site.

THE USE OF RAFT UNIT PAVEMENTS

The first recorded use of raft units was in the 1930's for airfield construction [2]. Since the 1960's raft unit use has rapidly increased, mainly in the area of port, airfields and industrial pavements. Three major reasons for this increasing use are:- a) 900kN axle loadings on container handling vehicles and 450kN container corner casting loads are not unusual, b) raft units are often used in areas where subgrade settlement is large. Raft units can be lifted out, the subbase relevelled and the raft units relaid, plus c) the rapidity with which a raft unit pavement can be constructed and be in service. In some cases within three to four hours.

A typical example might be where a dry culvert has failed and the road collapsed. Fill material would be pushed into the culvert and compacted to the appropriate level. Raft units would be placed on the fill material and the road back in use. Another advantage of raft units is that they can be stored for long periods before use. They would not be stored unused, but could be used initially as hard standing for a vehicle park, then moved to where required.

EXISTING RAFT UNIT DESIGN METHOD

The existing raft unit pavement design method was developed by Bull and is fully described in his book [2]. The design method is microcomputer-based and interactive with the pavement designer, who is able to enter and alter data at many points in the program. The method starts with the pavement parameter shown in Table 1. The program requests the required pavement life [number of passes] in terms of the heaviest wheel loading, the wheel configuration and the subgrade CBR. From this data, the program determines the design concrete stress in the form of a design pavement stress number [DPS] and the design subgrade stress, in the form of a design subgrade stress number [DSS]. The designer now has to input the concrete strength. The program then determines the applied pavement stress number [APS] and the applied subgrade stress number [ASS]. Provided the design numbers [DPS, DSS] are greater than the applied numbers [APS, ASS] respectively, the pavement design is satisfactory. The program will require further data, including the subgrade depth, and saturation. The granular subbase details are also requested. Alternatively a lean concrete subbase may be used. From the difference between the design values and the applied values, the pavement designer can determine the raft unit thickness and reinforcement requirements. The program will output the pavement life in terms of the raft unit serviceability limit state and the point at which subgrade settlement will require raft unit relevelling.

Table 1 Raft unit data input

Layer	Size m x m	Thickness mm	Young's Modulus	Poisson's Ratic
Raft Unit	2 x 2	150	34000MPa	0.15
Bedding	2 x 2	50	75MPa	0.25
Subbase	5 x 5	300	200MPa	0.25
Subgrade	5 x 5	600	3MPa	0.30

DESIGN EXAMPLE

An airfield pavement has to be repaired to allow 500 passes of a Tristar aircraft. The subgrade CBR is 3% and 450mm of granular subbase will be used beneath the 2m x 2m x 200mm thick raft unit. Table 2 gives the design method data input and output.

Table 2 Stages in raft unit pavement design

Stage No.	Data input		Data output
1	Tristar movements	500	DPS = 461
	Subgrade CBR =	3%	DSS = 178
2	Concrete	65MPa	
	Total load	1100kN	APS-DPS = 543-461 = 82
	4 wheels	1.78m x 1.32m	ASS-DSS = 126-178 =-52
3	Subgrade CBR =	3%	APS-DPS = 497-461 = 36
	Subgrade depth	5m+	ASS-DSS = 209-178 = 31
4	GSB CBR	20%	APS-DPS = 461-461 = 0
	GSB thickness	450mm	ASS-DSS = 181-178 = 3
5	Raft thickness	200mm	APS-DPS = 276-461 =-185
	Normal reinforcement		ASS-DSS = 173-178 =-5
6	Serviceability life		3859 movements
	Relevelling required after		555 movements

Stage 1 gives the required design values.

Stage 2 uses the concrete strength of 65 Mpa, and the main gear Tristar loading of 1100 kN on 4 wheels to predict the applied stress numbers. Subtracting the design stress number from the applied stress number shows that the applied pavement stress number is 82 in excess of the design stress number. This number must be reduced to zero.

Stage 3 The effect of the subgrade is assessed. The applied pavement stress number is too high by 36 and the applied subgrade number is too high by 31.

Stage 4 A 450mm thick subbase, compacted to a 20% CBR is used. The applied pavement stress number is the same as the design pavement number, but the applied subgrade number is too high by 3.

Stage 5 Because a 200mm thick raft unit is to be used, the applied pavement number reduces to 185 below the design pavement number. The applied subgrade number is 5 below the design subgrade number.

Stage 6 The raft unit design has achieved a value of 3859 Tristar movements before the raft units will require maintenance. After 555 Tristar movements the raft units will require relevelling due to subgrade displacement.

MODIFICATIONS TO THE DESIGN METHOD

The raft unit data of Table 1, equated to the known civil engineering properties of the most frequently used raft unit, the 2m x 2m x 150mm unit. The bedding layer related to the requirement of a level running surface associated with the allowable tolerances in raft unit thickness and subbase levels. Laboratory experiments had shown that to ensure the raft unit did not displace vertically to an excessive extent, the minimum subbase CBR and thickness had to be 20% and 300mm respectively.

Rigid pavement design had shown that for a single loaded wheel, three load positions must be considered [3]. Namely, the raft unit centre, the centre of one edge and at a corner. Because the raft unit cannot be considered as of infinite size, multiple wheels on a raft unit must be considered as multiple loaded points. They cannot be made equivalent to a single wheel load [4]. Thus, in this paper only single wheel loads are considered.

The original design method related all load induced stress to a centrally placed 10kN load. The vertical subgrade bearing stress [subgrade stress] was used to relate the number of load applications to the point where the raft unit pavement would require relevelling. The maximum principal concrete tensile stress [concrete stress] was used to determine the number of load applications to the raft unit serviceability and ultimate limit states [2]. Thus a raft unit pavement life expectancy could be established.

Finite element models with up to 1224 elements and 5285 degrees of freedom were used to refine the element mesh and provide asymptotic values for the subgrade and concrete stresses [2]. The results given in this paper relate to 848 elements and 3386 degrees of freedom. Each layer of Table 1 consisted of one layer of finite elements. The raft unit and the bedding layer consisted of 10 x 10 elements each. The subbase and subgrade layers consisting of 18 x 18 elements each.

The initial research and design method concentrated on the

stresses. In this paper, consideration will also be given to displacements. Furthermore, the pavement design technique in which each pavement layer is transformed into an equivalent layer through the use of equation 1 will be used [5]. For two layers to have a similar flexural stiffness, the term

$$[h^3 \times E]/[1 - \upsilon^2] \qquad [1]$$

is the same for each layer, where "h" is the layer thickness, "E" is Young's Modulus and "υ" the Poisson's ratio. The total depth under the raft unit is the same as in the finite element analysis, but a resulting single equivalent layer "E" value is determined.

Table 2a shows displacements, stresses and "E" values as the subgrade thickness is increased from 600mm to 1999mm. Rows 1 and 2 show the finite element calculated vertical displacements, directly beneath the centrally applied 10kN load. Row 1 showing the displacement on the upper surface of the bedding layer and Row 2, the displacement at the top of the subgrade layer. Row 3 gives the ratio of Row 2 to Row 1 in % and shows that between 95.6% and 97.8% of the vertical surface displacement takes place in the subgrade. In Row 4 the equivalent single layer "E" value is given. Both Rows 3 and 4 show the critical effect of low CBR subgrades and explain the large settlements encountered in many port pavements. The effect of increasing the subgrade thickness on the concrete stress is shown in Row 5. Research has shown that increasing the subgrade thickness beyond 1500mm has almost no effect on the concrete stress [6]. Row 6 and 7 show that increasing the subgrade depth reduces the subgrade stress and the vertical bedding stress.

Table 2a : Stress and displacement related to subgrade thickness

Subgrade thickness	mm	600	1000	1500	1999	Row No.
Bedding disp.	mm	0.1813	0.2400	0.2971	0.3484	1
Subgrade disp.	mm	0.1734	0.2323	0.2895	0.3396	2
Row 2 / Row 1	%	94.962	96.270	97.050	97.474	3
Equivalent "E"	MPa	26.681	16.273	11.280	8.9400	4
Concrete stress	MPa	0.4825	0.4890	0.4918	0.4928	5
Subgrade stress	kPa	1.100	0.867	0.708	0.615	6
Vert bed stress	kPa	4.08	3.93	3.84	3.83	7
Radius "a"	mm	883	900	910.5	911.6	8
"z" = 6a	mm	5298	5400	5463	5470	9
Disp. "S"	mm	0.3913	0.6299	0.8983	1.1318	10

BOUSSINESQ EQUATION

For a circular load uniformly distributed over a radius "a", the stresses, strain and deflection on the centre line of the load may be determined by using Boussinesq's equation [5]. For a surface stress of "σ_s" the applied stress at a depth "z" is found from equation 2.

$$\sigma_z = \sigma_o\{1 - \frac{1}{[1+(\frac{a}{z})^2]^3/2}\} \qquad [2]$$

Equation 2 shows that "σ_z" has a value even when "z" is very large. However, at a depth of "z" = "6a", "σ_z" has been reduced by 95.973%; thus in this paper that depth is used as the minimum depth of supporting material under the raft unit and the depth to which compaction is required. The surface stress "σ_z" at the top of the bedding layer, immediately beneath the raft is determined from the finite element analysis.

The bedding surface displacement "S" is significant in that it is directly related to the faulting, or vertical displacement between a loaded raft unit and an adjacent unloaded raft unit. "S" is found from equation 3, where the strain "ϵ" at a depth "z" is as shown in equation 4.

$$S = \int_o^H edh \qquad [3]$$

$$\epsilon_z = \frac{\sigma_o}{E}(1+v)\{\frac{(z/a)}{[1+(z/a)^2]^3/2} - (1-2v)\{\frac{z/a}{[1+(z/a)^2]^1/2} -1\}$$

[4]

Using a value of Poisson's ratio, υ = 0.4, and manually integrating equation. 4, "S" over the range "z" = 0 to 6a, is found to be :-

$$S = \frac{\sigma_o}{E} x\, 0.483\, x\, 6a \qquad [5]$$

The applied load was 10 kN and using the Row 7 stress values, the radius "a" of the loaded area can be determined [Row 8]. The value of "E" is that shown in Row 4. Row 9 gives the value of "z" for each subgrade thickness and Row 10 the displacement "S".

From Table 2a it is clear that increasing the subgrade thickness, and keeping the subgrade "E" value constant increases displacement, increasing concrete stress and the compactive depth "z". The equivalent "E" value reduces as, does the vertical bedding stress and the subgrade stress.

In effect this paper uses a linear elastic solution to solve a non-elastic problem. Consequently, the calculated displacement are only "best estimates" of the support deformation. However, a change in soil stress due to change in applied surface load causes a surface deformation or settlement. It is also observed that larger changes in stress produce larger settlements and experience indicates that this methodology provides satisfactory solutions [7].

FURTHER RAFT UNIT PAVEMENTS

The previous section showed that the subgrade CBR should be increased. In the following computer runs the subgrade the subbase CBR are set at 20%, [Young's Modulus = 200MPa]. Considering the depth effect "z", the combined subgrade and subbase thickness is made 5998mm. Initially a bedding layer thickness of 50mm was used. Following from site and laboratory tests, further alterations to the finite element model were made. The raft unit thickness was increased to 200mm. The bedding layer thickness later being reduced to 25mm. Materials other than sand are considered for the bedding layer.

Table 3 shows the results for the modified raft unit computer runs. Column 1 can be compared with column 1 of Table 2a. Table 3 shows that by increasing the raft unit thickness and the subgrade CBR value, the concrete stress reduces from 0.4825MPa to 0.2275MPa. The displacement "S" is reduced from 0.3913mm to 0.0612mm and the depth "z" is reduced from 5298mm to 4731mm. In reducing the concrete stress, larger loads can be carried. The reduction in "S" reduces faulting between raft units again allowing higher loads to be applied. The reduction in "z" reduces the required depth of compaction, thus increasing the speed of repair.

Comparing columns 1 and 2 of Table 3, it can be seen that increasing the combined subgrade/subbase thickness from 900mm to 5998mm reduces the concrete stress from 0.2275MPa to 0.2203MPa, increasing the equivalent "E" value from 191.314MPa to 198.619MPa and reducing "z" from 4731mm to 4314mm. However, the maximum vertical stress at the top of the bedding is increased from 5.12kPa to 6.16kPa.

Table 3 : Displacement and stresses

Bedding Thickness	mm	50	50	25	25	25	25
Bedding layer "E"	MPa	75	75	10	75	500	4000
Subbase+Subgrade	mm	900	5998	5998	5998	5998	5998
Equivalent "E"	MPa	191.3	198.6	198.4	199.3	200.9	207.1
Concrete stress	MPa	0.228	0.220	0.236	0.216	0.204	0.192
Bedding stress	kPa	5.12	6.16	4.31	7.47	10.90	25.00
Radius "a"	mm	788.5	719	859.4	652.8	540.4	356.8
"z"= 6a	mm	4731	4314	5156	3917	3242	2141
"S"	mm	0.0612	0.0646	0.0541	0.0709	0.0850	0.1249

One of the on-site sources of initial high settlement is the bedding layer. The sand was difficult to level and compact using normal civil engineering vibration techniques. The military engineering system uses a screed beam to give a level surface. Using the screed beam on the subbase, the surface can be levelled to more controlled tolerances. Thus, the 50mm thick sand bedding layer, could be reduced to 25mm. The significance of this reduction is shown in Columns 3, 4, 5 and 6 of Table 3.

Columns 3, 4 and 5 relate to a bedding layer with a CBR of 1%, 7.5% and 50% respectively. Column 6 relates to an asphalt concrete layer with a Young's Modulus of 4000MPa and a Poisson's ratio of 0.4. The direct result of reducing the bedding layer from 50mm [column 2] to 25mm [column 4], is to reduce the raft unit concrete stress from 0.2203MPa to 0.2155MPa, reduce the "z" value from 4314mm to 3917mm, and increase the equivalent "E" value from 198.619MPa to 199.306MPa. However, the stress at the top of the bedding layer is increased from 6.16kPa to 7.47kPa but as the stress increases more than the equivalent "E", the displacement "S" is increased from 0.0646mm to 0.0790mm.

The effect of increasing the Young's Modulus of the bedding layer, is to reduce the concrete stress from 0.236MPa to 0.192MPa and reduce the depth of compaction "z" from 5156mm to 2141mm. Increases were found in the equivalent Young's Modulus "E" from 198.431MPa to 207.091MPa, the vertical stress at the top of the bedding layer, from 4.31KPa to 25.00KPa and in the value of "S" from 0.0541mm to 0.1249mm. The effect of a 25mm thick asphalt bedding layer [column 6] is to increase the vertical bedding stress and the value of "S". However, the concrete stress is reduced as is the depth 'z' to which compaction is required.

LOADED AREA RELATED TO STRESS

Aircraft wheel contact sizes and tyre pressures are readily available [8]. For example, a Tristar with an all up mass of 245,850kg, has 45.65% of the mass on 1 main gear leg comprising dual tandem wheels with tyre pressures of 1.43MPa. This gives a single wheel contact area 439mm square. As the Tristar loading is very large, it is clear that a tyre contact area from a point load to in excess of 439mm square needs to be considered.

Table 4 Centrally loaded raft unit

Area loaded		200 x 200	400 x 400	600 x 600
Concrete stress	MPa	0.192	0.147	0.138
Bedding stress	kPa	25.0	14.4	11.67
Depth "z"	mm	2141	2821	3134
"S"	mm	0.1248	0.09747	0.08528

For a centrally placed load, the finite element computer program was run with three different loaded areas as shown in Table 4. The data input was as shown in the last column of Table 3. Table 4 shows that increasing the wheel contact area reduces the concrete tensile stress, reduces the bedding layer vertical stress and reduces the total faulting displacement. However, the depth effected by the load is increased from 2140mm to 3134mm, thus increasing the required depth of compaction.

Table 5 Corner loaded raft unit

Loaded Area		100 x 100	200 x 200	400 x 400	600 x 600
Concrete stress	MPa	0.6945	0.1098	0.0889	0.0772
Bedding stress	kPa	273.25	174.75	114.10	78.67
Depth "z"	mm	648	810	1002	1207
"S"	mm	0.4127	0.330	0.2667	0.2214

Table 5 shows the effect of increasing the loaded area for a load placed at a corner. For the corner placed loads, the centre of the loaded area moved towards the raft unit centre as the loaded area increased. It must be pointed out that the bedding layer was increased to 14 x 14 finite elements due to the corner loading.

Table 5 shows that increasing the loaded area reduces the bedding stress, the concrete stress and the faulting displacement, but the depth "z" required to be compacted must be increased.

DISCUSSION

It was implicit in the initial raft unit pavement design that civil engineering requirements, procedures and construction times would be observed. In the analysis and design of raft units used for emergency runway repair, it became necessary to consider a repair time of a few hours and a service life measured in days.

In civil engineering terms, faulting between a loaded raft unit and an adjacent unloaded raft unit was not particularly significant, as differences of 50mm could be easily accommodated due to the large diameter wheels of container handling equipment. In emergency runway repair, faulting of 30mm can cause expensive damage to an aircraft, thus it is necessary to reduce faulting between adjacent raft units, and between the newly laid raft units and the existing runway.

The depth of the void in the runway, is not within the control of the repair team. They must have at their disposal the knowledge and the means of reducing to a minimum the repair time. For this reason, it was necessary to provide indicators as to which areas of the existing raft unit design method should be further investigated. This is given in the next section.

CONCLUSIONS

In this paper the existing civil engineering design method

for raft unit pavements has been given, together with a design example. A discussion of certain aspects of the military rapid runway repair requirements have shown that a number of other raft unit design aspects need to be considered as follows:-

1. The existing raft unit design method can provide a basis for designing rapid runway repair using raft units.

2. Deep, weak subgrades must be improved in strength as they increase concrete stress, increase faulting and increase the required depth of compaction.

3. The subbase depth should be extended to the calculated required depth of compaction "z" and should have a minimum CBR of 20%.

4. The bedding layer should be reduced in thickness to 25mm and its CBR increased. It is recommended that a 25mm asphalt concrete layer be used as a bedding.

5. Use large wheel load contact areas as they reduce concrete stress, reduce bedding stress and reduce faulting, but the depth of compaction is increased.

6. Small contact area corner loads should be avoided as they produce high concrete stresses, high bedding stresses and large faulting displacement "S", but low values of compactive depth "z".

ACKNOWLEDGEMENTS

This work has been carried out with the support of Procurement Executive, Ministry of Defence [RARDE, Christchurch].

REFERENCES

1. D. Croney, "The design and performance of road pavements", Dept. of the Environment, TRRL, HMSO, London, 1977.
2. J.W. Bull, "Precast concrete raft units", Blackie and Son, Glasgow, UK, 1990.
3. H.M. Westergaard, "Stresses in concrete pavements computed by theoretical analysis", Public Roads, 7, No. 2, 1926, pp 25-53.
4. J.W. Bull and S.H. Salmo, "The use of the equivalent single wheel load concept for discrete raft-type pavements", Computers and Geotechnics, 3, 1987, pp 29-35.
5. P. Ullidtz and K.R. Peattie, "Pavement analysis by programmable calculation", Transportation Engineering Journal, Sept. 1980.
6. J.W. Bull and A. Singh, "A comparison between the stresses computed for finite sized precast concrete pavement units using Westergaard's equation and a numerical design method", awaiting publication in Computer and Geotechnics, 16p.
7. J.E. Bowles, "Foundation analysis and design", McGraw Hill, New York, 1988.

8. Property Services Agency [PSA], "A guide to airfield pavement design and evaluation, PSA, DoE, Croydon, UK, 1989
9. M.H.H. Ismail, "Research into and the validation of precast concrete pavement units, design method", Ph.D thesis, University of Newcastle upon Tyne, UK, 1990.

The Non-Linear Analysis of a Military Hangar

B.W. Golley
Department of Civil Engineering, University College, Australian Defence Force Academy, Campbell, ACT 2600, Australia

INTRODUCTION

The Australian Army frequently erects standard hangars which are used for a variety of purposes. The hangars are designed for ease of transportation and erection. The main frames, which are available in several spans, have cross sections as shown in Figure 1 and consist of six components which are bolted together in the field. The frames are generally lifted at two of the splices by crane, and are bolted to steel plates cast in concrete foundations. Elastic distortion of the frames during lifting is such that several personel are required to align the frame on the foundation plates.

In this paper, a procedure is described for the non-linear analysis of single loop and multi-loop plane structures. The technique is ideal for analysing this erection problem, taking into account sling length, effect of sling spreaders and location of lifting points.

NON-LINEAR PROCEDURE

Introduction

A shooting method has been developed by Harrison[1–3] for the large deflection analysis of a variety of single loop structures such as unguyed suspension towers, suspension cables and arches. The procedure has also been used for the analysis of carbon fibre suspension systems[4] and C springs[5]. Miller[6] has considered multiple supports on single loop structures and Coulter and Miller[7] have incorporated non-linear material properties. In Harrison's method, the loop is divided into elements connected at nodes. For geometric purposes, the elements are considered to remain straight between nodes. Shortening due to axial force, shrinkage and temperature change is simply incorporated in the analysis. Flexural effects are taken into account by an angular change between elements, the angle being obtained using the small deflection beam theory solution.

The solution of non-linear large deflection problems proceeds as follows. In the case of built in arches and frames, three actions (axial force, shear force and bending moment) are assumed at one end of the loop. Using a shooting method based on the straight elements, displacements (two translations and a rotation) at the second end corresponding to the assumed actions at the first end are determined. In general, the displacements at the second end are not zero, and the actions at the first end of the loop are adjusted using Newton-Raphson iteration and the shooting procedure is repeated. The procedure is repeated until the displacements at the second end of the loop are sufficiently small. Derivatives required in the procedure are obtained numerically by applying in turn small variations to the actions at the first end of the member and using the shooting procedure to determine the effect at the second end. The preceding description applies to single loop structures built in at both ends. The method may be applied to other boundary conditions, but requires separate programs or subroutines for the calculations.

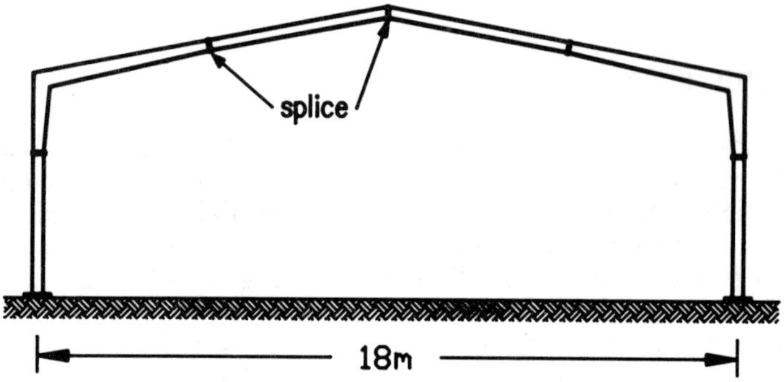

Figure 1. Typical Hangar Frame

The method discussed in this paper differs from Harrison's method in a number of respects.

1. The deformation of elements is obtained using linearised extensional *elastica* theory.

2. Derivatives used in the Newton-Raphson procedure are obtained in a semi-analytical manner.

3. A tangent stiffness matrix relating end actions and end displacements of the loop is determined. Elements of this matrix are obtained semi-analytically. Using the tangent stiffness matrix, any combination of stable boundary conditions at loop ends may be readily treated. In addition, multi-loop structures may be analysed in a straightforward way.

The method is similar in many respects to the procedure described by Qashu and DaDeppo[8] for the large deflection analysis of frames. In their procedure, the differential equations for each loop are integrated using Runge-Kutta forward numerical integration, whereas in this paper the same one-point boundary value problem is solved by considering structural deformation and equilibrium which is perhaps more familiar to engineers.

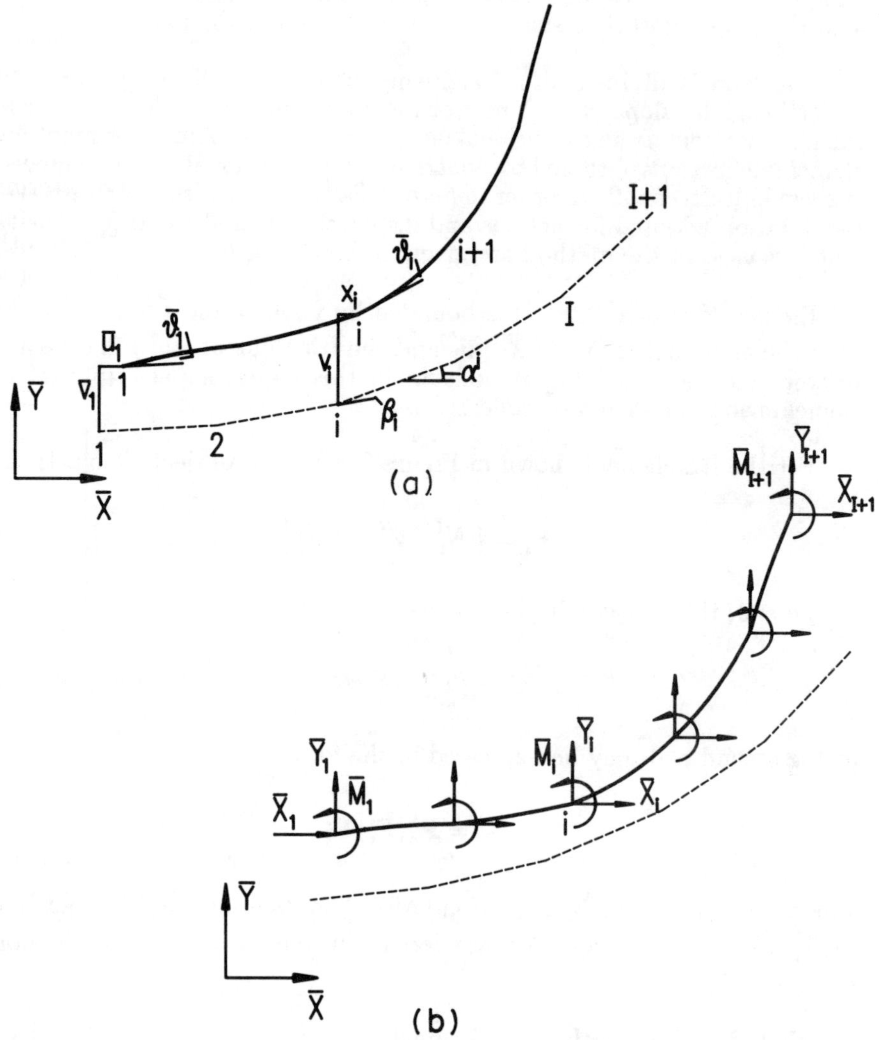

Figure 2. Typical Loop Showing Nodal Actions and Displacements

Solution of single loops

In this section, the determination of end actions and the tangent stiffness matrix of a single loaded loop with prescribed end displacements is discussed. The tangent stiffness matrix is the relationship between the differentials of actions and displacements at the ends of a loaded loop. The results are directly applicable to the solution of single loop structures, such as arches and portal frames, and form an essential part of the analysis of multi-loop structures, such as the hangar frame-sling system considered later.

Procedure A typical loop, forming either a complete structure or a component of a structure, is shown in the undeformed state by the dashed lines in Figure 2. It is assumed that the loop is made up of initially piecewise straight members, although curved members may be readily treated. Under load the loop deforms as shown by the solid lines in Figure 2.

The loop is divided into I elements which are sufficiently short that under load, the slope change between element ends is small. This permits the linearisation of the extensional elastica equations. The determination of element deflections obtained by linearising the extensional elastica equations is given in Reference 9. In order to permit the use of concise matrix notation, the notation adopted for actions and displacements and the transformation matrices used in the method are given in Appendix A.

Element i ($i = 1, 2, \dots I$) is bounded by nodes i and $i+1$. The angle α^i is the angle between the $\overline{X}$ axis and the ith element and β_i is the angle between elements meeting at node i. In this paper, superscripts refer to elements and subscripts to nodes.

For the ith element shown in Figure 3, which is subjected to actions

$$\mathbf{F}_i^i = \{\, N_i^i \quad V_i^i \quad M_i^i \,\}^T \tag{1}$$

at one end, the relative displacements

$$\mathbf{a}^i = \{\, u^i \quad v^i \quad \theta^i \,\}^T \tag{2}$$

at the second end may be expressed in the form

$$\mathbf{a}^i = \mathbf{P}_4^i(\mathbf{F}_i^i)\,\mathbf{F}_i^i \tag{3}$$

where $\mathbf{P}_4^i(\mathbf{F}_i^i)$ is given by Equation (A9). The functions $f_1(\mathbf{F}_i^i)$, $f_2(\mathbf{F}_i^i)$ and $f_3(\mathbf{F}_i^i)$ given in Equation (A9) are derived in Reference 9, and are functions of X_i^i and Y_i^i only.

Node i undergoes the displacements

$$\overline{\mathbf{a}_i} = \{\, \overline{u}_i \quad \overline{v}_i \quad \overline{\theta}_i \,\}^T \tag{4}$$

as shown in Figure 2(a). Concentrated actions may act at the nodes as shown in Figure 2(b). Nodal actions are designated by the vector

$$\overline{\mathbf{A}}_i = \{\, \overline{X}_i \quad \overline{Y}_i \quad \overline{M}_i \,\}^T \tag{5}$$

where $\overline{X}_i$, $\overline{Y}_i$ and $\overline{M}_i$ are as shown in Figure 2(b). At any stage of the non-linear iterative procedure adopted in this paper, end displacements $\overline{\mathbf{a}}_1$

and $\bar{\mathbf{a}}_{I+1}$ are specified. The determination of the corresponding end actions $\overline{\mathbf{A}}_1$ and $\overline{\mathbf{A}}_{I+1}$ and the tangent stiffness matrix $\mathbf{K}$ is carried out in a number of steps.

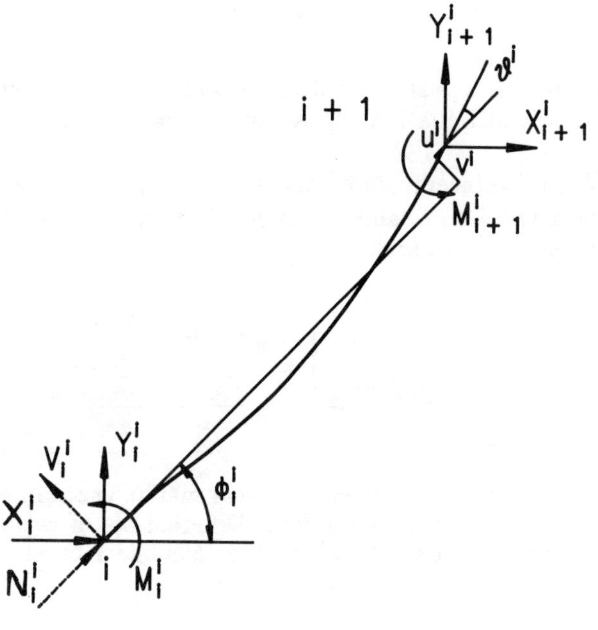

Figure 3. Single Element of Loop

Step 1 This step is similar to Harrison's procedure, but deflections are determined in a different manner and differential relationships are obtained in a semi-analytical way. Actions $\overline{\mathbf{A}}_1'$ at node 1 are assumed. Commencing with the prescribed displacements $\bar{\mathbf{a}}_1$ at node 1, a shooting procedure is used to determine corresponding displacements $\bar{\mathbf{a}}_{I+1}'$ at node $I+1$. In addition a matrix $\mathbf{J}_{12}'$ which contains derivatives of displacements at node $I+1$ with respect to actions at node 1 with the displacements $\bar{\mathbf{a}}_1$ remaining constant, namely

$$\mathbf{J}_{12}' = \left(\frac{\partial \bar{\mathbf{a}}_{I+1}}{\partial \overline{\mathbf{A}}_1^T} \right)' \tag{6}$$

is determined. The derivatives in Equation (3) follow the notation

$$\frac{\partial \bar{\mathbf{a}}_{I+1}}{\partial \overline{\mathbf{A}}_1^T} = \begin{bmatrix} \frac{\partial \bar{u}_{I+1}}{\partial \overline{X}_1} & \frac{\partial \bar{u}_{I+1}}{\partial \overline{Y}_1} & \frac{\partial \bar{u}_{I+1}}{\partial \overline{M}_1} \\ \frac{\partial \bar{v}_{I+1}}{\partial \overline{X}_1} & \frac{\partial \bar{v}_{I+1}}{\partial \overline{Y}_1} & \frac{\partial \bar{v}_{I+1}}{\partial \overline{M}_1} \\ \frac{\partial \bar{\theta}_{I+1}}{\partial \overline{X}_1} & \frac{\partial \bar{\theta}_{I+1}}{\partial \overline{Y}_1} & \frac{\partial \bar{\theta}_{I+1}}{\partial \overline{M}_1} \end{bmatrix} \tag{7}$$

In general, the calculated displacements $\bar{\mathbf{a}}_{I+1}'$ do not equal the specified displacements $\bar{\mathbf{a}}_{I+1}$ and actions at node 1 are corrected using Newton-Raphson iteration. The corrections $\Delta\overline{\mathbf{A}}_1'$ are obtained by solving

$$\mathbf{J}_{12}'\Delta\overline{\mathbf{A}}_1' = \mathbf{a}_{I+1} - \mathbf{a}_{I+1}' \tag{8}$$

The procedure is repeated until the errors in displacements, given by the right hand side of Equation (8), are sufficiently small.

Step 2 With displacements $\overline{\mathbf{a}}_1$ and actions $\overline{\mathbf{A}}_1$ determined as described in Step 1, the actions $\overline{\mathbf{A}}_{I+1}$ and additional matrices of derivatives $\mathbf{J}_{11}$, $\mathbf{J}_{21}$ and $\mathbf{J}_{22}$ are determined, where

$$\begin{bmatrix} \mathbf{J}_{11} & \mathbf{J}_{12} \\ \mathbf{J}_{21} & \mathbf{J}_{22} \end{bmatrix} = \begin{bmatrix} \dfrac{\partial \overline{\mathbf{a}}_{I+1}}{\partial \overline{\mathbf{a}}_1^T} & \dfrac{\partial \overline{\mathbf{a}}_{I+1}}{\partial \overline{\mathbf{A}}_1^T} \\ \dfrac{\partial \overline{\mathbf{A}}_{I+1}}{\partial \overline{\mathbf{a}}_1^T} & \dfrac{\partial \overline{\mathbf{A}}_{I+1}}{\partial \overline{\mathbf{A}}_1^T} \end{bmatrix} \tag{9}$$

Step 3 In this step, the tangent stiffness matrix is obtained using matrices determined in the previous section. Differentials of actions and displacements at node $I+1$ are related to those at node 1 by

$$\begin{Bmatrix} d\overline{\mathbf{a}}_{I+1} \\ d\overline{\mathbf{A}}_{I+1} \end{Bmatrix} = \begin{bmatrix} \mathbf{J}_{11} & \mathbf{J}_{12} \\ \mathbf{J}_{21} & \mathbf{J}_{22} \end{bmatrix} \begin{Bmatrix} d\overline{\mathbf{a}}_1 \\ d\overline{\mathbf{A}}_1 \end{Bmatrix} \tag{10}$$

from which we obtain

$$\begin{Bmatrix} d\overline{\mathbf{A}}_1 \\ d\overline{\mathbf{A}}_{I+1} \end{Bmatrix} = \begin{bmatrix} \mathbf{K}_{11} & \mathbf{K}_{12} \\ \mathbf{K}_{21} & \mathbf{K}_{22} \end{bmatrix} \begin{Bmatrix} d\overline{\mathbf{a}}_1 \\ d\overline{\mathbf{a}}_{I+1} \end{Bmatrix} \tag{11}$$

where

$$\mathbf{K}_{11} = -\mathbf{J}_{12}^{-1}\mathbf{J}_{11} \tag{12a}$$

$$\mathbf{K}_{12} = \mathbf{J}_{12}^{-1} \tag{12b}$$

$$\mathbf{K}_{21} = \mathbf{J}_{21} - \mathbf{J}_{22}\mathbf{J}_{12}^{-1}\mathbf{J}_{11} \tag{12c}$$

$$\mathbf{K}_{22} = \mathbf{J}_{22}\mathbf{J}_{12}^{-1} \tag{12d}$$

The matrix

$$\mathbf{K} = \begin{bmatrix} \mathbf{K}_{11} & \mathbf{K}_{12} \\ \mathbf{K}_{21} & \mathbf{K}_{22} \end{bmatrix} \tag{13}$$

is the 6 × 6 tangent stiffness matrix containing derivatives of end actions with respect to end displacements.

SHOOTING PROCEDURE

Given the displacements and actions at node 1 in a loop, it is required to determine displacements and actions at node $I+1$. In addition, the derivatives given in Equation (9) are required. Commencing with the specified values of displacements and actions at node 1 of element 1, displacements at node 2 are determined using Equation (3). Actions at node 2 of element 1 are then determined by considering the eqilibrium of that element. By considering equilibrium and compatibility at node 2, values of displacements etc. at node 2 of element 2 are evaluated and the procedure is repeated for all elements. Differential relationships are also included in the shooting process.

Prior to commencing the iteration for element i, the following are known:

$$\phi_i^i,\ \mathbf{A}_i^i,\ \bar{\mathbf{a}}_i,\ \mathbf{T}_{11}^i,\ \mathbf{T}_{12}^i,\ \mathbf{T}_{21}^i,\ \mathbf{T}_{22}^i$$

where ϕ_i^i is the angle shown in Figure 3 and the matrices $\mathbf{T}_{11}^i$ etc. relate differentials by the equation

$$\begin{Bmatrix} d\bar{\mathbf{a}}_i \\ d\mathbf{A}_i^i \end{Bmatrix} = \begin{bmatrix} \mathbf{T}_{11}^i & \mathbf{T}_{12}^i \\ \mathbf{T}_{21}^i & \mathbf{T}_{22}^i \end{bmatrix} \begin{Bmatrix} d\bar{\mathbf{a}}_1 \\ d\bar{\mathbf{A}}_1 \end{Bmatrix} \tag{14}$$

We also write for convenience

$$d\bar{\theta}_i = [\,\mathbf{t}_1^i \quad \mathbf{t}_2^i\,] = \begin{Bmatrix} d\bar{\mathbf{a}}_1 \\ d\bar{\mathbf{A}}_1 \end{Bmatrix} \tag{15}$$

where $\mathbf{t}_1^i$ and $\mathbf{t}_2^i$ are the third rows of $\mathbf{T}_{11}^i$ and $\mathbf{T}_{12}^i$ respectively. At the completion of the iteration for element i, the corresponding values for element $i+1$ are required.

The calculations indicated in Equations (16) to (24) are performed for $i = 1, 2, \ldots, I$.

$$\mathbf{F}_i^i = \mathbf{P}_2(\phi_i^i)\,\mathbf{A}_i^i \tag{16}$$

$$\mathbf{a}^i = \mathbf{P}_4(\mathbf{F}_i^i)\,\mathbf{F}_i^i \tag{17}$$

$$\bar{\mathbf{a}}_{i+1} = \bar{\mathbf{a}}_i + \mathbf{b}_1^i(\phi_i^i) + \mathbf{P}(\phi_i^i)\,\mathbf{a}^i \tag{18}$$

$$\phi_{i+1}^{i+1} = \phi_i^i + \theta^i + \beta_{i+1} \tag{19}$$

$$\mathbf{A}_{i+1}^i = -\mathbf{A}_i^i + \mathbf{P}_3^i(\mathbf{a}^i, \phi_i^i)\,\mathbf{A}_i^i \tag{20}$$

$$\mathbf{A}_{i+1}^{i+1} = -\mathbf{A}_{i+1}^i + \mathbf{A}_{i+1} \tag{21}$$

$$\mathbf{R}_1^i = \mathbf{P}_7(\mathbf{F}_i^i)[\mathbf{P}_8(\phi_i^i)\,\mathbf{A}_i^i\,\mathbf{t}_1^i + \mathbf{P}_2(\phi_i^i)\,\mathbf{T}_{21}^i] \tag{22a}$$

$$\mathbf{R}_2^i = \mathbf{P}_7(\mathbf{F}_i^i)[\mathbf{P}_8(\phi_i^i)\,\mathbf{A}_i^i\,\mathbf{t}_2^i + \mathbf{P}_2(\phi_i^i)\,\mathbf{T}_{22}^i] \tag{22b}$$

$$\mathbf{T}_{11}^{i+1} = \mathbf{T}_{11}^i + \mathbf{b}_2^i(\phi_i^i)\,\mathbf{t}_1^i + \mathbf{P}_1(\phi_i^i)\,\mathbf{R}_1^i + \mathbf{P}_5(\phi_i^i)\,\mathbf{a}^i\,\mathbf{t}_1^i \tag{23a}$$

$$\mathbf{T}_{12}^{i+1} = \mathbf{T}_{12}^i + \mathbf{b}_2^i(\phi_i^i)\,\mathbf{t}_2^i + \mathbf{P}_1(\phi_i^i)\,\mathbf{R}_2^i + \mathbf{P}_5(\phi_i^i)\,\mathbf{a}^i\,\mathbf{t}_2^i \tag{23b}$$

$$\mathbf{T}_{21}^{i+1} = \mathbf{T}_{21}^i - \mathbf{P}_9^i(\mathbf{A}_i^i, \phi_i^i)\,\mathbf{R}_1^i - \mathbf{P}_{10}^i(\mathbf{a}^i, \phi_i^i)\,\mathbf{A}_i^i\,\mathbf{t}_1^i - \mathbf{P}_3^i(\mathbf{a}^i, \phi_i^i)\,\mathbf{T}_{21}^i \tag{24a}$$

$$\mathbf{T}_{22}^{i+1} = \mathbf{T}_{22}^i - \mathbf{P}_9^i(\mathbf{A}_i^i, \phi_i^i)\,\mathbf{R}_2^i - \mathbf{P}_{10}^i(\mathbf{a}^i, \phi_i^i)\,\mathbf{A}_i^i\,\mathbf{t}_2^i - \mathbf{P}_3^i(\mathbf{a}^i, \phi_i^i)\,\mathbf{T}_{22}^i \tag{24b}$$

After performing the calculations for $i = I$, we obtain

$$\mathbf{J}_{11} = \mathbf{T}_{11}^{I+1} \tag{25a}$$

$$\mathbf{J}_{12} = \mathbf{T}_{12}^{I+1} \tag{25b}$$

$$\mathbf{J}_{21} = -\mathbf{T}_{21}^{I+1} \tag{25c}$$

$$\mathbf{J}_{22} = -\mathbf{T}_{22}^{I+1} \tag{25d}$$

from which the tangent stiffness matrix $\mathbf{K}$ is obtained using Equation (12). The actions at node $I + 1$ are also determined as

$$\overline{\mathbf{A}}_{I+1} = \mathbf{A}_{I+1}^I \tag{26}$$

TIE ELEMENT

The end actions and tangent stiffness matrix of a tie element corresponding to specified end displacements may be determined analytically in a straightforward manner. This type of element, which neglects flexure, has four degrees of freedom and was found desirable to treat the wire rope sling.

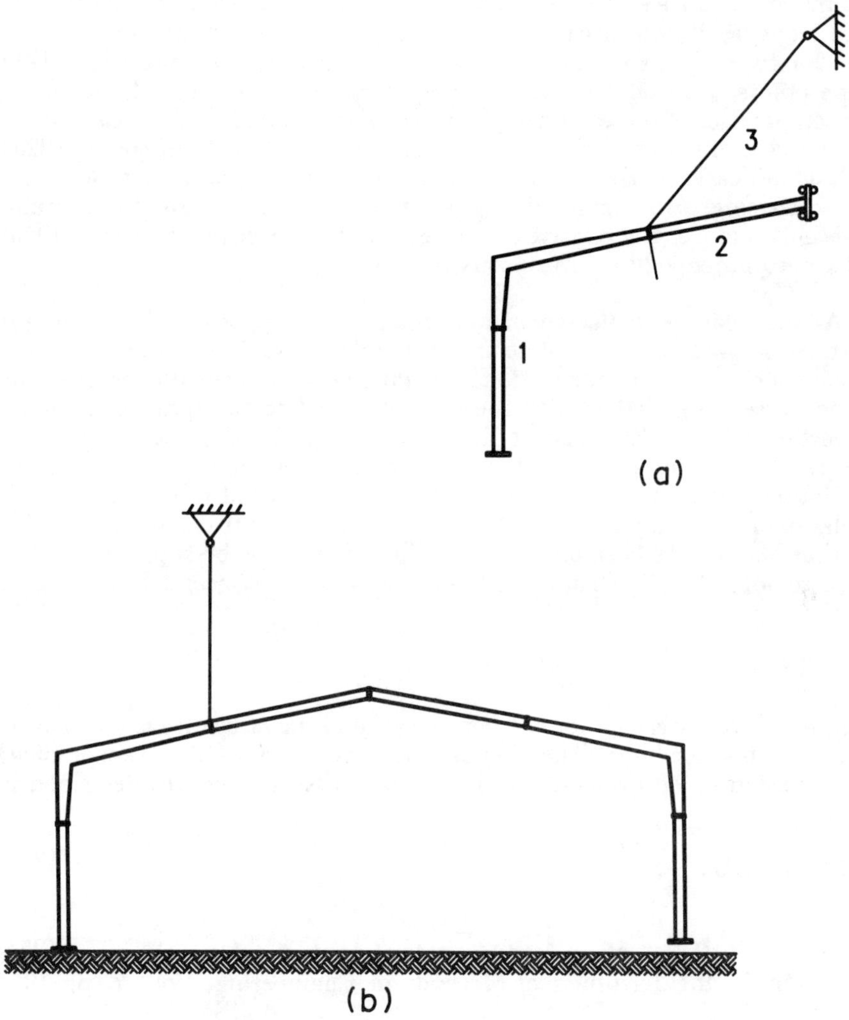

Figure 4. Frame Configurations Analysed

SOLUTION OF MULTI-LOOP STRUCTURES

The structure is divided into loops, each of which may only be connected to another loop at the ends. Supports and free ends are designated master nodes, as are nodes where loops are connected. With known values of displacements and estimated values for unknown displacements at master nodes, end actions and tangent stiffness matrices are determined for each loop as described earlier. Standard procedures are adopted to assemble the global tangent stiffness matrix and equilibrium error vector, and estimated displacements are corrected until errors in equilibrium at master nodes are sufficiently small.

HANGAR ANALYSIS

To simulate lifting a single hangar frame, a half of the frame was divided into three loops, as shown in Figure 4(a). Loop 1 was divided into 15 elements, loop 2 into 8 elements and a single element represented the sling. The lifting points were selected as the splice locations, following the standard erection practice of these frames. The analysis indicated that the base plates moved together 42 mm. This was less than the reported closure, and further analysis indicated horizontal forces of 1.3 kN were required to provide the reqired spacing. Moving the lifting points about 2 m towards the columns essentially removes the relative displacement of the base plates during lifting and is recommended for future construction.

As the relative displacement was smaller than reported, the frame was also analysed with one base plate on the foundation and the frame supported at one splice only, a configuration thought likely to occur during erection. In the worse case, that is with the sling attached to the splice nearest the supported column, the relative displacement of the base plates was 136 mm, requiring a force of over 4 kN to spread the second base plate to the foundation plate. With the sling attached to the second splice, the relative displacement was only 10 mm. The recommendation is to ensure that during erection, should one base plate be fixed first, the frame be supported at the splice remote from the supported column while the second base plate is positioned.

CONCLUSIONS

The procedure described for the non-linear analysis of frames is a relatively simple method for large deflection analysis. Erection problems experienced in the past may be overcome by following simple procedures described in the paper.

REFERENCES

1. Harrison, H.B. "Post-buckling Analysis of Non-uniform Elastic Columns", Int J Numerical Methods in Engineering, Vol. 7, pp 195–210, 1973.

2. Harrison, H.B. "Non-linear Analysis of Unguyed Towers and Stacks", Civil Engg Trans IE Aust, Vol. CE 17, No 1, 1975.

3. Harrison, H.B. Structural Analysis and Design, Pergamon, 1980.

4. Harrison, H.B. "On Grossly Deformed Arches and Carbon Fibres", Proc Eleventh Australasian Conf on the Mechanics of Structures and Materials, Auckland, New Zealand, pp 307–310, 1988.

5. Ng, S.H. "Analysis of Spline Springs Under Large Deformations", MEngSc Thesis, University of Sydney, 1988.

6. Miller, R.E. "Numerical Analysis of a Generalised Plane Elastica", Int J Numerical Methods in Engineering, Vol. 15, pp 325–332, 1980.

7. Coulter, B.A. and Miller, R.E. "Numerical Analysis of a Generalised Plane 'Elastica' with Non-linear Material Behaviour", Int J Numerical Methods in Engineering, Vol. 26, pp 617–630, 1980.

8. Qashu, R.K. and DaDeppo, D.A. "Large Deflection and Stability of Rigid Frames", J of Engineering Mechanics, ASCE, Vol. 109, No 3, pp 765–780, 1983.

9. Golley, B.W., "A Procedure For the Gross Deflection Analysis of Plane Structures", Research Report, Department of Civil Engineering, University College, University of NSW, 1990.

APPENDIX A MATRIX TERMS

The following transformation matrices are used in this paper. In addition, the notation used for actions and displacements is summarised below.

$$\mathbf{a} = \{u \quad v \quad \theta\}^T \tag{A1}$$

$$\mathbf{A} = \{X \quad Y \quad M\}^T \tag{A2}$$

$$\mathbf{b}_1^i = \{l^i(\cos\phi - \cos\alpha^i) \quad l^i(\sin\phi - \sin\alpha^i) \quad 0\}^T \tag{A3}$$

$$\mathbf{b}_2^i = \{-l^i \sin\phi \quad l^i \cos\phi \quad 0\}^T \tag{A4}$$

$$\mathbf{F} = \{N \quad V \quad M\}^T \tag{A5}$$

$$\mathbf{P}_1(\phi) = \begin{bmatrix} \cos\phi & -\sin\phi & 0 \\ \sin\phi & \cos\phi & 0 \\ 0 & 0 & 1 \end{bmatrix} \tag{A6}$$

$$\mathbf{P}_2(\phi) = \begin{bmatrix} \cos\phi & \sin\phi & 0 \\ -\sin\phi & \cos\phi & 0 \\ 0 & 0 & 1 \end{bmatrix} \tag{A7}$$

$$\mathbf{P}_3^i(\mathbf{a},\phi) = \begin{bmatrix} 0 & 0 & 0 \\ 0 & 0 & 0 \\ -(l^i+u)\sin\phi - v\cos\phi & (l^i+u)\cos\phi - v\sin\phi & 1 \end{bmatrix} \tag{A8}$$

$$\mathbf{P}_4^i(\mathbf{F}) = \begin{bmatrix} -\frac{l^i}{EA} & 0 & 0 \\ 0 & f_2(\mathbf{F}) & f_1(\mathbf{F}) \\ 0 & -f_1(\mathbf{F}) & f_3(\mathbf{F}) \end{bmatrix} \tag{A9}$$

$$\mathbf{P}_5(\phi) = \begin{bmatrix} -\sin\phi & -\cos\phi & 0 \\ \cos\phi & -\sin\phi & 0 \\ 0 & 0 & 0 \end{bmatrix} \tag{A10}$$

$$\mathbf{P}_6^i(\mathbf{F}) = \begin{bmatrix} 0 & 0 & 0 \\ V\frac{\partial f_2(\mathbf{F})}{\partial N} + M\frac{\partial f_1(\mathbf{F})}{\partial N} & V\frac{\partial f_2(\mathbf{F})}{\partial V} + M\frac{\partial f_1(\mathbf{F})}{\partial V} & 0 \\ -V\frac{\partial f_1(\mathbf{F})}{\partial N} + M\frac{\partial f_3(\mathbf{F})}{\partial N} & -V\frac{\partial f_1(\mathbf{F})}{\partial V} + M\frac{\partial f_3(\mathbf{F})}{\partial V} & 0 \end{bmatrix} \tag{A11}$$

$$\mathbf{P}_8(\phi) = \begin{bmatrix} -\sin\phi & \cos\phi & 0 \\ -\cos\phi & -\sin\phi & 0 \\ 0 & 0 & 0 \end{bmatrix} \tag{A12}$$

$$\mathbf{P}_9(\mathbf{A},\phi) = \begin{bmatrix} 0 & 0 & 0 \\ 0 & 0 & 0 \\ -X\sin\phi + Y\cos\phi & -X\cos\phi - Y\sin\phi & 0 \end{bmatrix} \tag{A13}$$

$$\mathbf{P}_{10}^i(\mathbf{a},\phi) = \begin{bmatrix} 0 & 0 & 0 \\ 0 & 0 & 0 \\ -(l^i + u)\cos\phi + v\sin\phi & -(l^i + u)\sin\phi - v\cos\phi & 0 \end{bmatrix} \tag{A14}$$

Retro-Fitting of Expedient Protection for Buried Structures

M.C.R. Davies, A.B. Ismail

School of Engineering, University of Wales College of Cardiff, U.K.

ABSTRACT

Buried structures, such as basements and field defences, often lack sufficient protection against loading resulting from the propagation of blast waves through soil. The installation of a barrier can offer some protection to the structure by attenuating the intensity of the stress wave. Such barriers may be installed rapidly to provide expedient protection to existing structures or temporary defensive excavations. By considering the transmission and diffraction of stress waves by protective barriers, a theoretical assessment of the use of protective barriers to reduced blast loading on buried structures is conducted.

INTRODUCTION

The current method of design against blast loading for buried structures is to provide a structure which can absorb the blast loading. Occasionally sacrificial elements, such as blast slabs, are included in the design. The provision of a strengthened structure can often be highly expensive or difficult to construct. However, all too frequently buried structures, such as basements and trenches, lack sufficient protection against blast.

An alternative to strengthening the structure is to provide a means of reducing the level of stress which reaches the structure by attenuating the amplitude of the blast pressure wave. Such means of protection may be included in the design of new structures but, more importantly, may be retro-fitted to existing facilities affording expedient protection against ground blast. Even in situations when structures have been specifically constructed to withstand blast loading, levels of vibration may be detrimental to sophisticated equipment, such as computers, located in basements and means of reducing this shock loading are desirable.

In the paper the theoretical aspects of a proposed method of retro-fitted expedient protection using a barrier are considered.

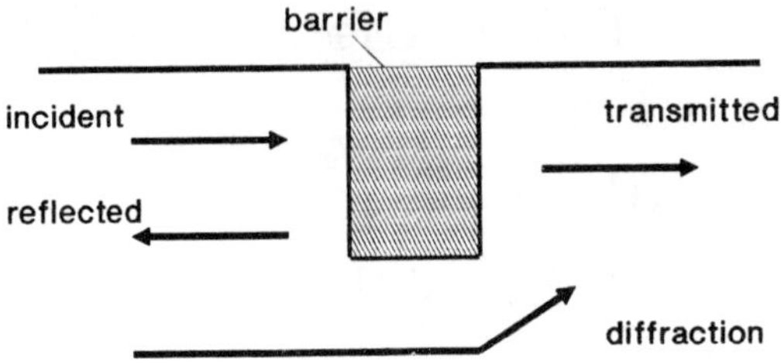

Figure 1. Body waves resulting from the interception of a propagating P wave with a barrier.

Stress wave attenuation using barriers

When a body wave travelling through soil encounters a barrier of infinite width and depth part of the energy of the wave may be transmitted through the barrier with the remainder of the energy being reflected by the barrier. However, in practice the barrier will be of finite dimensions allowing energy to be diffracted around the sides or base of the barrier, as shown in Fig. 1. As will be shown below, the proportion of the energy in the wave being transmitted through or diffracted around the extremities of the barrier - and hence the efficiency of the barrier as a means of protection - is governed by the nature of the body wave and by the dimensions and material properties of the barrier.

It is well established from the theory of wave propagation in an elastic medium, e.g. Timoshenko and Goodier [1], that the most efficient barrier to prevent the transmission of elastic propagating waves is a void. Energy is reflected back into the medium since it cannot be transmitted across the void. Woods [2] conducted a series of field trials to investigate the attenuation of surface waves, such as might result from the vibration of machines, using open trenches. The tests revealed that high degrees of surface wave attenuation could be achieved using an open trench provided the trench was constructed to a sufficient depth (in this case at least 1.2 times the wavelength of the surface wave) in order to reduce stress wave diffraction. When considering the protection of buried structures or basements from body waves resulting from buried explosives it may not be practical to surround the structure with a void which may require support or become inundated with water. A moat around a structure might also result in undesirable restriction of access to the facility. A barrier constructed from a material which provides attenuation without some of these disadvantages might be more desirable.

WAVE PROPAGATION IN SOIL

Dynamic stress waves

The energy released from an explosion is propagated through the soil by body waves. These are compressive (P) waves, in which the particle velocity of the soil is parallel to the direction of propagation, and shear (S) waves, in which the particle velocity is perpendicular to this direction. As a result of its greater velocity the P wave will arrive at the structure before the S wave. A contained blast will generally result in the localised loading of an element of a structure, and the normal loading from the P wave is considered potentially more damaging than the shear stresses imposed by the S wave (this is unlike dynamic loading resulting from earthquakes, in which the large ground movements caused by the S wave affect the complete structure and are considered more damaging). It is therefore important that in order to be effective barriers should attenuate the P wave component of the blast wave.

In an elastic soil it may be shown (e.g. Richart et al [3]) that the normal stress, σ_0, resulting from a propagating P wave is a function of the mass density of the soil, ρ, the velocity of the wave through the soil, c, and the particle velocity, v, viz:

$$\sigma_0 = \rho\, c\, v \tag{1}$$

The values of the mass density and velocity of propagation terms in Equation 1 depend on the soil type through which the wave is propagating. The product ρc is known as the characteristic impedance, R.

Transmission of stress waves

At the boundary between two media (i.e. the soil/barrier interface) continuity dictates that the particle velocity, v, in both materials must be the same. Thus the sum of the particle velocities of the incident wave, v_I, and the reflected wave, $-v_R$, will be equal to the particle velocity of the transmitted wave, v_T, i.e.

$$v_I - v_R = v_T \tag{2}$$

The stresses either side of the interface must be in equilibrium hence,

$$\sigma_I + \sigma_R = \sigma_T \tag{3}$$

where σ_I, σ_R, σ_T are the incident, reflected and transmitted normal stresses respectively. Using Equation 1 the particle velocities in Equation 2 may be expressed in terms of the appropriate stress and characteristic impedance for the materials on each side of the interface. Submitting these into Equation 3 yields the relationship between the stress transmitted from medium 1 into medium 2, viz

$$\sigma_T = \frac{2\,\sigma_I\,R_2}{R_2 + R_1} \tag{4}$$

where R_1 and R_2 are the characteristic impedances for medium 1 and medium 2, respectively. Equation 4 indicates that if the characteristic impedance of medium 2 is less that of medium 1 the stress wave will be attenuated.

Intensity transmission coefficient, C_T

The energy associated with a wave may be quantified by calculating its intensity, I, which is the average flow of energy through an area. For a plane harmonic wave the intensity is given by (e.g. Sears, [4])

$$I = \frac{\sigma_{max}{}^2}{2\,\rho\,c} \tag{5}$$

where σ_{max} is the peak pressure amplitude. The ratio of the intensities of the incident and transmitted waves is termed the intensity transmission coefficient, C_T, i.e.

$$C_T = \frac{I_T}{I_I} \tag{6}$$

TRANSMISSION OF STRESS WAVES THROUGH A BARRIER

A propagating wave travelling from the source of disturbance (i.e. a blast) to a structure protected by a barrier travels through three media; (1) the soil between the disturbance and the barrier, (2) the barrier, and (3) the soil between the barrier and the structure. It may be shown (e.g. Stumpf, [5]) that the ratio of the intensity of the wave incident on the barrier and the wave transmitted into medium 3 from the barrier, C_T, is

$$C_T = \frac{4\,(R_1/R_3)}{[\{1+(R_1/R_3)\}^2.\cos^2 k_2\,t]+[\{(R_1/R_2)+(R_2/R_3)\}^2.\sin^2 k_2\,t]} \tag{7}$$

where t is the thickness of medium 2 (the barrier) and R_1, R_2 and R_3 are the characteristic impedances of the three media. The term k_2 is the wave number of the propagating stress wave, which has a frequency f and wave length λ, in medium 2, through which it travels at velocity c_2, i.e.

$$k_2 = \frac{2\pi}{\lambda} = \frac{2\pi f}{c_2} \quad (8)$$

Generally the soil will be the same either side of a barrier installed to protect a structure (i.e. media 1 and 3 are the same) and Equation 7 becomes

$$C_T = \frac{4}{[4\cos^2 k_2 t]+[\{(R_1/R_2)+(R_2/R_1)\}^2.\sin^2 k_2 t]} \quad (9)$$

It follows from the consideration of the transmission of stress above that in order for the barrier to attenuate the intensity of the propagating wave its characteristic impedance, R_2, should be less than that of the surrounding soil. This may be achieved by selecting a material for the barrier which has a lower density, ρ, than the surrounding soil or causes a reduction in the propagation velocity, c, of the stress wave.

material	mass density, ρ kg/m^3	propagation velocity, c m/s
soil	2000	1500
polystyrene	40	1820
water	1000	1481
air (at 20°C)	1.2	343

Table 1. Typical values of material constants

Table 1 gives typical values of ρ and c for soil and three potential barrier materials, polystyrene, water and air (at 20°C). A comparison of the attenuation achieved from barriers made from these three materials can be made by substituting the appropriate values into Equation 9. Figure 2 displays the comparison of the variation in intensity transmission coefficient, C_T, with normalised thickness of barrier, t/λ; where λ (= c/f) is the wavelength of the propagating wave. The diagram clearly indicates that, as would be expected from the discussion above, an air void provides the most efficient barrier allowing negligible transmission of stress. By contrast, water displays only very little attenuation at high normalised thicknesses. As the

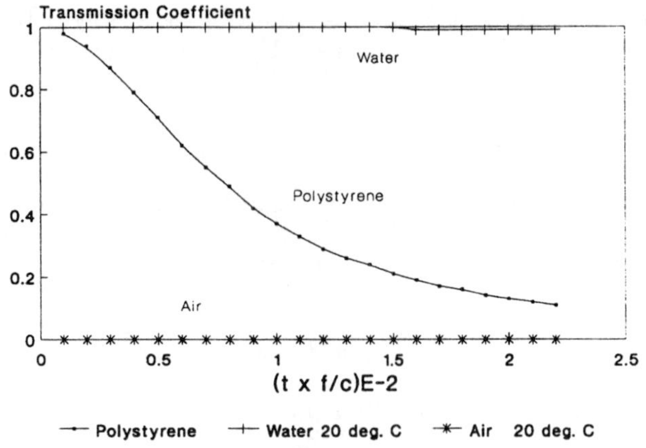

Figure 2. Variation in intensity transmission coefficient with normalised thickness of barrier - for barriers constructed from water, polystyrene and air (at 20°)

normalised thickness of the barrier is increased Fig. 2 indicates the attenuation provided by a polystyrene barrier increases.

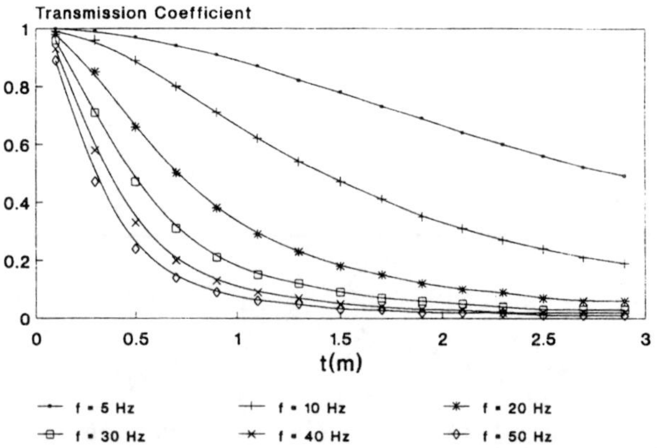

Figure 3. Variation in intensity transmission coefficient with thickness of polystyrene barrier for waves of different frequency

The effect of the frequency (and hence wavelength) of the propagating wave on the predicted value of the intensity transmission coefficient is shown in Fig. 3. In addition to decreasing with thickness of barrier, the intensity transmission coefficient may also be seen to decrease as the blast frequency increases. For example, the value of intensity transmission coefficient predicted for a barrier of 1m thickness is 0.9 for a frequency of 5Hz but reduces to 0.1 if the propagating wave frequency is 50Hz. It is therefore essential that the characteristics of the typical blast wave in the soil are known if the attenuation provided by the barrier is to be correctly predicted.

DIFFRACTION OF STRESS WAVES

Diffraction is the term applied to the bending of waves around an obstacle. The earliest attempt to model this phenomena observed in light waves was conducted in the early nineteenth century by Fresnel; after whom this analytical approach to diffraction is named. Fresnel postulated that a wave front may be divided into zones, each of which obeys Huygens principle, i.e. each contributes to the intensity ahead of the wave front. The combined "interference" of the zones results in the intensity of the wave. If a wave propagating from a source, S, Fig. 4, meets a barrier, the contribution of the wave intercepted by the barrier is eliminated and the intensity at point P results from the contribution of the zones which comprise the lower part of the wave only. Since only half the wave front is effective at point P the amplitude of the wave (the peak stress, σ_{max}, for a propagating P wave) will be 50% of the amplitude of the complete wave front. It can be seen from Equation 5 that under these conditions the ratio of the intensity, I, at point P to the intensity of the wave at point P if there were no barrier present, I_0, is 0.25. The ratio $I:I_0$ is termed the relative intensity.

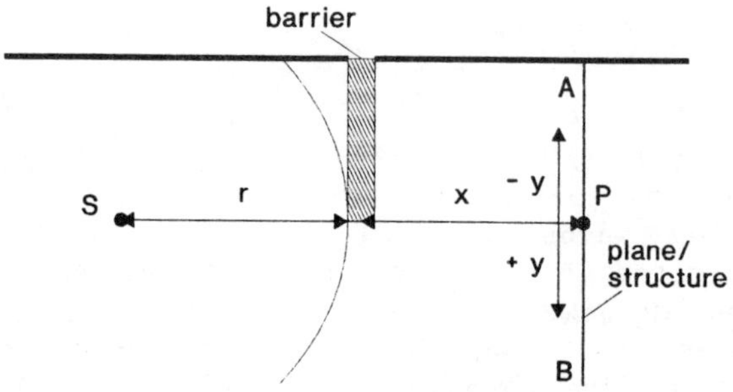

Figure 4. Diffraction of a propagating wave by a barrier

In the zone behind the barrier (the "geometrical shadow") the zones of the lower part of the wave front continue to contribute to the intensity, thus giving the impression that the wave is bending around the barrier.

Fresnel showed a consequence of the curvature of the wave front is that the size of each zone, and thus its contribution to the intensity of the wave in the geometrical shadow, is reduced as the distance from the zone to the plane AB, Fig. 4, increases. Therefore, the intensity of the diffracted wave front on plane AB decreases with distance above the base of the barrier.

Herein the analytical theory of "Fresnel diffraction" developed for light, e.g. Sharman [6], is used to predict the intensity, I of a wave front on a plane AB (i.e. the structure), Fig. 4. For a source at S (i.e. an explosion), of wave length λ, located a distance r from the barrier, the intensity, I, at elevation y on plane AB, a distance x from the barrier, is given by,

$$I = \frac{1}{2} \frac{r.x.\lambda}{(r + x)} \{[\tfrac{1}{2} + x(v)]^2 + [\tfrac{1}{2} + y(v)]^2\} \qquad (10)$$

where

$$x(v) = \int_0^v \cos \frac{\pi}{2} v^2 \, dv \qquad (11a)$$

$$y(v) = \int_0^v \sin \frac{\pi}{2} v^2 \, dv \qquad (11b)$$

and v is a geometric parameter, i.e.

$$v = y\,[2\,(r + x)(r.x.\lambda)]^{\frac{1}{2}} \qquad (12)$$

For large values of r relative to x Equation 12 simplifies to

$$v = y\,[2/(x.\lambda)]^{\frac{1}{2}} \qquad (13)$$

Predictions of relative intensity (I/I_0) at different locations on plane AB are presented in Fig. 5. for propagating waves of different frequencies. Results for planes located at x = 1.0m and x = 10.0m from the barrier are presented in Fig. 5 (a) and Fig. 5 (b) respectively. Both plots show that, as has been discussed above, at the elevation of the base of the barrier, y = 0, the relative intensity is 0.25. Above this level (-ve y) the intensity drops off more rapidly the closer the plane is located to the barrier. In field tests of "passive" isolation barriers Woods [2] observed an increase in surface wave amplitude with distance behind the barrier. This behaviour is consistent with the predicted Fresnel diffraction behaviour. As the frequency of the P wave is decreased, resulting in an

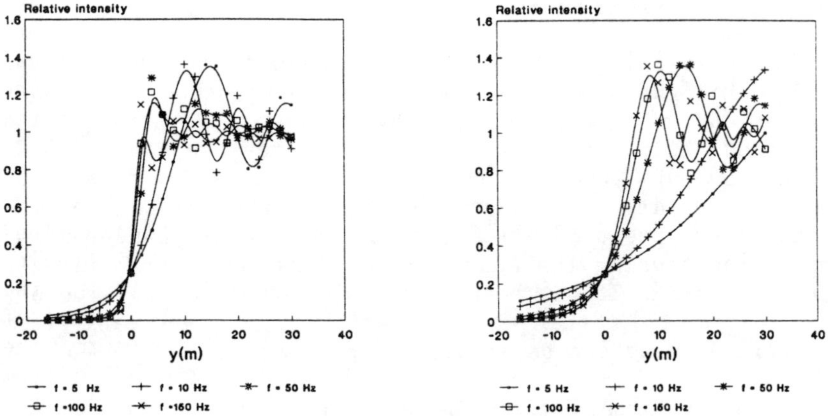

(a) x = 1.0m (b) x = 10.0m (see Fig. 4)

Figure 5. Relative intensity against distance along y axis

increase in the wavelength, Fig. 5 indicates that the relative intensity in the geometrical shadow is increased. However, this never exceeds the value of 0.25 at y = 0.

At elevations below the barrier (+ve y) a series of maxima and minima are observed, the peaks of which reduce with increasing y. This behaviour results from the constructive and destructive interference of the strips; each of which is $\lambda/2$ out of phase. The peaks reduce with the value of relative intensity, converging on unity - the intensity of on unobstructed wave - as the influence of the barrier reduces with increasing y. The distance, y, for convergence increases with wavelength.

These observations indicate that below the barrier there is a magnification of P wave intensity. Therefore barriers should be constructed to at least the depth of the structure to be protected. Since the level of relative intensity decreases with distance above the base of the barrier, improved protection can be obtained from barriers which extend to a greater depth that the structure.

PROTECTIVE BARRIERS IN SOIL

The above analyses indicate that, provided the theories of sound and light wave propagation are applicable to P wave propagation in soil, the energy transmitted through and diffracted below protective barriers may be quantified. Attenuation of stress waves in soil is also caused by geometric damping. The process of geometric damping results as the waves spread out from the source of disturbance during propagation causing the energy to be dissipated over an ever increasing area. It has been shown from experimental studies, e.g. Davies [7], that P wave attenuation resulting from geometric damping can be accurately predicted.

It is important that the effects of geometric damping are included when considering the protection afforded to a buried structure by a barrier. Geometric damping will result in the attenuation of the intensity of a P wave arriving at a barrier from a source. It will also reduce the intensity of both transmitted and diffracted components of the wave as they travel on to the structure. Experimental studies, using centrifuge models, are currently being conducted to investigate this phenomenon and thus validate the above analyses for application in geotechnical engineering.

The consideration of P wave transmission through and diffraction below barriers herein, indicates that attenuation may be achieved if a barrier constructed from a material with a low characteristic impedance is installed to the depth of the base of the structure. As has been previously discussed, a void would provide the greatest protection but presents problems of excavation support and the possibility of inundation with water (which would negate any benefits gained). Expanded closed cell foams, such as polystyrene, appear to be the most appropriate protective barrier material. This may be installed as pre-formed blocks or mixed in-situ. Since the density of expanded foam is considerably less than its constituent components, from a logistical stand point, the latter method of installation would be more appropriate in the case of rapidly assembled military structures.

It is envisaged that for the protection of the majority of military structures such barriers may be rapidly installed using conventional plant. However, the use of techniques employed for the excavation of deep trenches for diaphragm walling could be used for the installation of protective barriers to greater depths, such as for the protection of deep basements.

CONCLUSIONS

Applying the theories for the propagation of sound and light waves to the propagation of P waves in soil, a theoretical assessment of the use of protective barriers to reduced blast loading on buried structures has been conducted. The major conclusions of this study are

1. The intensity of stress transmitted through a barrier is a function of the relative intensities of the soil and the material forming the barrier, the thickness of the barrier and the frequency (or wavelength) of the P wave.

2. Diffraction of P wave energy results in loading of the structure in the geometrical shadow of the barrier and an amplification of intensity a some locations below the base of the barrier. The relative intensity of the diffracted wave is a function of the location of the structure from the barrier and the frequency of the wave.

3. Materials having a low characteristic impedance are the most efficient for use as protective barriers. It is recommended that barriers should extend to the full depth of the structure.

4. The intensity of a propagating P wave is attenuated by geometric damping. This effect should be considered when predicting the intensity of P wave arriving at a structure protected by a barrier.

5. Protective barriers may be rapidly installed to provide expedient protection against the effects of blast waves on existing buried structures.

ACKNOWLEDGEMENT

The financial support of the Royal Armaments Research and Development Establishment, Christchurch, U.K. for the research in dynamic soil/structure interaction at University of Wales College of Cardiff is gratefully acknowledged.

REFERENCES

1. Timoshenko, S. and Goodier, G.N. Theory of Elasticity, McGraw-Hill, New York, 1951.

2. Woods, R.D. Screening of Surface Waves in Soils, Journal of Soil Mechanics and Foundations Division, ASCE, Vol. 94, SM4, 1968, 951-979.

3. Richart, F.E., Hall, J.R and Woods, R.D. Vibrations of Soils and Foundations, Prentice-Hall, Englewood Cliffs, New Jersey, 1970.

4. Sears, F.W. Mechanics, Wave Motion and Heat, Addison-Wesley, New York, 1958.

5. Stumph, F.B. Analytical Acoustics, Ann Arbor Science Publishers, 1980.

6. Sharman, R.V. Vibrations and Waves, Butterworths, London, 1963.

7. Davies, M.C.R. Physical Modelling of Impact Loading of Buried Structures, in Structures Under Shock and Impact (Ed P.S. Bulson), Proceedings of the 1st Int. Conf. on Structures Under Shock and Impact, Cambridge, Massachusetts, Elsevier, Amsterdam, 1989, 431-442.

SECTION 2: RAPIDLY ASSEMBLED AND ACTIVE CIVIL STRUCTURES

Temporary Seating Assemblies

J.F. Dickie, S.J. Gibbs

Engineering Department, University of Manchester, Oxford Road, Manchester M13 9PL

ABSTRACT

The principal forms of temporary seating assemblies possess decks with retractable supports or decks supported by temporary structure. All of the seating deck support structures whether modular,system scaffold or standard tube and fitting possess joint slackness. Dynamic loading results from audience behaviour; of particular significance are synchronised spectator sway and spectator rhythmic dancing. The design significance of these loading forms is examined using a computer program that incorporates the nonlinearities introduced by joint slackness. Current code recommendations are also considered using additional evidence from both field and laboratory studies.

INTRODUCTION.

There is a multitude of events that spectators wish to observe and hear. The spectrum of arenas ranges very widely. A single purpose facility may be a concert hall or a golf course. A multi-purpose facility will 'stage' any form of event sporting or otherwise and may be covered or in the open air. A common requirement for the vast majority of events is the provision of seating.

A stage can create sightlines for audiences standing or seated on a common level. The concern herein relates to events where the spectators observe events below them. Sightlines,comfort and safety are the principal considerations whether within a public hall,cultural centre,school gymnasium or ancient monument. Baseline requirements are present in these areas,economics simply governs the distancing of the facility from the minimal provision.

Fig.1 Retractable seating. (Audience Systems Ltd.)

Fig.2 Demountable seating.
(Arena Promotional Facilities Ltd.)

Fig.3 Demountable seating upon system scaffolding.

Fig.4 Storage of retractable seating.

Space management may be achieved using the flexibility of telescopic or demountable systems. The former is illustrated in Fig.1 and is used indoors. Folding,telescopic or retractable seating is definable as a structure for tiered seating of persons and whose overall shape and size may be reduced without being dismantled for purposes of moving or storing. Demountable systems are shown in Figs.2 and 3 and may be used both indoors and outdoors. Fig.2 illustrates a modular system and Fig.3 shows a stand erected using system scaffold.

SEATING GEOMETRY

Minimum requirements are dictated by the Guide to Fire Precautions in Existing Places of Entertainment and the Like Premises,[14], and the Guide to Safety at Sportsgrounds,[13]. The latter is normally referred to as the Green Guide. No provisions in the Green Guide are mandatory however it is widely used by local authorities as the authoritative document. Means of escape and hall/auditorium capacity have a particular bearing on overall arrangements for indoor seating. Distance from exit/place of safety is particularly significant. There should in all cases be an unobstructed seatway of at least 305mm measured between the back of the one seat and the front of the seat immediately behind it.

The seating assigned to each person shall be not less than:-

a. 760 mm deep where backs are provided to the seats or 600 mm deep where backs are not provided; and
b. 500 mm wide where arms are provided to the seats or 450 mm wide where arms are not provided.

The number of seats in a row should not exceed:-

a. 7 seats where there is a gangway at one end only except that up to a maximumum of eleven seats may be permitted if the 305 mm seatway is increased by 25 mm for each additional seat over 7; and
b. 14 seats where there is a gangway at each end. However larger rows of up to 22 seats with a gangway at each end may be allowed if the seatway is increased by 25 mm for each additional pair of seats and/or odd seat over 14.

These indoor provisions taken from Reference [14] are more onerous than the Green Guide which should be applied outdoors.

The maximum local seating density permitted is 3.60 persons/m^2 and would apply to bench seating. The provision of backs reduces this figure to 2.92 persons/m^2 and where arms are also provided the figure is further reduced to 2.63 persons/m^2. It is of interest to note that an individual can stand within a rectangular space of 460 mm x 280 mm which is equivalent to 7.77 persons/m^2.

Disagreements occur as to the definition of the radial passageways between seating. Are they stairs or gangways? The Oxford dictionary in defining a gangway refers specifically to

" a passage,especially between seats".

The Green Guide adopts this definition to apply to tiered seats. A gangway permits a greater rake and relaxes guarding requirements. Gangways need not contain intermediate landings; provisions on the number of steps and turns that apply to stairways are not appropriate. Nevertheless the designer must recognise that stairs are a significant hazard to individual users. Annual figures for the UK alone are approximately 600 fatalities.

Gangways should be positioned so as to provide ready access to exits and should have a slip resistance surface. No spectator in any part of the seated area should have to travel more than 30 m, measured along the line of seating and the gangway, from their seat to an exit or place of safety. Gangways should be of sufficient width for the seats served and should be not less than 1.1 m wide. The gradient should not exceed that achieved by step dimensions for stairways that are stated in Reference [13] as

"The preferred going should be 305 mm but not less than 280 mm; the preferred height of the risers should be 150 mm but not more than 190 mm."

In essence the seating and gangway requirements dictate static loadings that are globally in the order of 3 persons/m^2. Should an audience remain at their seats whilst others queue to gain access to the gangway local density can be 6 persons/m^2.

BARRIER PROVISION.

BS 6180:1982, Protective barriers in and about buildings (currently under revision) gives recommendations for the general design and construction of temporary and permanent protective barriers to be provided in and about buildings and places of assembly,such barriers being positioned and designed to protect persons from various hazards and to restrict or control the movements of persons.

Various barrier positions are indicated in Fig.5. Section 181 of the Green Guide states that any stepped side gangway (i.e. with viewing accommodation on one side only),position A, should be provided with a hand rail fixed at a height of not less than 840 mm (measured vertically from the line of the nosing). The rail should be designed for 3.4 kN/m width where at right angles to the direction of flow and 2.2 kN/m width where parallel to the direction of flow. Section 122 of the Green Guide concerning stairs is applicable. Where guarding is next to open wells or spaces,position B, such guarding should conform to Building Regulations but should be not less than 1.1 m height (measured vertically from the nosing of the step). The guarding should have a strength equivalent to that of crush barriers of the weaker type, i.e. 3.4 kN/m width design force. Section 183 is also applicable. Those sections of balustrading at the foot of gangways or stairways,position F, should be as strong as crush barriers of the stronger type. The author interprets this as 5 kN/m width.

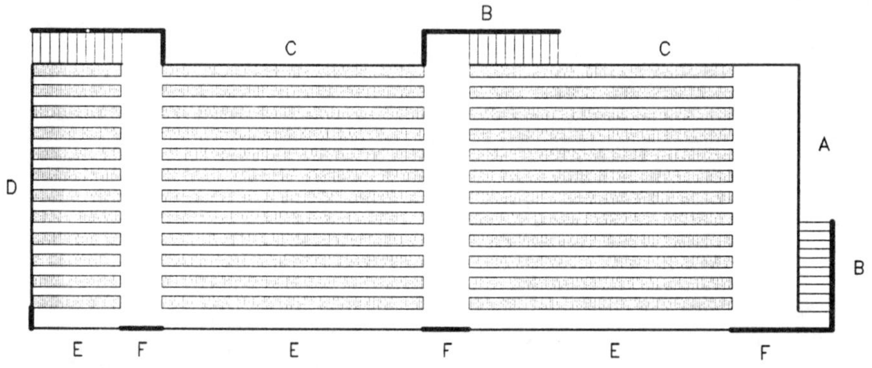

Fig.5 Perimeter guardrails.

Guard rail provision is also necessary for spectators occupying perimeter seats as indicated at positions C and D in Fig.5. BS 6180 does refer to spectators sitting within a specific distance from a guardrail however it is effectively directed at theatre balconies. The ISE guide relating to the design of demountable grandstands,[15], recommends that guardrails in these positions should be designed for 1 kN/m width. Height requirements for rear barriers, position B in Fig.5, are not adequately covered by design codes. The tendency of spectators is to sit on the rear guard rails even where adequate stewarding is present. Local authorities are increasingly demanding higher rear guardrails following the case of

JONES -v- HARVEY GOLDSMITH

Whilst attending a Rod Stewart pop concert at Earls Court Exhibition Centre,Mr Colin Jones stood up during the course of the concert and leaned back. Immediately behind him was a dark curtain. The curtain hid a drop of about 9.1 m to a concrete floor below. Colin's fall from the back of the balustrade resulted in very serious injuries. The plaintiff's solicitors, Russell-Cooke, Potter and Chapman contended that the Defendants failed to comply with the Greater London Council's (GLC) code of practice for pop concerts and with the GLC's recommendation for balustrades which required a minimum height of 42 in (1067 mm). Consequently they contended that the seating was unsafe by reason of the height of the balustrade which was below the minimum recommended height and furthermore did not take into account the excitable nature of audiences attending a pop concert. It was also contended that there was a breach of common duty of care under the Occupiers Liability Act 1957 and a breach of the District Surveyors duty under Section 26 of the London Building Acts 1939. Finally it was contended that the GLC were negligent in failing to ensure that there were sufficient attendants to maintain public safety.

The plaintiff was awarded £625,000 agreed damages.

An outstanding issue concerns the loading criteria for the higher rear guardrails likely to result as a consequence of this litigation.

STATIC LOADING.

Basic loading requirements are provided by BS6399, Reference [7]. The occupancy class is Public Assembly and the live loading for assembly areas with fixed seating is 4 kN/m^2 and 5 kN/m^2 in gangways. International practice adopts similar values for basic floor loadings. Given that local densities could be 6 persons/m^2, a bodyweight of 75 kg gives a loading of 4.5 kN/m^2. Limit state design requires a load factor of 1.6 consequently the actual loading is adequately covered. The Standard states explicitly that imposed load values whilst allowing for small dynamic effects do not allow for dynamic loads due to crowds. It is also recognised that the magnitude of the dynamic load effect depends upon the response of the structural system.

An additional floor loading requirement of BS6399 specifies a concentrated loading of 3.6 kN. It is essential to recognise that all demountable stands used indoors will exceed this value unless special provision is made to spread individual leg loadings. Older wooden flooring is particularly at risk.

DYNAMIC LOADING.

Transient loadings occur from spectator movements such as rising or sitting. The UK provision is contained in BS 5973,[5]. There is a requirement that bracing should be provided in both directions to accommodate 5% of the vertical imposed loading. This is to account for horizontal surges. The German Standard, DIN 1055, adopts a similar figure.

The US provision is contained in Reference [1] and requires:-

a.	Horizontal line load parallel to seats.	360 N/m
b.	Perpendicular to seats.	150 N/m

Spectator loadings were extensively examined by Tuan and Saul,[17]. Their sample comprised of 22 individuals of average weight 70 kg with a range of 50-98 kg. A complete range of spectator movements was considered

Sitting
- standing up suddenly
- swaying sideways
- swaying back and forth

Standing
- jumping
- continuous random jumping
- falling into a seat
- swaying sideways
- swaying back and forth
- flexing at the knees
- periodic jumping.

The statistically reduced statically equivalent horizontal loading components suggested for design were as follows :-

a.	Parallel to rows of seats	156 N/seat.
b.	Perpendicular to rows of seats	200 N/seat.

For seats 460 mm in width these values become 336 N/m and 438 N/m respectively.

Tuan and Saul also obtained a maximum force of 120 N for an individual swaying at a frequency of 0.64 Hz.

Ebrahimpour and Sack,[9], extended this work using groups of four people performing in unison. For four persons swaying at a frequency of 1 Hz a maximum force of 712 N was obtained equivalent to 178 N/person. The increase may be attributed to group synergy.

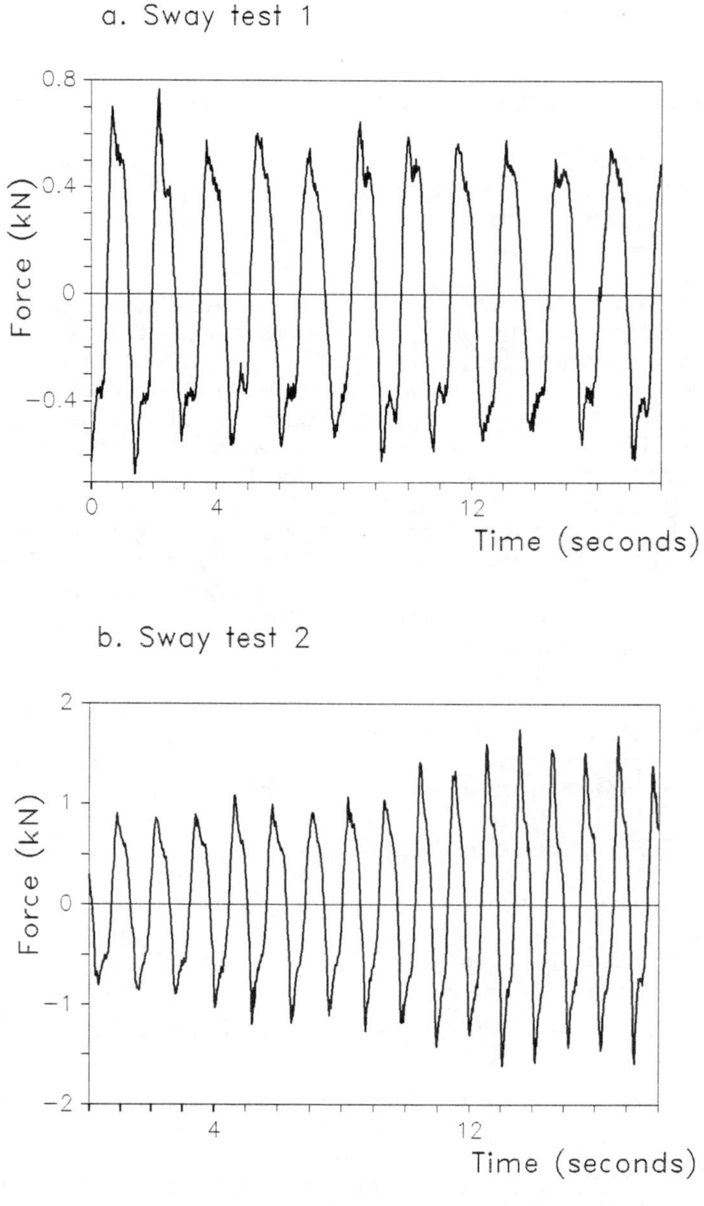

Fig.6 Sway forces.

The authors have measured dynamic forces induced by six people swaying on a single bay of seating in a laboratory study. The testing equipment involved a bank of six seats mounted on four triaxial load cells linked to a microcomputer with a sampling rate of 50 Hz. Two forms of sway test were carried out:-

1. Moderate sway 0.6 Hz
2. Energetic sway 0.9 Hz

Fig.6 shows typical test records obtained by the authors.

For test form 1 the maximum force of 766 N corresponds to a force of 128 N/person.
For test form 2 the maximum force of 1745 N corresponds to a force of 291 N/person.

The latter value is substantially higher than values adopted in current design practice and would need to be considered at popular music/rock concerts where energetic synchronous movements could occur.

Few building codes impose limitations upon the natural frequency range of the structure to avoid resonance. One exception is Commentary A of the Canadian Building Code 1977. This recommends that fundamental frequencies of less than 5 Hz should be avoided in the design of floors for active assembly. The ISE document,[15],includes the following statement

"If the fundamental frequency is less than 4Hz analysis must be carried out to estimate the effects of low forcing frequencies. Should the frequency be between 4 Hz and 6 Hz special consideration should be given to the proposed use of the stand. A value of 6Hz would in normal circumstances be acceptable."

THEORETICAL CONSIDERATIONS.

The solution of Eqn.1 is required.

$$M\ddot{u} + C\dot{u} + Ku = R \qquad (1)$$

M = Mass matrix; K = Stiffness matrix; C = Damping matrix;
u = Displacement vector; R = Vector of externally applied forces.

A suitable model must reflect the nature of the excitation,damping,and the nonlinearities of the system. Where structures are joint dominated it is essential to have a realistic joint model. To predict the influence of joints on response,detailed consideration of joint properties is necessary.
These are schematically represented in Fig.7.

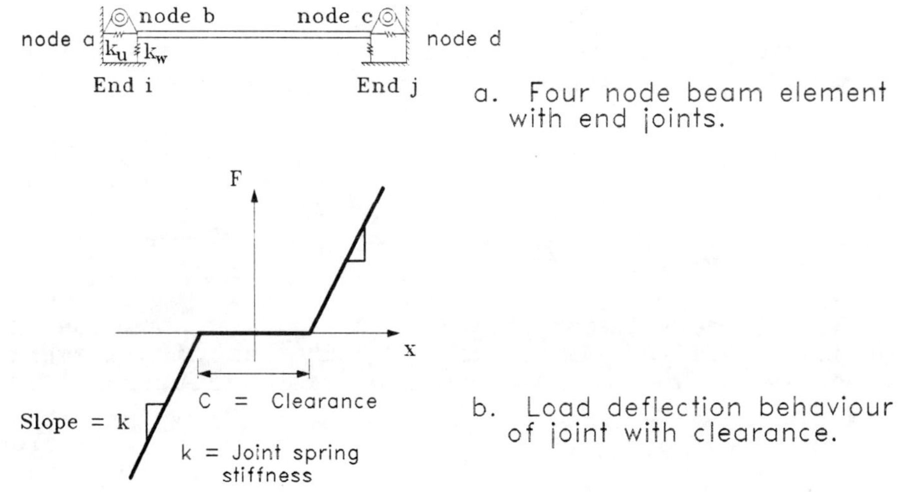

Fig.7

The solution of Eqn.1 can be obtained by direct step-by-step integration. However the presence of joint clearance or slackness in member connections introduces non-linearities. The slackness introduces a dead band in the member stiffness characteristics as schematically illustrated in Fig.7. The dead band may be present simultaneously or singly in the axial,the rotational or the transverse stiffnesses of an element. The theoretical development given in Reference [10] adopts the Wilson-Θ method within a non-linear time history analysis. The member end displacements are calculated for each member to establish whether one or more dead bands have opened or closed during the current time step. To avoid large errors it was found necessary to ascertain the transition point and change the stiffness matrix within the time step.

The method advances the solution from some known state at time t to the new solution at time $t+\delta t$ by first linearly extraploating to a hypothetical solution at time $t+\tau$ where $\tau = \Theta\delta t$.

Eqn.1 can be written

$$M\ddot{u}_{t+\tau} + C\dot{u}_{t+\tau} + K_t(u_{t+\tau} - u_t) = R_{t+\tau} - F_t^e \qquad (2)$$

where F_t^e is the internal resisting force vector at time t. Assuming that the acceleration varies over the time interval τ enables the equation to be written as

$$(K_t + \frac{6}{\tau^2}M + \frac{3}{\tau}C)(u_{t+\tau} - u_t) = R_{t+\tau} - F_t^e + M(\frac{6}{\tau}\dot{u}_t + 2\ddot{u}_t) + C\left(2\dot{u}_t + \frac{\tau}{2}u_t\right) \quad \text{.........(3)}$$

which may be solved for $u_{t+\tau}$.

The increase in computational effort as a consequence of the dead band is illustrated by the displacement response of the simple two member frame illustrated in Fig.8.

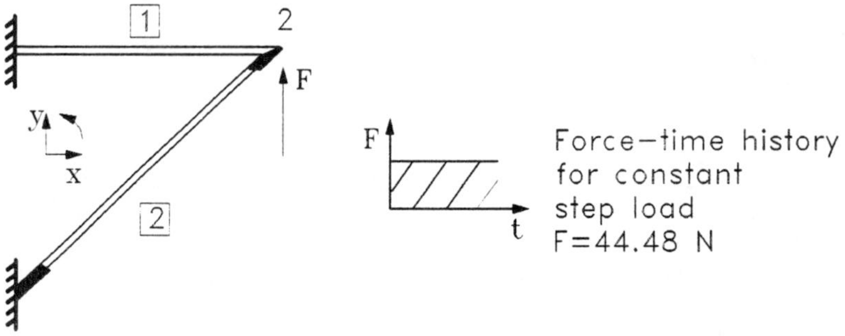

Fig.8 Two member frame.

Member 1 has rigid joints whilst member 2 has a non-linear axial joint stiffness at each end. Fig.9a shows the vertical displacement at node 2 as a consequence of the applied force shown in Fig.8. The horizontal displacement of node 2 is shown in Fig.9b. The joint chatter is very noticeable and complicates the integration path required to ascertain the motion.

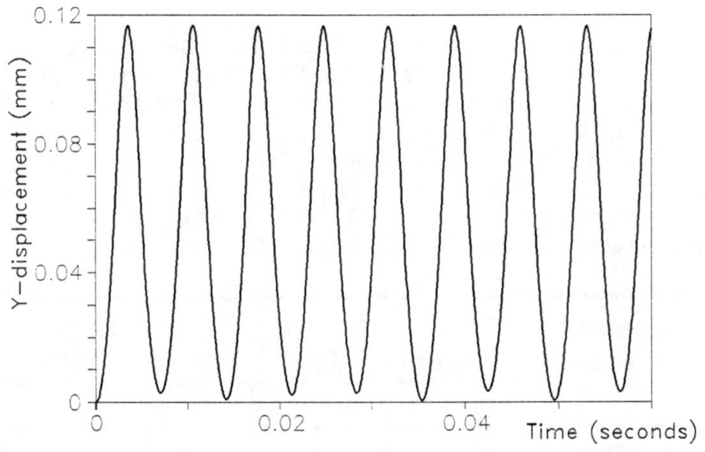

(a) vertical displacement

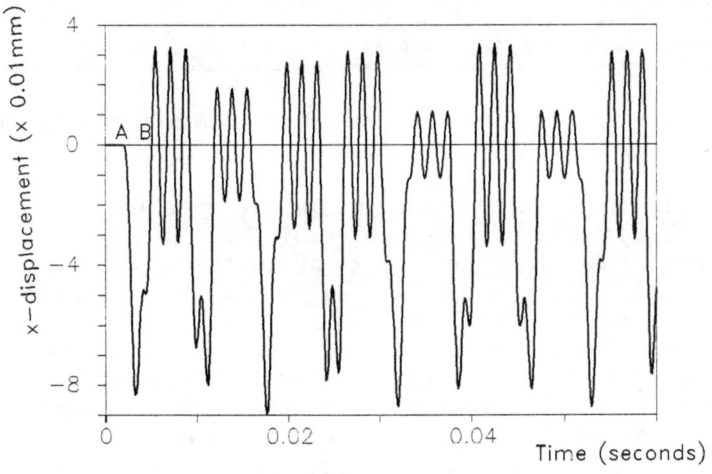

(b) horizontal displacement

Fig.9 Nonlinear analysis of two-member frame.

Linear dynamic (eigenvalue) analyses of this structure for

a. member 2 with zero axial stiffness
b. member 2 with an axial joint stiffness of k_u = 17.5 kN/mm

were also carried out to assist explanation of the results. The frequencies and mode shapes are given in Table 1.

There is very little initial movement in the x-direction until the dead band in the diagonal member has been taken up (point A, Fig.9b). This is because the horizontal displacement in the first mode is very small in comparison with the vertical displacement when the axial stiffness in member 2 is zero. The axial force in member 2 and hence in member 1 is zero so the frame will not vibrate in mode 2 until the mass has gained some momentum in the x-direction.

Frequency/ Mode shape	k_u = 0 (member 2)		k_u = 17.513 kN/mm (member 2)	
	Mode 1	Mode 2	Mode 1	Mode 2
Frequency (Hz)	8.2	608	198	648
X-displ.,	-.000013	0.78	-0.091	0.77
Y-displ.,	0.78	.000017	0.77	0.091
Ratio x/y	1/61,648	46,337	1/8.5	8.5

Table 1. Frequencies and mode shapes for 2-member frame

The high amplitude response corresponds to the vertical direction frequency of 140 Hz. The amplitude is higher because the ratio of horizontal to vertical displacement for mode 1 is higher when the axial joint stiffness of member 2 is zero.

Each time the dead band in the diagonal member opens (e.g. at point B), a lower amplitude frequency component of 610 Hz is seen oscillating about zero. This corresponds to the second mode when the axial stiffness in member 2 is zero. The frame responds in this way because at this stage there is zero axial force in member 2. The momentum which the mass has gained therefore keeps it vibrating at its natural frequency in the x-direction and the y-response has no effect on it until the dead band closes again. The behaviour indicates that the higher frequency components have an effect on the analysis and can interfere with the fundamental frequency response.

Sinusoidal loading is the primary interest however impulsive loadings have also been examined using the theoretical work. Damping must be used in a sinusoidal load anlysis for a steady state solution to be obtained. At the beginning of the analysis for a lightly damped system,transients are present. The solution time must therefore be long enough to allow these to die out so the steady state solution may be found. An expression for the number of cycles, n, taken to reach the steady state response is derived by Newland, [17].

$$n = \frac{1}{2\pi\zeta}\ln\left(\frac{20}{(1-\zeta^2)^{\frac{1}{2}}}\right)$$

It is clear from the foregoing theoretical outline that the numerical problems are relatively severe. Fortunately the sway motion of the seating forms considered herein is essentially two dimensional in nature. Their modular format and discrete panel bracing shown in Fig.10 affords simplification.

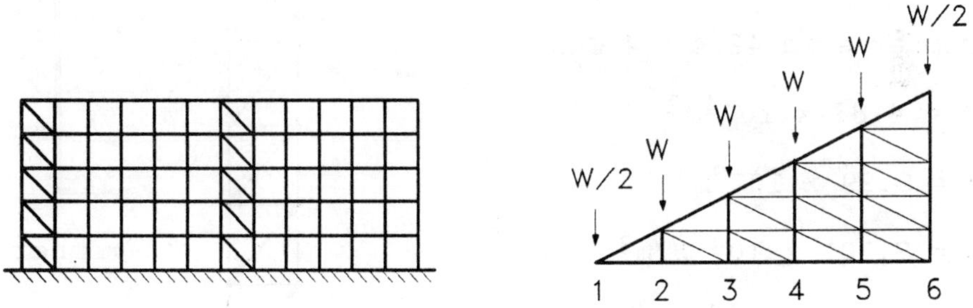

Fig.10 Demountable scaffold stand.

The lateral stiffness of the frame would be increased significantly were the bracing panels to be adjacent to one another. However it is normal practice to distribute the bracing uniformly throughout the structure. This is in effect extending the principle in section 190 of the Green Guide that failure of any one support member or connection should not render the structure unstable,cause local or give rise to progressive collapse.

The lowest frequencies observed in practice were seating decks supported by system scaffold systems. Typical slackness in the diagonal bracing members of a scaffold frame could be a few millimetres. The lowest natural frequency of such a stand is usually in the side to side sway direction. The front to back sway frequency is higher because the height decreases towards the front of the frame. A support frame is shown in Fig.10. Where there is a lack of plan bracing or deck continuity rear sections are readily excitable in sway.

A typical stand would be braced every six bays so that a 6-bay unit would need to be analysed to give the correct frequency. This would require considerable computational effort which may be avoided by analysing a single bay with six times the added mass. This will give nearly the same frequency as the 6-bay unit because minimal increase in stiffness is gained by the extra unbraced bays. It may be shown that the analysis of a single-bay stand with an added mass of six-times that of a single bay is equivalent to that of a 6-bay stand with one-bay braced.

The modular nature enables the fundamental characteristics of a support frame similar to that illustrated in Fig.10 to be obtained from consideration of a plane frame of the form shown in Fig.11. The inner frame,5, will possess the lowest frequency as it has twice the tributary area of rear frame. A damping ratio of 7% of critical was used throughout as a typical value for a steel structure with bolted joints. The diagonal members have an axial dead band at each end.

Member size 42.4 x 4 CHS

$A = 4.83 \times 10^{-4} m^2$

$I = 8.99 \times 10^{-8} m^4$

$\rho = 7.701 \times 10^4\ N/m^3$

$E = 2.06 \times 10^{11} N/m^2$

Bay width = 1.4 m

Storey height = 1.9m

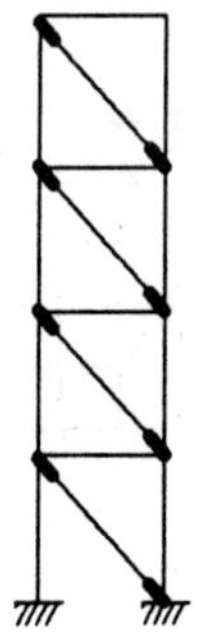

Fig.11

The diagonal members also possessed

Axial joint stiffness $k_u = 4.2x10^2 kN/m$

Rotational joint stiffness $k_\phi = 10^{-10} Nm/rad$

The rotational stiffness of $k_\phi = 10^{-10} Nm/rad$ was used to represent a pinned joint in the diagonal members. If the rotational joint stiffness had been zero analysis problems would arise due to all diagonal terms in the member stiffnesses becoming zero when the joint dead bands open.

An eigenvalue analysis ascertained the natural frequency in the absence of dead bands. A time history analysis using a constant step load of 12% of the vertical loads was carried out for dead band sizes ranging between 0.5 mm and 4 mm at each end of the diagonal members. The results were as follows:-

Dead band mm	Δt (x10^{-4}s)	$\Delta t/T$ (x10^{-4})	Amplitude (x10^{-2}mm)	Frequency Hz
0.0	-	-	-	4.54
0.5	1.5	5.1	0.77	3.40
1.0	2.0	5.9	1.19	2.97
2.0	2.0	4.9	2.00	2.43
4.0	2.0	3.7	3.56	1.87

Table 2. Variation of frequency with dead band size for stand subjected to a constant step load.

The frequency is considerably reduced by the presence of the dead band.

The same structure was analysed under sinusoidal loading with an added mass of 9 people of 70 kg per bay. This gave a mass of 315 kg per leg per bay. A time history analysis was carried out for dead band sizes ranging up to 1 mm. The damping ratio used in all cases was 7% of critical. The value of the mass damping coefficient for Rayleigh (or proportional) damping,α, was varied to keep the damping ratio constant at each natural frequency. Initial analyses were carried out to find the approximate frequency for a particular dead band size. This enabled α to be determined. The natural frequency was then found more accurately using the correct damping ratio.

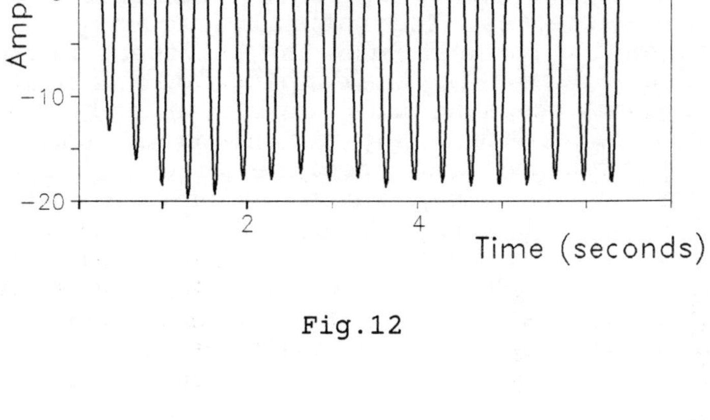

Fig.12

Dead band (mm)	α	Δt ($x10^{-4}s$)	Frequency (Hz)	Amplitude (mm)	Frequency (Hz)
0.1	3.74	4.70	4.10 4.20 4.30 4.40	12.9 14.1 14.4 13.2	4.26
0.2	3.61	4.90	4.00 4.10 4.20 4.30	13.8 14.9 14.5 12.9	4.13
0.5	3.25	5.40	3.60 3.70 3.76 3.77	15.4 16.3 17.2 14.0	3.76
1.0	2.73	6.45	2.90 3.00 3.10 3.15	15.5 17.7 18.0 8.0	3.10

Table 3. Variation of frequency with dead band size for stand subjected to a sinusoidal load.

A time step size to period ratio $\Delta t/T$ of 2 x 10^{-3} was used in each case to give an accurate result. Table 3 gives the results and Fig.12 gives a typical displacement characteristic.

The natural frequency of a stand with a joint dead band characteristic depends upon the size of the dead band, the magnitude of the force and the damping ratio. The frequency of the stand decreases with increasing dead band size,as expected,because the stiffness of the structure decreases. The natural frequency decreases with decreasing force because the stiffness is reduced. The effect of decreasing the damping ratio is to increase the frequency because the amplitude and hence the stiffness is greater. Any reduction of the natural frequency is significant because it could reduce the frequency into the range which can be excited by crowd motion. However the most important consideration from the designer's viewpoint is the increase in amplitude and hence stress. The typical increase in amplitude for 1 mm dead bands is approximately 40% between 2Hz and 3 Hz. For 2mm dead band the increase is 74% at a frequency of 2Hz.

PRACTICAL CONSIDERATIONS.

Stands at popular music concerts are excited by significant percentages (>50%) of the occupants dancing in the gangways and seat aisles. This form of activity is good natured and not overly exuberant.

A large number of stands have been examined at various venues. Dynamic testing is described in references [4] and [19]. Standard procedures were adopted by the authors and low-g accelerometers were used. These high sensitivity accelerometers are designed for measurement of very low level vibrations down to $2x10^{-5}$ ms^{-2} at frequencies extending from 0.3 Hz to 1 kHz. Signals were recorded on tape and where mains power was available a spectrum analyser was also employed.

Fig.13 concerns a free standing retractable stand 'seating' approximately 500 people at a Stevie Wonder concert. The stand was truncated at the front giving a trapezoidal side elevation. The stand was adequately braced against sidesway motion however the front to back motion was considerable. Whilst the spectator activity as described appears random it is driven by the beat of the music. The time history of the force amplitude is approximately a half-sine wave and requires higher frequency components than 2Hz to describe it. The spectrum shown in Fig.13 was recorded during music with a beat of 2.2 Hz. The multiples are clearly seen and there are four predominant frequencies in the range 0-8Hz. Although the lowest natural frequency was in evidence when the audience activity was low the significant frequency at higher levels of audience activity is the second natural frequency. The maximum motion was observed at 2.6 Hz. The problems created by front to back sway in this instance could be alleviated simply by tying the structure in.

A similar spectra was recorded on a demountable stand of the form shown in Fig.3 at a Phil Collins concert and this is shown in Fig.14.

No appreciable movement occurred as the nature of the jointing system created considerable damping.

For the large number of stands that have been examined the lowest natural frequency obtained for an empty stand was 1.7 Hz. The stand was outdoors and vibrations were occurring under wind excitation. The added mass of 5000 persons would reduce this to a value of approximately 1Hz. The stand was used at the Jean Michel Jarre concert held in the Docklands 1987,fortunately audience activity was minimal. The fact that multiples of the music beat are reflected in the loading spectrum and that the music beat will vary means that a stand will almost certainly be excited at its natural frequency at some point during a concert.The

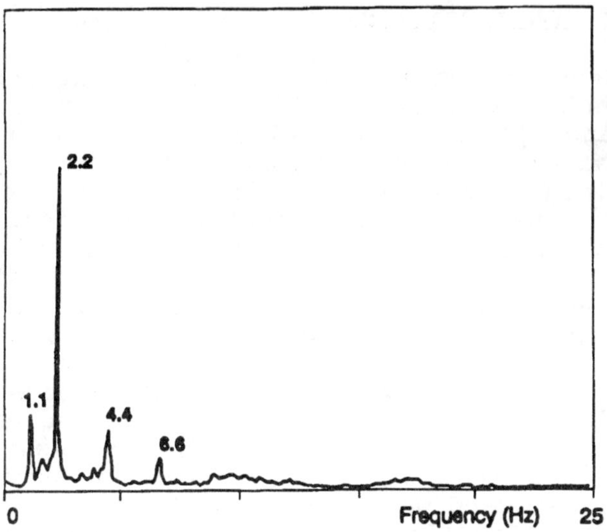

Fig.13 Frequency spectrum, Retractable stand.

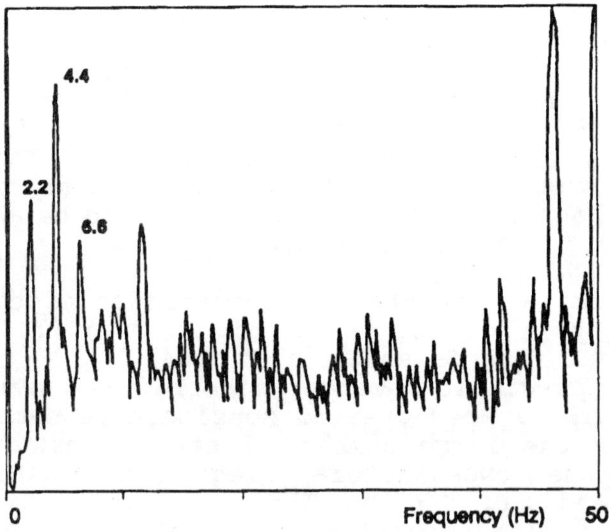

Fig.14 Frequency spectrum, Demountable stand.

degree of structural motion is a function of the music beat and does not necessarily reflect the degree of audience activity.

CONCLUSION

Retractable and demountable seating stands possess low natural frequencies. Joint slackness makes a significant contribution to this aspect of seating assemblies. In the presence of a low natural frequency careful consideration is required as to the use of this structural form to avoid undue excitation as a consequence of spectator behaviour. Current design codes are not sufficient to adequately account for audience activity that may be concerted.

REFERENCES

1. ANSI/NFPA.102. Assembly seating,tents and membrane structures,1986.

2. Bathe,K.J., and Wilson,E.L. Stability and Accuracy Analysis of Direct Integration Methods, Earthquake Engineering and Structural Dynamics, Vol.1,1973, pp. 283-291.

3. Belvin,W.K., Modelling of Joints for the Dynamic Analysis of Truss Structures, NASA Technical Paper, May,1987,Langley Research Centre,Hampton,Virginia.

4. Broch,J.T., Mechanical Vibrations and Shock Measurements, Bruel and Kjaer, Denmark 1984.

5. BS5973:1981. Access and working scaffolds and special scaffold structures in steel.

6. BS6180:1982. Protective barriers in and about buildings.

7. BS6399:Part 1:1984. Loadings for buildings.

8. The Building Regulations, HMSO, 1985.

9. Ebrahimpour, A., and Sack, R.L., Modelling Dynamic Occupant Loads, Journal of Structural Engineering, ASCE, Vol.115, No. 6, June, 1989, pp.1476-1496.

10. Gibbs, S.J., Dynamic response of loosely jointed temporary grandstand seating. PhD thesis, University of Manchester, 1991.

11. Greater London Council,1978. Code of practice for pop concerts.

12. Health and Safety Executive. Guidance for health and safety at pop concerts and related events. (To be published).

13. Home Office,1990. Guide to Safety at Sportsgrounds.

14. Home Office,1990. Guide to Fire Precautions in Existing Places of Entertainment and the Like Premises.

15. Institution of Structural Engineers,(1987). Safety considerations for the design and erection of demountable grandstands.

16. National Building Code of Canada, Commentary A, Supplement No.4.

17. Newland,D.E., Mechanical Vibration Analysis and Computation. Longman Scientific and Technical, 1989.

18. Park, K.C., Practical Aspects of Numerical Time Integration. Computers and Structures, Vol. 7, 1977, pp. 343-353.

19. Randall,R.B., Frequency Analysis. Bruel and Kjaer, Denmark,1987.

20. Russell-Cooke,Potter and Chapman. Private correspondence.

21. Tuan, C.Y., and Saul, W.E., Loads due to Spectator Movements,Journal of Structural Engineering,ASCE,Vol.111, No.2,Feb., 1985, pp. 418-434.

Active Control of Bridge Structures

C.J. Burgoyne(*), W.H. Chan(**)
(*) Cambridge University Engineering Department
(**) Ove Arup Partnership, Hong Kong

ABSTRACT

The paper describes a possible replacement for the traditional 'Bailey Bridge' type of structure. The system consists of Vierendeel Truss panels, which can easily be joined together, and minimises the secondary moments normally associated with such trusses by the use of an actively controlled prestressing system.

Straight prestressing tendons pass from end-to-end of the structure along the mid-line of the Vierendeel panels, and long stroke jacks are connected between the tendon and the top joints between the panels. As a load passes over the bridge, these jacks are extended to counteract the bending effects within the panels, and to increase the force in the prestressing tendon, thus ensuring that the trusses are subjected to primarily axial forces and that all joints remain closed.

The requirements of such a system to deal with large moving loads are considered, such as might be encountered in military applications or as temporary works on site. The case of a 25m span bridge carrying a 100 Tonne HB-type vehicle at a speed of 0.25m/s is considered to determine the requirements of the mean prestressing force, jack stroke and hydraulic power capacity. The principles of the control system are considered, as are the nonlinear effects of the varying inclination of the jacks and the large deformations of the prestressing tendons.

INTRODUCTION

Structures are traditionally regarded as carrying their load in a fixed manner. The distribution of dead load forces is controlled by the erection sequence with, in special cases, redistribution being induced at that time by the application of prestressing forces to stays or cables, or by jacking at the supports. The live load distribution is simply that which results from the normal behaviour of the structure; the only control that the designer exerts over the distribution comes from the relative sizing of the members.

The implications of this are so fundamental to normal structural design that they pass unremarked by most engineers, but the effect is that structures have to be designed for the maximum range of loads that can be applied throughout the life-time of the structure. The penalty in terms of weight and cost can be large, especially in structures in which the extreme live load forms a high proportion of the total load on the structure.

The use of active, or reactive, structures, which change their configuration in response to the applied loads, offers a way of overcoming that inefficiency, but at the penalty that the fixed structure becomes a machine, with all the attendant difficulties of control and maintenance. Certain types of structures conventionally have maintenance personnel in attendance; these include large structures, such as major river crossings, and temporary bridges, such as those used by the military or civil engineers. It is to this last class of structures that this paper is addressed.

It is perhaps worthy of note that not all 'structures' have evolved in the conventional way. The function of many of the muscles in animal and human bodies appears to be to reduce the stresses in bones [1]; the resulting costs in terms of muscle fibre, control mechanisms and energy have presumably been found to be made worthwhile, in evolutionary terms, by the consequential reduction in bone size that this allows.

The work presented here is largely the result of a third year project [2] carried out at Imperial College, inspired by ideas from Prof Sir Alan Harris, and his contribution to this work, and his permission to publish this paper in this form, are gratefully acknowledged.

STRUCTURE USED IN DESIGN EXAMPLE

The structure used as the basis for this study is a demountable Vierendeel truss bridge, with a span of 25m, designed to carry an HB vehicle of total weight 100 Tonnes. This load is assumed to be evenly distributed on 4 axles, with spacings of 1.8m, 6m and 1.8m. For calculating the power requirements of the system, it is assumed that this vehicle is travelling fairly slowly (0.25m/s = 0.5mph). This is probably reasonable given that such a large load would be under fairly careful control while crossing a bridge of this type. When presenting results which depend on the position of the vehicle on the bridge, the position of the *centre* of the vehicle will be given, measured relative to the left hand end of the structure. Thus, some part of the vehicle will be on the bridge with this position varying from -4.8m to 29.8m. The vehicle is assumed to be travelling from left to right on the bridge, in cases where the order of application of the forces is important.

The prestressing tendon is assumed to be a Type G Parafil rope, which consists of a core of high modulus (126kN/mm^2) Kevlar 49 yarns, contained within a polymeric sheath. The design strength of such tendons (1930 N/mm^2) compares well with conventional steel tendons, and there are advantages such as light weight and resistance to corrosion which would be important in a demountable structure. Such tendons are now being widely considered for use as prestressing tendons in concrete structures [3], and also as stay cables in bridge and tension structures [4]. Although the numerical results would be slightly different if steel tendons were used in the present case, the broad conclusions relating to active prestressing would be unaffected.

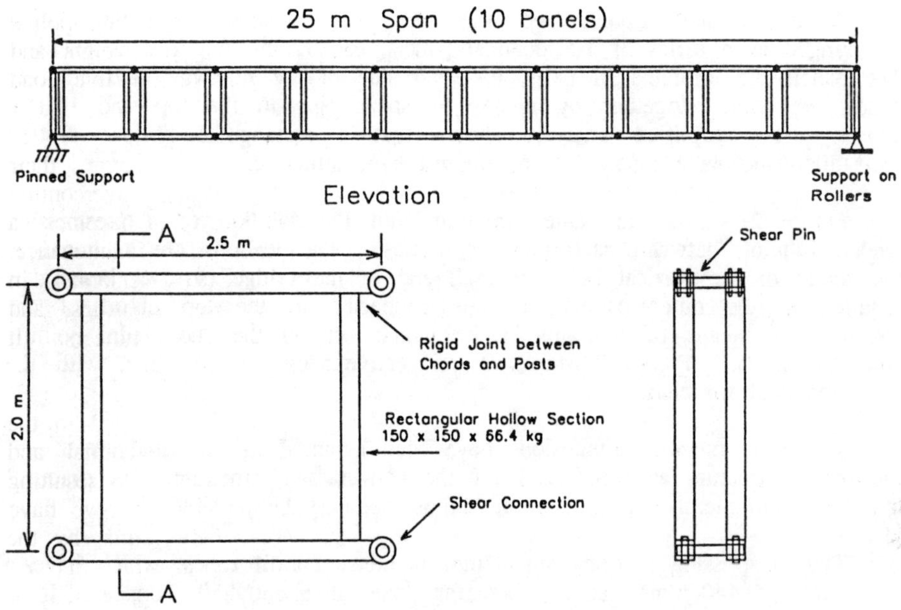

Figure 1. *Unprestressed demountable Vierendeel truss.*

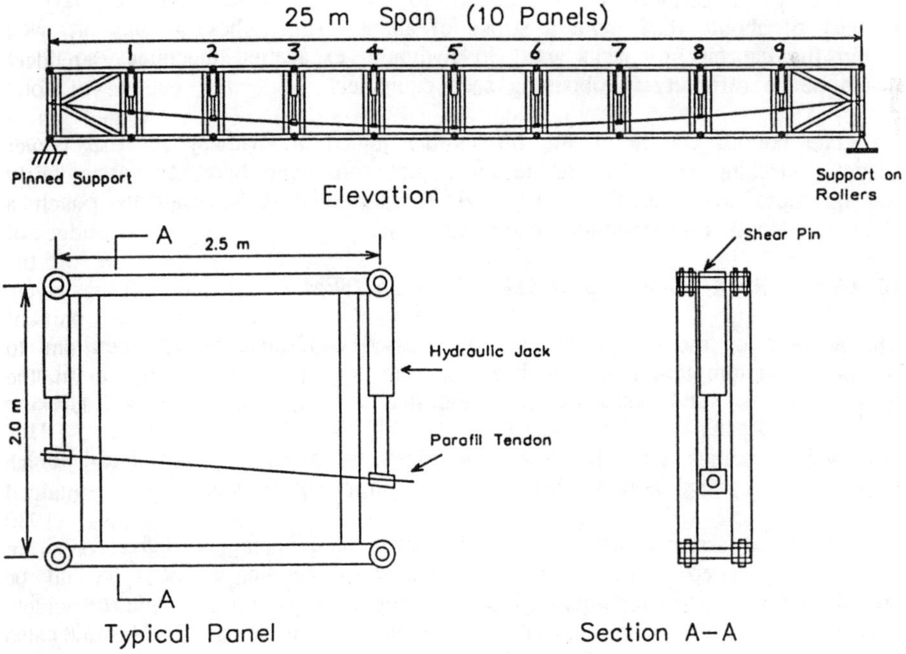

Figure 2. *Truss prestressed with Parafil tendon. Note the addition of K bracing in end panels and jacks to act on the prestressing cable.*

A conventional, unprestressed, Vierendeel truss is shown in Figure 1. It is arranged as a series of 10 identical panels, each 2.5m long and 2.0m high. The panels are arranged in pairs, on each side of the roadway, and adjacent panels are joined together by means of shear pins in the top and bottom chords. A simplified analysis shows that a rectangular hollow section 150*150*66.4kg is adequate for the unprestressed structure.

Figure 2 shows the same structure with the addition of a prestressing tendon running between each pair of trusses. The tendons are anchored at the centre of the vertical posts at each end of the bridge (so 'K' bracing is required in the outer panels), and is connected to the top chord of the structure by means of hydraulic jacks which act on the shear pins joining adjacent panels. Figure 3 shows a perspective view of one bay, with the jacks removed for clarity.

The same member dimensions have been assumed for the analysis of the prestressed structure as were used for the conventional structure. In practice, a reduction in member weight would almost certainly be possible.

The prestressing tendons are 87mm diameter Parafil ropes, which have a core area of 4400mm^2, and a breaking load of about 850 Tonnes. It is assumed that each tendon has an initial prestress of 160 Tonnes. These figures have been obtained on the basis that the deflection of the tendon below the neutral axis under live load is to be limited to about 40% of the depth of the truss.

The jacks connecting the tendon to the top chord need to have a capacity of about 80kN, with a stroke of about 0.8m. These figures are well within the capacity of jacks used in hydraulic excavating machines, so there would be no difficulty in obtaining such equipment.

The bottom chords of the trusses are joined at roadway level by cross beams. Decking units (whose design is not considered here) are assumed to rest on these cross beams, so no load is transmitted to intermediate positions along the length of the bottom chord members.

BEHAVIOUR OF THE UNPRESTRESSED SYSTEM

The analysis of the bridge in its unprestressed configuration has been carried out on the assumption that the shear force in each panel is equally distributed between the top and bottom chords, and that there are points of contraflexure within each panel at the mid-point of both the chords and posts. This effectively assumes that the posts are rigid when sharing the shear forces between the top and bottom chords, but the error will be small.

On these assumptions, the structure takes up a complex deformed shape (shown exaggerated in Figure 4), with clear local bending in both the chords and the posts. The maximum global bending moment on a single truss (i.e. one of four), is 1055kNm, and the maximum shear force is 123kN, under the action of the 100 Tonne HB vehicle.

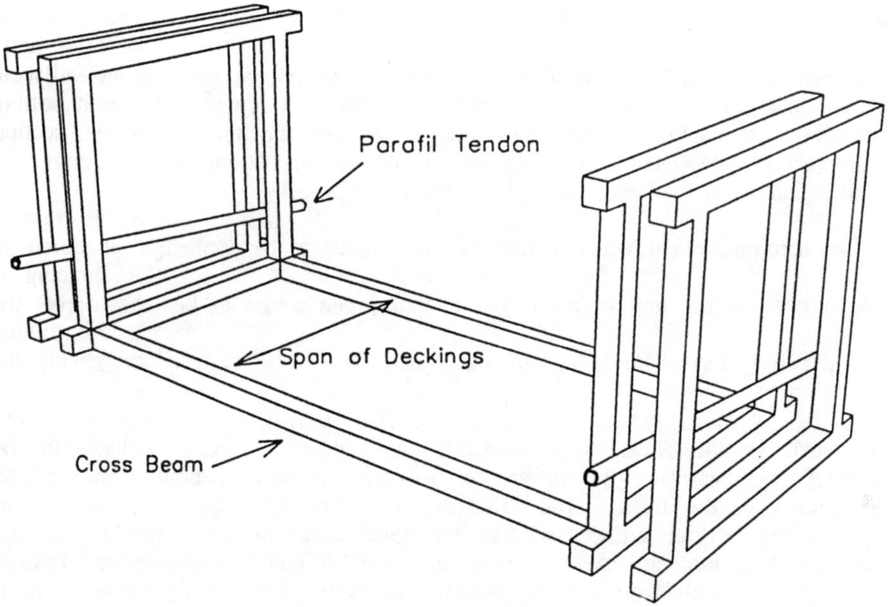

Figure 3. *Typical unit of truss. Shear pins and hydraulic jacks have been omitted for clarity.*

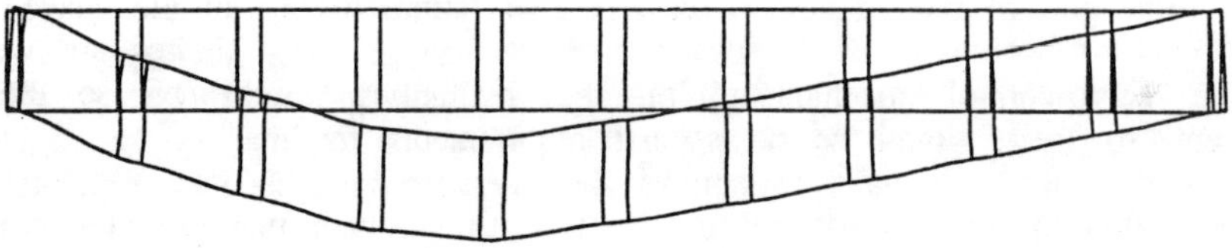

Figure 4. *Deformation of unprestressed Vierendeel truss, exaggerated to show secondary effects in panels.*

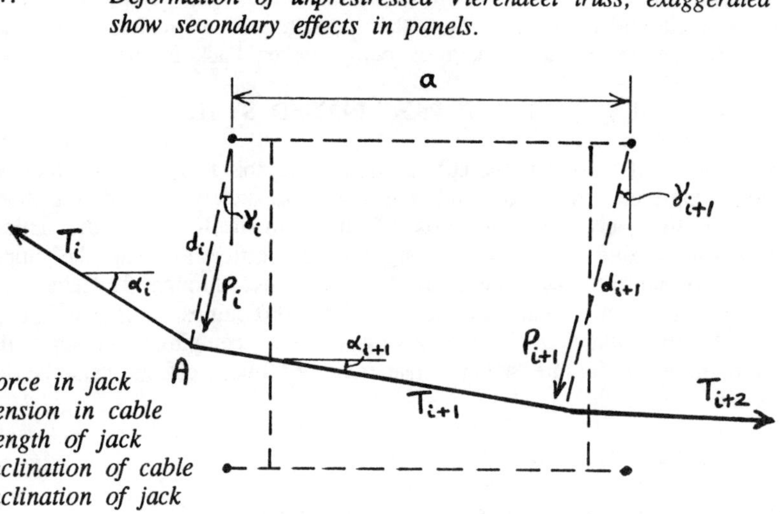

P_i *Force in jack*
T_i *Tension in cable*
d_i *Length of jack*
α_i *Inclination of cable*
γ_i *Inclination of jack*

Figure 5. *Geometry of deformed cable profile.*

ACTIVE CONTROL SYSTEM

The aim of the active control system will be to relieve the Vierendeel truss of as much of the secondary loads as possible. This will be done by extending the jacks to exert upwards forces on the top chord joints that counteract the downwards loads being applied to the bottom chord through the cross beam. To that end, the system must comprise:-

(a) a sensing system which determines the forces to be applied,

(b) a computation unit to evaluate the appropriate action to be taken, and

(c) a servo mechanism to control each jack.

The sensing system is most logically provided by monitoring the difference in the shear force between two adjacent panels, which can be obtained by measuring the difference between the local bending moments in the chords of the truss. The difference between strain gauge readings from the top and bottom surfaces of the horizontal chord members would give the local bending moment in that truss, and the difference in moment between two adjacent panels would be proportional to the required applied force.

The computation unit could be very simple, and would not need to be a digital system, since analogue readings of the strain gauges could be used directly. However, it might be sensible to feed the results through a computer system of some sort to monitor the behaviour of the system and to provide some sort of redundancy in the event of failure of one of the gauges.

The servo-control mechanism for the hydraulics would be standard equipment, so there would be no special requirements for that system. The system would, however, have to control the *pressure* in each jack separately, and allow flow of oil into or out of the jack to maintain that pressure. As will be shown later, there will be occasions when the force in the jack is increasing while the flow of oil is back towards the pump; this occurs when the structure is unloading after the passage of a vehicle, and the energy stored in the prestressing tendon is being passed back to the jacking system.

STATIC ANALYSIS OF THE PRESTRESSED SYSTEM

The method of analysis adopted to determine the behaviour of the prestressed system considers the truss and the cable separately, linked together by the forces in the jack. The analysis of the truss itself, under the action of the active prestressing force is not significantly different from the unprestressed case, since all the loads are applied to the truss through the node points, and the deflections are small so there are no difficulties with non-linear effects. However, the behaviour of the tendon is more complex, and since deflections of the order of 1m are being considered, non-linear effects must be taken into account.

The analysis must allow for:-

(a) the inclination of the tendon to the horizontal,

(b) the eccentricity of the tendon from the neutral axis of the structure,

(c) the inclination of the jack to the vertical, which results in some horizontal forces being applied to the truss, and

(d) the axial stiffness of the tendon.

It is assumed that there is no slip between the jack and the tendon, that the jack is free to pivot about its point of attachment to the top chord, and that the response of the jacking system is instantaneous. The implications of this last assumption will be considered later in terms of the power requirements for the hydraulic system, and the response time of the control system.

Consider the system in Figure 5.

At each clamp (e.g. A), vertical equilibrium gives

$$T_i \sin\alpha_i - T_{i+1} \sin\alpha_{i+1} - P_i \cos\gamma_i = 0$$

while horizontal equilibrium gives

$$-T_i \cos\alpha_i - P_i \sin\gamma_i + T_{i+1} \cos\alpha_{i+1} = 0$$

and geometrical compatability gives

$$d_{i+1} \cos\gamma_{i+1} - d_i \cos\gamma_i = \{a + d_i \sin\gamma_i - d_{i+1} \sin\gamma_{i+1}\} \tan\alpha_{i+1}$$

The force in the tendon can be related to the geometrical changes by

$$\frac{a + d_i \sin\gamma_i - d_{i+1} \sin\gamma_{i+1}}{\cos\alpha_{i+1}} = \left[\frac{1 + \frac{T_{i+1}}{EA}}{1 + \frac{T_0}{EA}} \right] a$$

By setting up a set of equations of this form, and specifying the vertical component of the force in each jack, a set of non-linear simultaneous equations is produced, which can be solved by any suitable method, such as multi-dimensional Newton Raphson iteration.

BEHAVIOUR OF THE PRESTRESSED SYSTEM

The cable profile, for the vehicle in 7 different positions, is shown in Figure 6. Maximum cable deflections of about 0.76m occur with the vehicle near midspan. The force in each tendon increases from its initial value of 160 Tonnes, up to a maximum of 286 Tonnes; the total tension force is about 5.8 times the weight of the vehicle, which is fairly typical for prestressed concrete bridges. Both the top and bottom chords of the truss are in compression, with most of the bending being carried by the cable.

The jack forces do not completely cancel out the bending moments and shear forces in the bridge, since the jacks pivot about the top chord of the

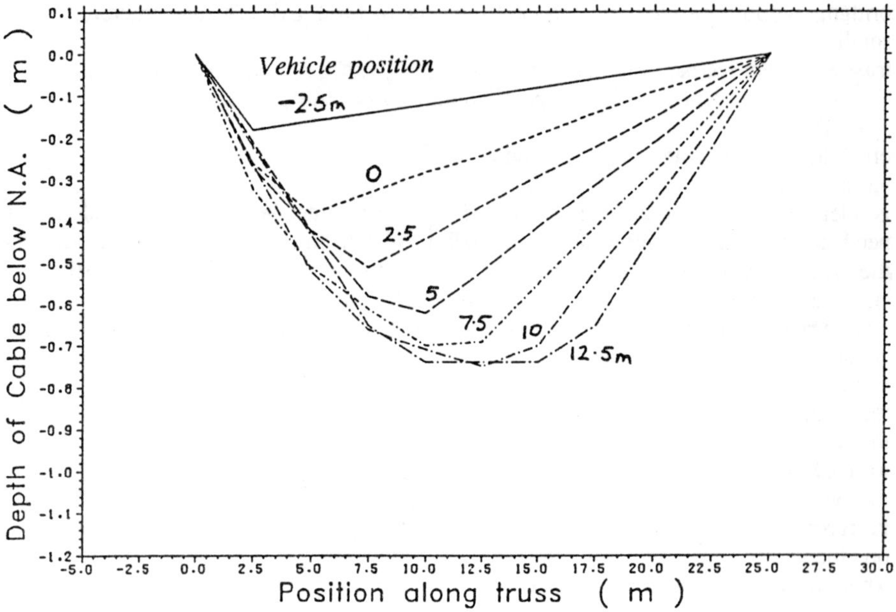

Figure 6. *Cable profile for different positions of HB vehicle.*

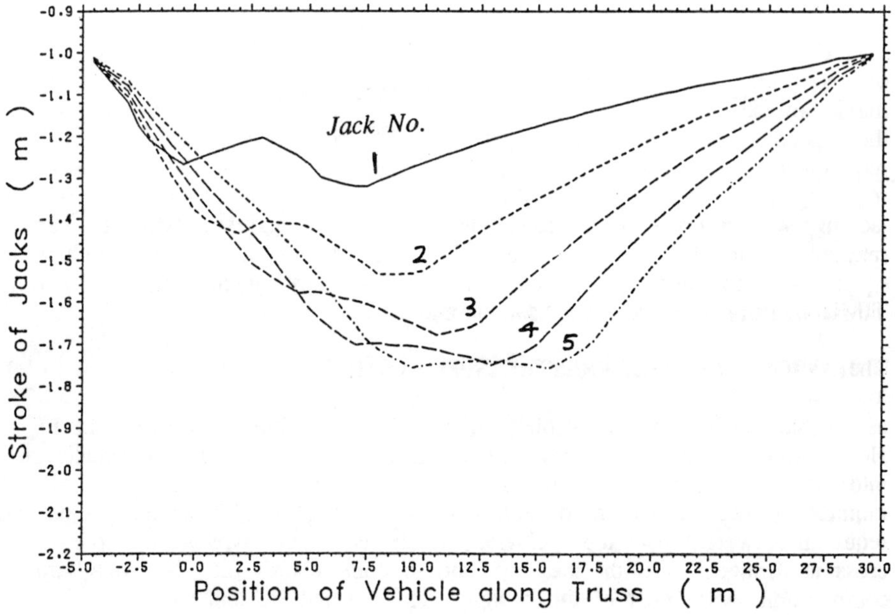

Figure 7. *Stroke variation in jacks as vehicle passes over the truss.*

bridge, while the load is applied to the bottom chord. These effects are small, and most of the local bending moments associated the Vierendeel trussses are eliminated.

The stroke of the various jacks (measured from the top node), is shown on Figure 7, against the position of the centre of the vehicle, while the force variation in one jack (jack 3), is shown in Figure 8. From these figures it is clear that the control system, which would be controlling jack force, would need to be able to allow the jack extension to vary at the same time, since the cable profile would be affected by the actions of the other jacks.

The total power requirement for each jack can be calculated by considering the movement of the jack and the force in the jack, and the results for jacks 1-5 are shown in Figure 9. The power scale needs to be multiplied by the vehicle speed to get true power; the figure also relates to the jack power applied to each half of the truss. For these jacks, which are to the left of the mid-span region, net positive work is done by the jacks as the vehicle moves on to the bridge, while only a small amount of the work is recovered as the vehicle moves off. However, for the corresponding jacks on the other side, the situation would be reversed, and the total amount of work done in the bridge as a whole should, at least in theory, be recoverable.

This observation leads to the intriguing prospect that a power pack could charge up a hydraulic accumulator, which could be used to power the jacks; suitable one-way valves could then be used to recharge the accumulator as the vehicle moved off the bridge. This system would be energy efficient, but more importantly, would not need a large source of hydraulic power. However, the design of the hydraulic system, and also the control system, would need to be carefully worked out in detail.

One important parameter for the design would be the rate of application of the jack force. This is dependent on the speed of the vehicle; the maximum rate of loading in any one jack is about 31.4*v kN/s, where v is the speed of the vehicle. For $v = 0.25$ m/s, the maximum power requirement in any one jack is 3.25 kW. However, not all jacks would be requiring power at any one time, as can be seen from Figure 9, so the jacking system would not need to power all 18 jacks simultaneously at this rate.

TIME DEPENDENT CONSIDERATIONS

The system described so far is capable of effectively transferring the whole load of the heavy vehicle to the cable, while the actual bridge structure acts as a reaction frame under axial force and distributes purely local loads. However, there are two principal time dependent effects that need to be taken into account in a practical application. There will be a delay between monitoring the out-of-balance shear forces in adjacent panels, and applying the required correcting force. Similarly, in a real situation, the vehicle may not cross at a speed that the pump system can cope with.

Two numerical simulations have been carried out to illustrate these points. In each case, the vehicle speed, the sampling interval and the force capacity (expressed as a maximum change in jack force per second) are

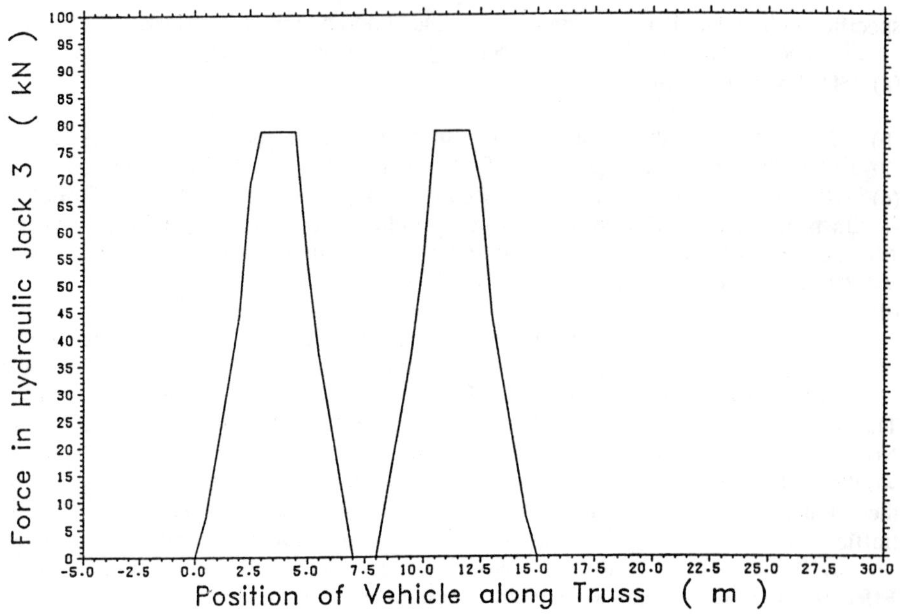

Figure 8. *Force variation in jack 3 as vehicle passes over the truss.*

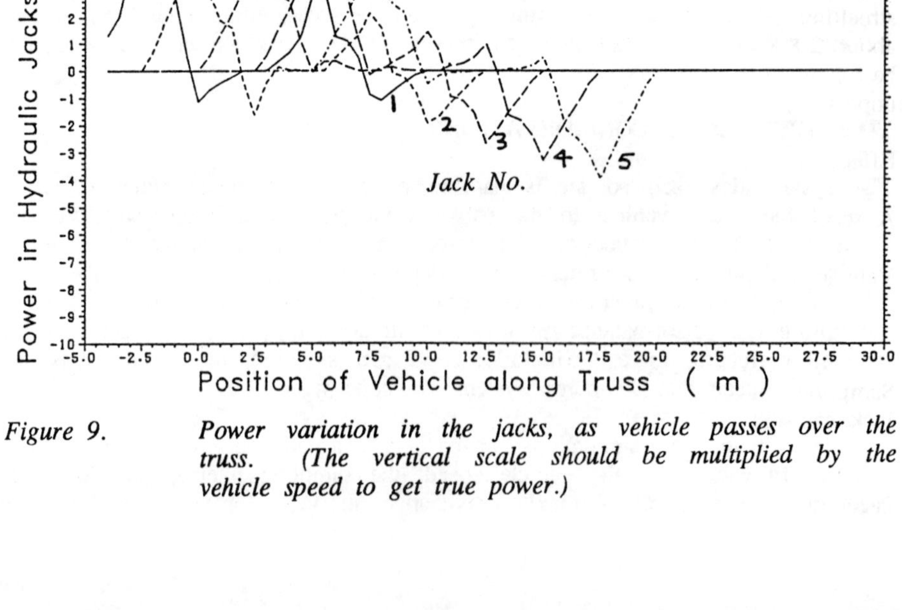

Figure 9. *Power variation in the jacks, as vehicle passes over the truss. (The vertical scale should be multiplied by the vehicle speed to get true power.)*

specified. The following procedure has been adopted:-

(a) The difference in the shear force in adjacent panels is calculated.

(b) The desired new force to be applied at the next time step is determined.

(c) If the jack force capacity is high enough, this force is applied to the system. If not, the maximum force change that the system can apply is determined, and this value used instead.

(d) The new cable profile, and the corresponding force distribution in the truss, at the next time step are then calculated.

The whole process is repeated for the complete passage of the vehicle. The response of the system lags behind the requirement, and is thus *reactive.* No attempt has been made to build predictive elements into the system, which would make the system truly *active*, although it should be possible to refine the model to make it predictive, especially for a structure carrying one-way traffic.

Effect of varying sampling interval

A simulation has been carried out with the following parameters:-

Vehicle velocity	:	0.25 m/s
Sampling interval	:	0.5 s
		1.0 s
		2.0 s
		4.0 s
Jack capacity	:	15 kN/s

Figure 10 shows the resulting bending moment in the truss for one position of the vehicle (at 17.5 m), for the different sampling intervals chosen. This position is the worst position for out-of-balance forces, since it corresponds to the position where the vehicle axles start to move off the structure. At an interval of about 0.5 s, the maximum bending moment is about 20 kNm, which increases to about 100 kNm at 4.0 s. These should be compared with a maximum bending moment of 1055 kNm in the unprestressed structure.

Effect of increasing vehicle speed

A second simulation has been carried out varying the vehicle speed.

Vehicle velocity	:	0.125 m/s
		0.25 m/s
		0.5 m/s
		1.0 m/s
Sampling interval	:	1.0 s
Jack capacity	:	15 kN/s

At 0.5 m/s, the speed just about matches the jacking capacity, but at the faster speed, it is to be expected that the system will not be able to respond

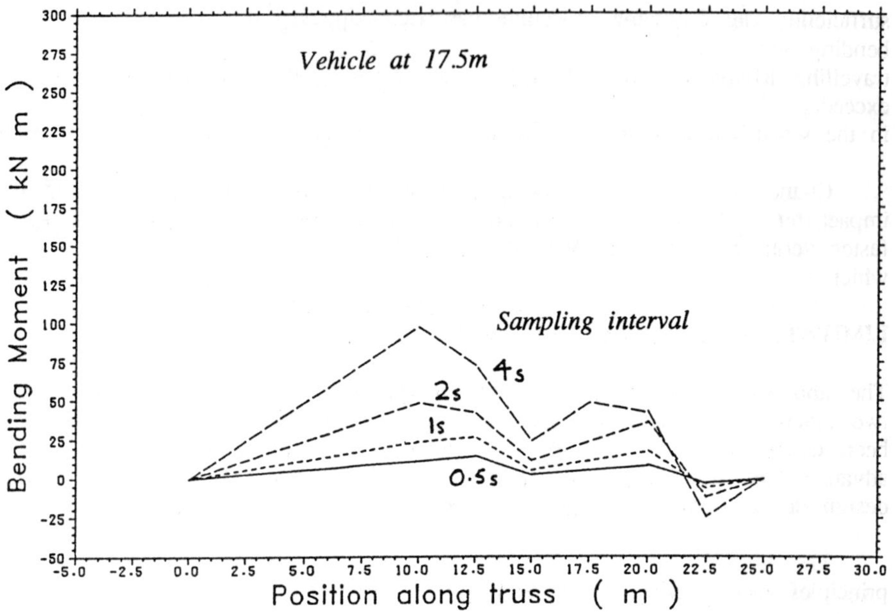

Figure 10. ***Bending moment in the truss, as a function of the sampling interval of the control system, for the centre of the vehicle at 17.5m from left support.***

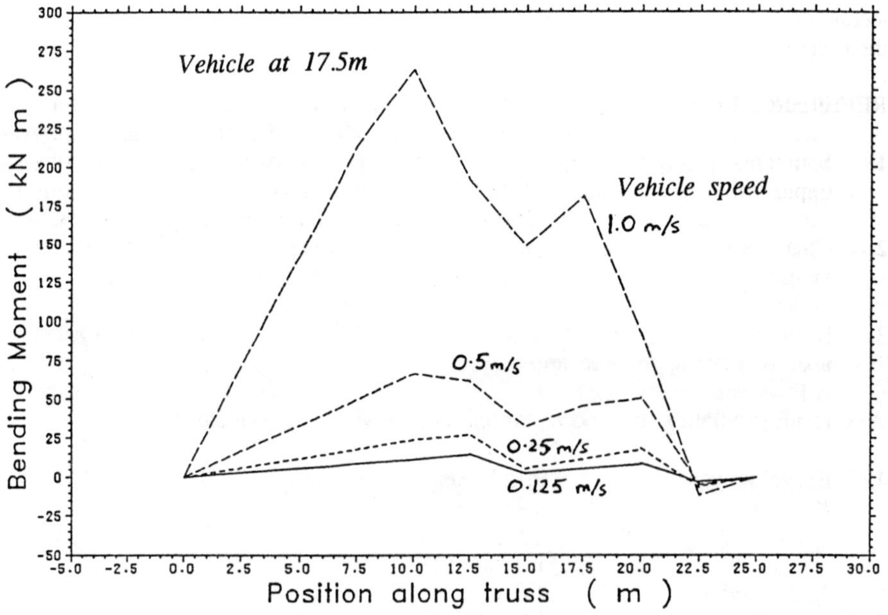

Figure 11. ***Bending moment in the truss, as a function of vehicle speed, for the centre of the vehicle at 17.5m from left support.***

sufficiently quickly. This is illustrated in Figure 11, which again shows the bending moment in the truss. The response is reasonable if the vehicle is travelling at lower speeds, but once the capacity of the bridge to respond is exceeded, the bending moment increases dramatically, reaching about 270 kNm in the worst condition (illustrated in the figure).

Changing the speed of response of the system would have a small impact on reducing the bending moments for the slower vehicles, but for the faster vehicle, where the response of the system is governed by the rate at which the jacks can apply force, no such improvement could be made.

LIMITATIONS OF THE ANALYSIS

The analysis presented here is limited. The truss has been considered as a two-dimensional structure, and detailed design of the flooring system has not been carried out. No attempt has been made to redesign the truss to take advantage of the reduced bending moment in the system, nor has detailed design of the hydraulic system been undertaken.

The absence of these considerations, however, does not invalidate the principles described here. The idea of measuring shear forces in adjacent panels of a Vierendeel truss, and minimising these by applying upwards forces to the structure which are reacted against a prestressing cable, remains valid.

CONCLUSIONS

The paper has demonstrated that active control of bridge structures offers a way of enhancing the load carrying capacity of the bridge. All of the techniques described in the paper make use of existing technology, although their combined application to this type of problem may be new.

REFERENCES

1. Saumarez R.C., An analysis of action of intercostal muscles in human upper rib cage. *J. Appl. Physiol.* **60**(2), 690-701, 1986

2. Chan W.H., Active Control of Bridge Structures, Undergraduate Research Project, Dept of Civil Eng., Imperial College, London Univ., 1988

3. Burgoyne C.J., Properties of polyaramid ropes and implications for their use as external prestressing tendons, *External Prestressing in Bridges*, ed. A.E.Naaman and J.E.Breen, American Concrete Institute, SP-120, 107-124, Detroit, 1990.

4. Burgoyne C.J., Polyaramid ropes for tension structures, *1st Int. Oleg Kerensky Memorial Conference on Tension Structures*, London, 1988.

The Pantographic Deployable Mast: design, structural performance, and deployment tests

A.S.K. Kwan, S. Pellegrino

Department of Engineering, University of Cambridge, Trumpington Street, Cambridge,CB2 1PZ, U.K.

ABSTRACT

This paper presents recent experimental work on the new Pantographic Deployable Mast, a three-dimensional pantograph which is deployed and prestressed by two active cables. Many prestressed passive cables stiffen the mast, when it is fully deployed. The design and manufacture of a 1200 mm high prototype of the Pantographic Deployable Mast are described, including a joint of a novel type between the rods of the pantograph. The fully-deployed prototype has been tested under both lateral and axial loads, and its stiffness measured. Deployment tests have been conducted: in 90% of the cases, the height of the mast has been within ±0.2 mm of the mean height, while the r.m.s. error on the position of the joints at the top has been 0.4 mm.

Keywords: deployable mast, cable, structural mechanism.

1. INTRODUCTION

Deployable/collapsible masts feature regularly in space missions, mainly to guide the deployment of solar arrays or to support scientific experiments. Some deployable masts have already been fully developed and tested in space, but larger and yet stiffer masts, will be needed for future applications. The challenge, now, is to identify suitable structural concepts.

Some of the existing deployable masts deform elastically during folding, and the strain energy thus accumulated is gradually released during deployment. The Astromast [1] and the Simplex Mast [2] are sophisticated braced trusses with a triangular cross section. During folding, their battens take a helical shape and coil into a canister. The STEM [3] and the CTM [4] are open- and closed-section tubes, respectively, which can be flattened and rolled up, rather like a steel measuring tape. Concepts for deployable masts which remain essentially stress-free during folding have also been proposed, e.g. Dornier's telescopic tube [5] or the box truss [6], whose braced modules have some foldable members. Typically, such assemblies are deployed by slowly activating their mechanisms, and are locked in their final position by latching devices. In general, masts of the first type tend to have rather straightforward connections between elements, resulting in simple, lightweight structures. However, the presence of a deformation mode, through which they can be collapsed, tends to affect adversely their stiffness. Masts of the second type tend to require mechanically elaborate connections between members, and are usually much heavier. Although for entirely different reasons, they also tend to lack

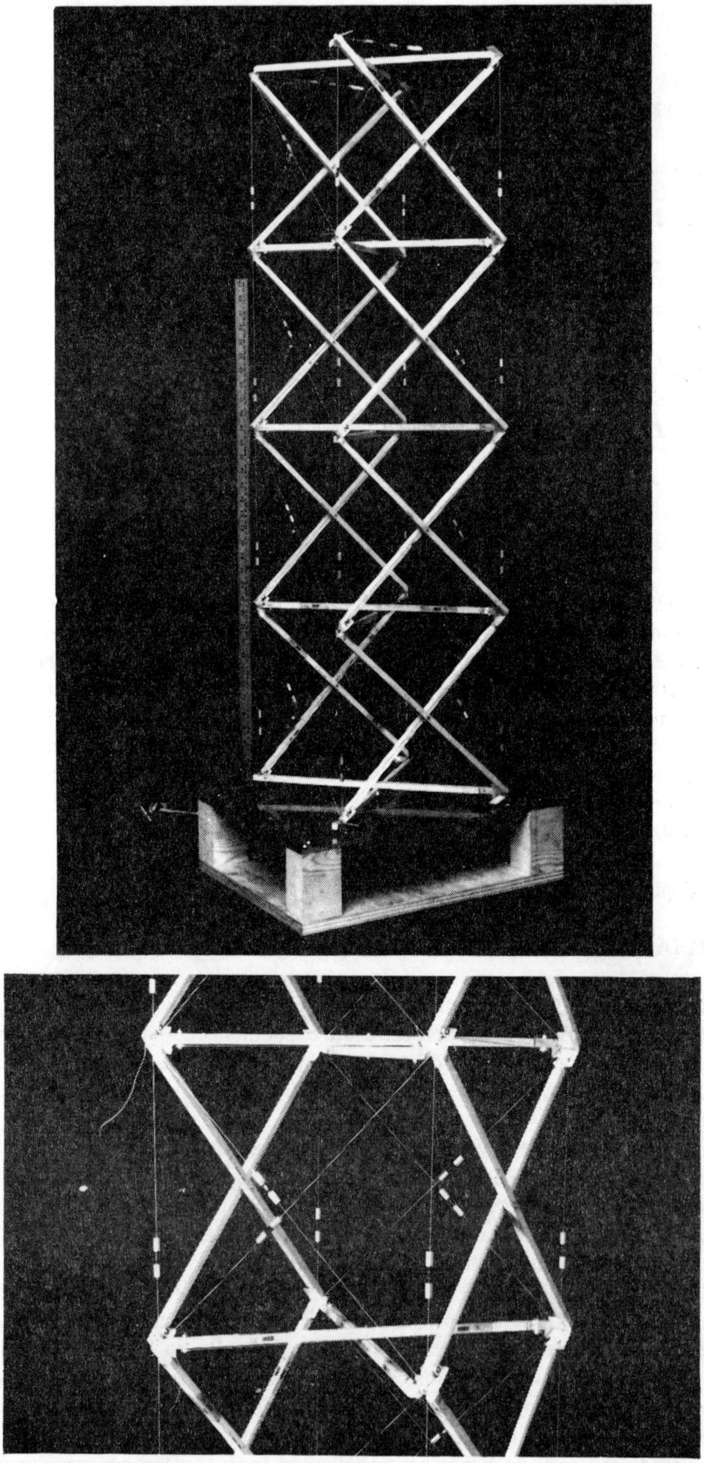

(a)

(b)

Fig. 1. (a) View of fully-deployed PDM. (b) An intermediate unit of the mast. Note that a small turnbuckle is included in each cable segment.

stiffness: for small loads, which are most usual in a gravity-free environment, their performance is dominated by the small mechanical tolerances at the joints. However, this can be improved by introducing a small amount of prestress in each connection, after deployment has been completed.

In [7] we have proposed a new concept, the Pantographic Deployable Mast (PDM), shown in Fig. 1, which combines the advantages of the two types of mast described above, but avoids many of their disadvantages. The PDM is particularly intended for those applications which require high stiffness. The key component of the PDM is a *three-dimensional pantograph*, formed by a set of rods connected at the ends and in the middle. Figure 2c shows the bottom unit of this pantograph, consisting of four pairs of coplanar rods, connected in the middle by a pin which can transmit in-plane forces and out-of-plane couples only. The end connection between rods allows them to rotate freely, in their own plane. The PDM shown in Figs 1 and 2 has eight coplanar rods in the vertical plane $x - z$, and eight more in a plane parallel to it; they are connected by five pairs of rods in equidistant horizontal planes. The pantograph is deployed by winding-in two *active cables*, running over pulleys which are located near the joints of the pantograph. The route followed by one of the active cables is shown in Fig. 2b: each kink requires a pulley. The second active cable starts at joint 8 and follows the symmetrical route. The active cables have *maximum* length when the PDM is folded and *minimum* length when it is fully deployed. The pantograph is folded by winding-in an additional active cable, which goes directly from the tip of the mast to its base. This cable, without which the mast cannot be operated in a gravity-free environment, is not shown in any of the figures. When the pantograph is fully deployed, at a deployment angle $\theta = 90°$, the short, *passive cables* which connect the ends of each vertical beam pair (vertical passive cables) or brace each pantograph unit (diagonal passive cables), suddenly become taut, as shown in Fig. 2d. At this point, a small, additional winding-in of the active cables imposes a state of prestress onto the mast. Now all of the cables are in a state of pretension, and therefore are able to carry any tensile or compressive force changes induced by the external loads, thus behaving much in the same way as pin-jointed rods. Of course, the compressive axial forces

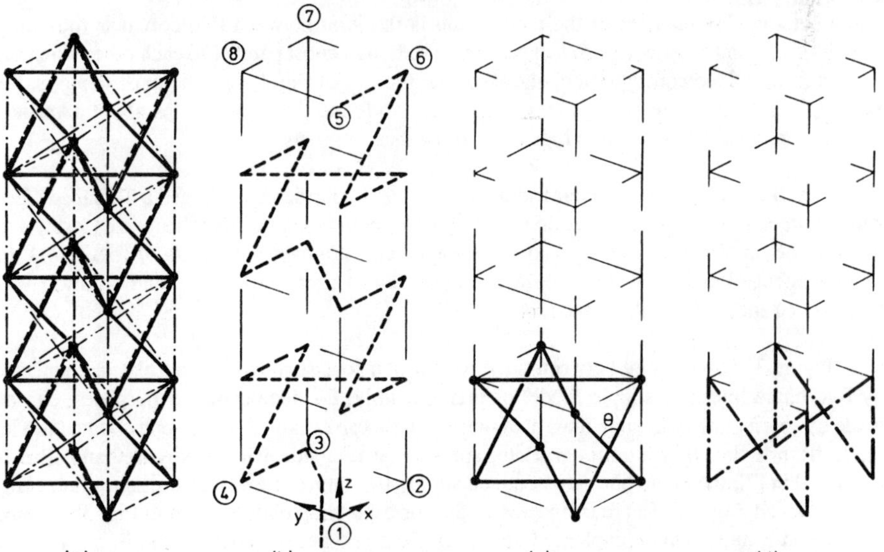

Fig. 2. (a) Schematic view of four-unit PDM. (b) Route followed by one of the two active cables. (c) Bottom unit of pantograph, consisting of four pairs of rods connected at the ends. A centre pin connects two rods in the same pair. The bottom two rods are connected by the same pin to a base plate, see Fig. 4. (d) Eight passive cables for the pantograph unit shown in (c).

must be smaller than the initial pretension. An additional effect of this state of prestress is to remove joint backlash. When the passive cables are slack, the PDM has an internal 'free' mechanism which allows it to fold down to $\theta = 180°$, as discussed in Section 4. However, when the PDM is fully deployed and prestressed, its internal mechanism suddenly disappears, due to the introduction of many extra members which are fully able to carry loads. The resulting *braced truss* is remarkably stiff.

An analytical investigation of the PDM has been reported in [7]. In this paper, we report on the measured performance of the four-unit prototype shown in Fig. 1. The layout of the paper is as follows: Section 2 briefly describes the design and manufacture of the prototype; Sections 3 and 4 describe its performance when fully deployed and during deployment. Section 5 concludes the paper.

2. DESIGN AND MANUFACTURE OF A FOUR-UNIT PDM

The prototype shown in Fig. 1 is the first complete PDM which has been built and tested. When fully deployed, it has a square section of 300×300 mm^2, and a height of 1200 mm. The corresponding (minimum) length of each active cable is 5.6 m, approximately. The PDM folds down to a small bundle, of approximate size $430 \times 50 \times 150$ mm^3; the corresponding (maximum) length of each active cable is 6.6 m. The main design aim for this model was to verify the feasibility of the proposed concept by finding solutions to the two main problems which had been identified from one- and two-unit simpler models, namely, the fairly complex nature of the joints of the pantograph, and the possibility that significant losses of tension in the active cables might occur, due to friction. For this prototype, comparatively little effort has gone into the other components, e.g. the rods of the pantograph or the deployment motor, which explains their fairly crude simplicity. Apart from ensuring the integrity of the PDM under a fairly low prestress and small external loads, see Section 3, its components have not been optimised in any way.

2.1 Pantograph

The most complex element of the pantograph is the joint between two coplanar rods, in a vertical plane, and a horizontal rod. These three rods are almost parallel to each other when the mast is in its folded configuration, shown at the top-left of Fig. 8. As the mast deploys, the two coplanar rods rotate about a horizontal axis, while the third one rotates about a vertical axis. The joint between the three rods must allow these motions.

The first, rather straightforward solution which was considered envisaged a T-shaped piece with the three rods hinged to it. A double-unit pantograph based on this scheme was tested and found to be satisfactory. However, due to general concern over the effects of friction on the large number of hinges which would be required in larger masts, a second solution was investigated and, in the end, adopted.

Figure 3 shows a joint between three rods which consists of five small plates, connected by five hinge lines. A simple model of this assembly, for demonstration purposes, can be made by scoring the five hinge lines on a piece of light cardboard. A more sophisticated model made from a flat, thin aluminum plate, sprayed on one side with a polyurethane film is described in [7]; the hinge lines are obtained simply by cutting grooves in the aluminium plate. Rods 1, 2 and 3 of Fig. 3a are connected to the corresponding plates shown in Fig. 3b, hence rods 1 and 3 must remain coplanar. If we fix the angle between rods 1 and 3, clearly, the five-plate assembly poses some restrictions on the position of rod 2. If we reduce the angle between rods 1 and 3, rod 2 is drawn in towards them, and the inverse: to a certain extent, the movements of the three rods are thus synchronised. Actually, the particular arrangement shown in Fig. 3a is not suitable for the PDM: when rods 1 and 3 are at 90°, rod 2 - which should form an angle of 45° with the plane of rods 1 and 3 - is, instead, at $\cong 65°$. We discovered this only after assembling the prototype. To obtain the correct angles in the fully folded and fully deployed configurations, the five-plate joint should not be flat, but with an angular defect of $\cong 40°$. An assembly of five plates connected by five concurrent hinges has two *internal mechanisms*, the first mechanism is that required for folding the pantograph, but the second, unwanted mechanism allows rod 2 to rotate quite freely about a horizontal axis bisecting rods 1 and 3. If properly supported, a complete pantograph structure can suppress this motion.

In the PDM shown in Fig. 1, each five-plate assembly has been produced as a single piece, by injection moulding, using Nylon 66. Light hardwood rods with square cross section of 10×10 mm^2 have been used as the main pantograph elements. Their Young's Modulus is E = 4600 N/mm^2.

(a)

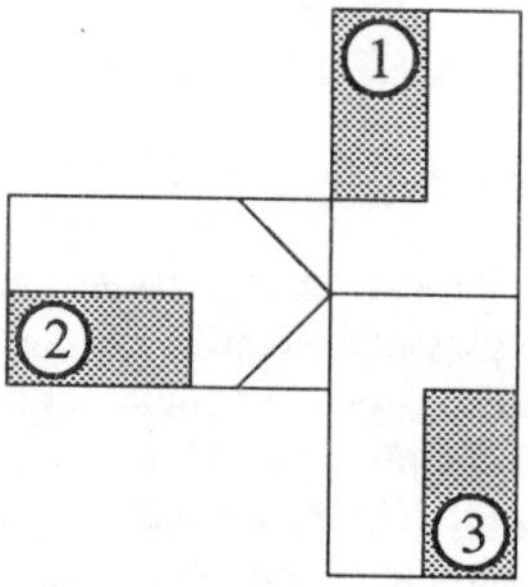

(b)

Fig. 3. Five-plate joint of PDM. In (a), note the three pulleys for the active cables. (b) Flat five-plate joint, which results, incorrectly, in an angle of 65° between rod 2 and the plane containing rods 1 and 3, when $\theta = 90°$.

2.2 Active and passive cables

Standard Kevlar cable, Kevlar 49 yarns covered with a polyethylene sheath, has been used throughout. The axial stiffness of this cable is AE = 8400 N and its tensile strength is just over 300 N. Each active cable runs over 16 pulleys and, to minimise the variation in tension, all pulleys are mounted on small ball-bearings. Each passive cable incorporates a small turnbuckle, to adjust its length.

2.3 Base and deployment mechanism

The centre pin which connects the two rods at the bottom of the mast is fixed to a base plate and, beneath, a small DC motor turns two threaded drums which wind the active cables. This arrangement is shown in Fig. 4.

3. PERFORMANCE OF FULLY-DEPLOYED MAST

In this section, we report on the structural performance of the fully-deployed and prestressed PDM, under a range of static load conditions. In each case, joint displacements and cable forces have been measured. While the general level of prestress in the PDM is regulated by the active cables, the precise value of the pretension in a particular passive cable depends crucially not only on its length but also on the lengths of the passive cables in its vicinity. To achieve - as far as possible - uniform tension in all passive cables, before the structural tests started their lengths were adjusted using the turnbuckles. However, small fluctuations in the cable tensions

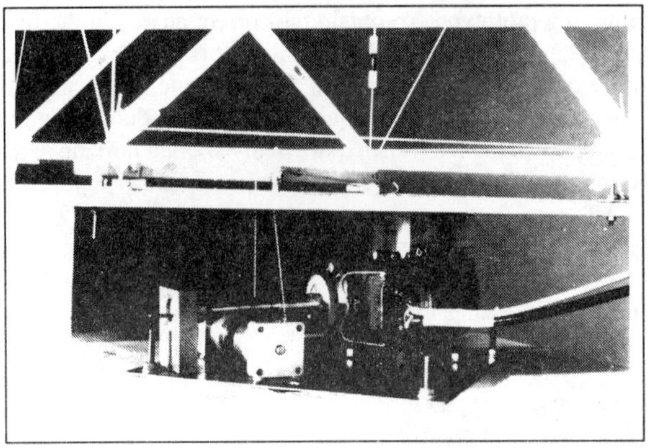

Fig. 4. Base of PDM, note the drum which winds the two active cables.

still occur: a typical set of tensions in the horizontal segments of an active cable, measured soon after deployment, is from bottom to top, 26.6, 26.6, 25.8, 23.2, 21.7, 20.0, 22.0 N; the tension in the last segment could not be measured. The corresponding forces in the passive cables are in the region of 12 N. Although this represents a fairly uniform state of stress, a 25% reduction in force over only four units of a long PDM would be certainly unacceptable. However, it should be possible to achieve rather significant improvements by fairly minor design changes; for instance, in the present model the active cables tend to rub against the central pins between rods, which is clearly undesirable.

Figures 5-7 show the response of the mast to external loads applied to the top joints. The following load conditions have been considered.

(i) Two equal forces at joints 6 and 7 in the x-direction, Fig. 5a, and at joints 7 and 8 in the y-direction, Fig. 5b. The response of the mast is fairly linear and predictable; numerical predictions based on the analysis described in [7] are also shown, for comparison. The stiffness, for symmetric loads at the tip of the mast, is 1.3 N/mm in the x-direction and 1 N/mm in the y-direction. Hence, in stiffness terms, bracing rods and diagonal bracing cables are almost equivalent.

(ii) Four equal forces at joints 5, 6, 7 and 8 in the negative z-direction, Fig. 5c. The axial stiffness of the mast is 25 N/mm.

(iii) A single load at joint 6 in the x-direction, Fig. 6. The torsional stiffness of the mast is 24 Nm/rad.

(iv) Figure 7 shows the response of the PDM to a load cycle (two equal forces at joints 7 and 8, in the y-direction). In the 30 minutes delay between the end of the first load cycle and the beginning of the second, the deflection was reduced by 0.2 mm.

Probably, the most considerable source of non-linearity has been the gross deformation of the Nylon plates at the joints of the pantograph, which had to bend during deployment to accomodate the angular mismatch discussed in Section 2.1. Obviously, this deformation is not allowed for by the computer model. A further source of non-linearity in the response of the mast is the slackening of some cables, apparent in Fig. 5c; this would be eliminated simply by increasing the prestress level.

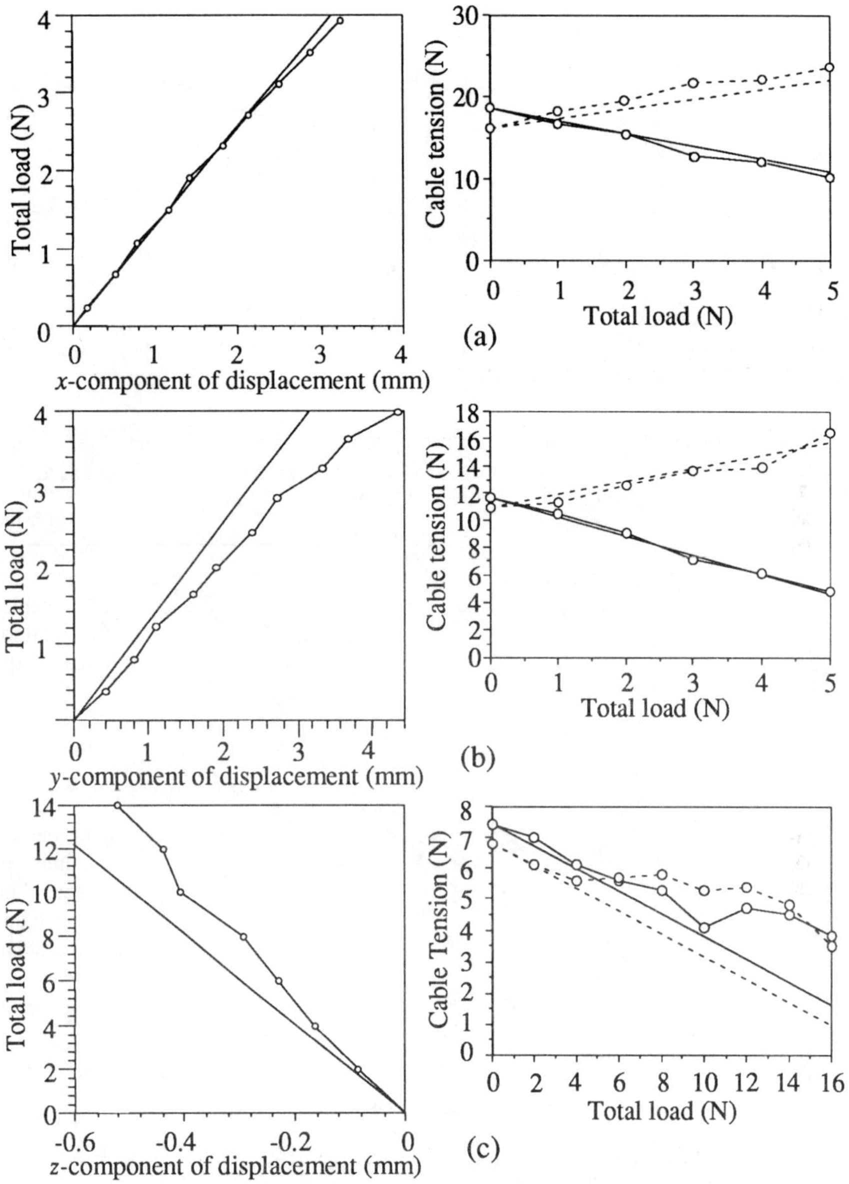

Fig. 5. Performance of fully-deployed PDM. The load/deflection and load/tension responses were measured during separate tests, in fact the load steps are different. Experimental values are marked by circles, while unmarked lines indicate theoretical predictions. (a) Response to two equal forces in the x-direction, at joints 6 and 7. The mean measured deflection, in the x-direction, of the two loaded joints and the mean tensions in the bottom vertical passive cables are given. The solid lines and the broken lines refer to the cables connected to joints 2 and 3, and joints 1 and 4, respectively. (b) As (a), but for loads at joints 7 and 8, in the y-direction. The mean deflection of the two loaded joints and the mean tensions, in the vertical passive cables connected to joints 3, 4 (solid lines) and 1, 2 (broken lines), are given. (c) Response to four equal forces in the negative z-direction, at joints 5, 6, 7, and 8. The mean axial deflection and the mean tensions in the vertical passive cables connected to joints 1 (solid lines) and 2 (broken lines) are given.

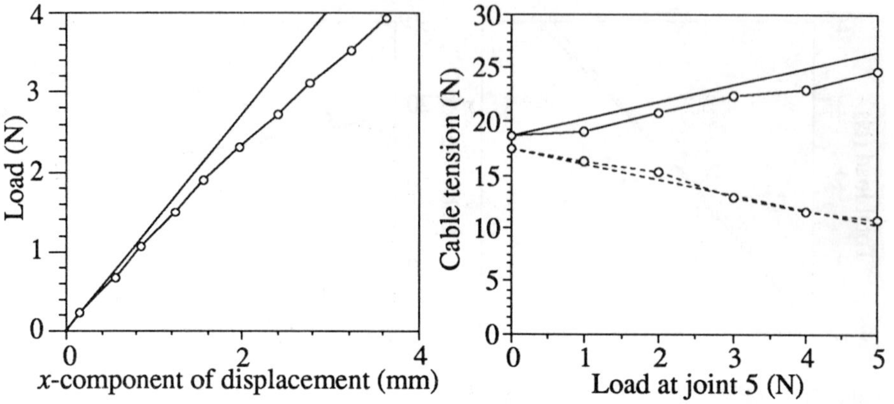

Fig. 6. Response to a single force, in the x-direction, at joint 6. The x-deflection of joint 6 and the mean tensions in the bottom vertical cables (solid lines correspond to cables connected to joints 2 & 3, and broken lines to cables connected to joints 1 & 4) are shown.

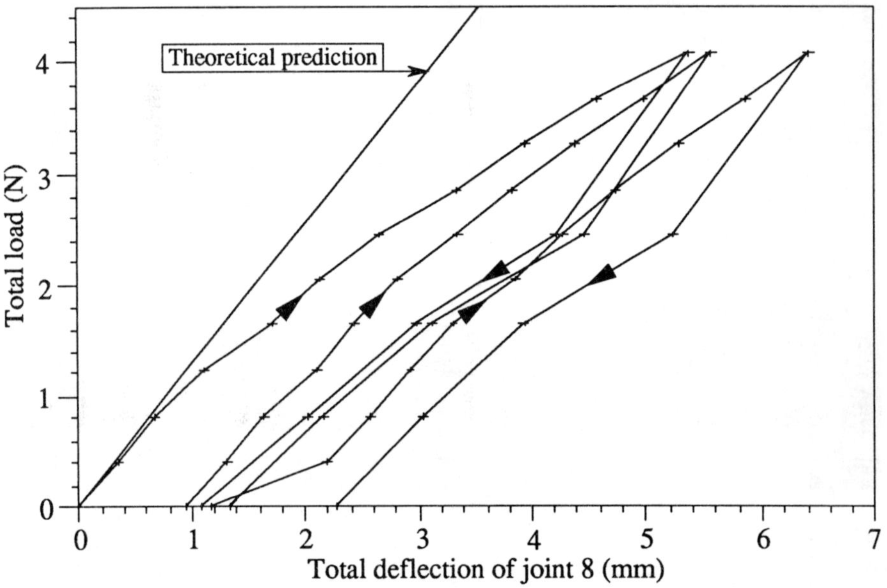

Fig. 7. Deflection of the PDM during a load cycle. The total deflection of joint 8, due to equal forces in the y-direction at joints 7 and 8, is shown.

4. DEPLOYMENT TESTS

The deployment sequence of the PDM is shown in Fig. 8, while the corresponding active cable length is shown in Fig. 9. Deployment starts from the fully-folded configuration, top-left of Fig. 8, which corresponds to a deployment angle θ slightly lower than 180°. At this stage, each of the two active cables has a total length of about 6.5 m.

Note that the length of the active cable becomes maximum, and hence stationary, at a deployment angle θ = 180°, see Fig. 9, which of course would be difficult in practice to

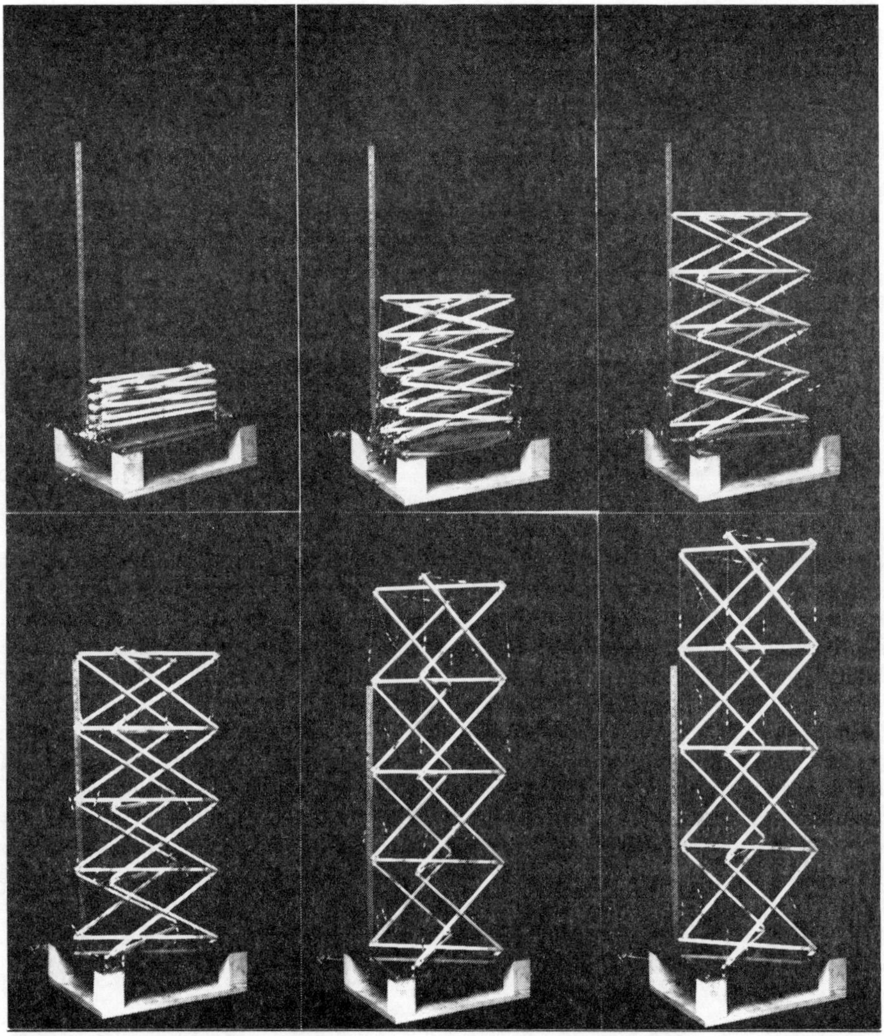

Fig. 8. Deployment sequence of PDM. The 1 m rule on the side of the mast is fixed to the base, which is 130 mm high.

achieve but, if - with some special joints - the mast could be folded down to $\theta = 180°$, it would be practically impossible to re-deploy it only by shortening the active cables. In practice, though, even the small slopes in the region of $\theta = 170°$ could give rise to problems, a small amount of friction would be sufficient to prevent re-deployment. Our prototype never got stuck. Nonetheless, in future, it would be wise to include some soft springs near the joints.

As the two active cables are shortened by equal amounts, the deployment of the PDM proceeds smoothly, as shown in Fig. 8, until the fully-deployed configuration, bottom-right in the figure, is reached. At this stage, each active cable is shortened by a further 25 mm, thus inducing the prestress level indicated in Section 3. It has been found that, during deployment, and particularly near mid-height, the prototype is very floppy indeed: its top joints can sway almost freely ± 100mm in the y-direction, which is not braced by beam pairs and hence relies on the stiffness of the joints. Clearly, the joint design shown in Fig. 3 is too flexible; it should be replaced, for the top and bottom joints of the PDM at least, with a stiffer joint.

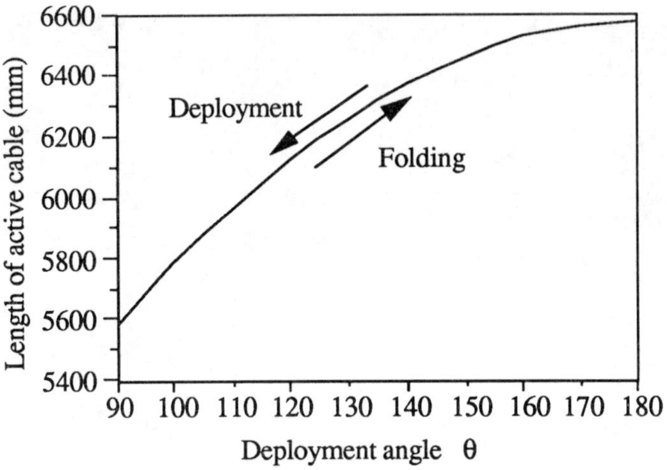

Fig. 9. Active cable length vs deployment angle.

Remarkably, though, when deployment is complete the transition from a mechanism to a stiff structure is entirely successful.

As well as stiffening the PDM by pretensioning the 32 passive cables, prestress has the further effect of tightening up any slightly loose connections. The end result is that, if the mast is folded and re-deployed, it comes back to almost exactly the same position. We have folded and re-deployed the PDM twenty-one times and, after each cycle, measured the position of joints 5 and 8 using the technique described in the Appendix. The motor was switched off when a mark on an active cable became aligned with a reference mark. Figure 10 shows the measured error in the x- and y-coordinates of these joints, i.e. the difference between any measured value and the mean. In both x- and y-directions, all but one of the measured points are within ± 0.6 mm from the mean. Regarding the z-coordinates, Fig. 11a, all but one points are within ± 0.3 mm from the mean, which is quite remarkable, given the fairly crude nature of the model we have tested. The corresponding values of the r.m.s. total error, indicating a length accuracy of better than 0.1%, are given in Fig. 11b.

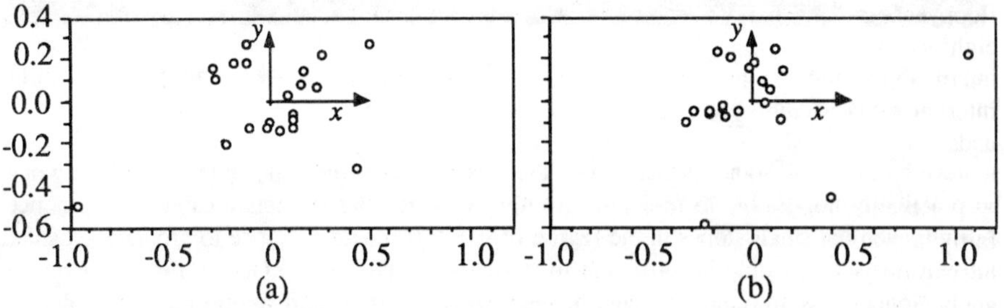

Fig. 10. Error in x- and y-coordinates of joint 5, plot (a), and joint 8, plot (b), when the mast is folded and re-deployed twenty-one times. All values are in mm.

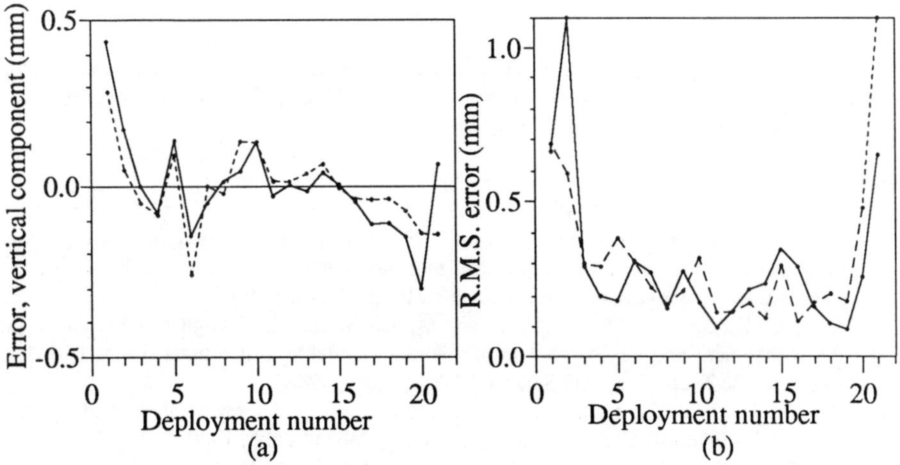

Fig. 11. Error in z-coordinate and r.m.s. error of joints 5, solid lines, and 8, broken lines, during the deployment tests.

5. CONCLUSIONS

The overall performance of the four-unit PDM described in this paper has been satisfactory, thus demonstrating that the concept of a pantographic deployable mast is indeed feasible. We are particularly pleased with the high repeatability of its fully deployed shape, which makes it suitable for applications that require high accuracy, high stiffness, and high ratio of deployed to folded volume. We are confident that, simply by making all components more carefully, an even higher shape accuracy could be achieved.

Many points which need further development have emerged. We are currently investigating some alternative schemes, still based on the same principles of the PDM, but with different layouts of rods and cables, which should result in a more stable deployment process. The total number of rods, which are the main contibutors to the mass of the mast, will remain essentially unchanged. Although the five-plate joint has proved to be very reliable and easy to make, its undesired internal mechanism has resulted in an unexpected and excessive floppyness of the pantograph which, during deployment, would be unacceptable. In future, the eight joints at the top and bottom ends of the pantograph will not be of the same type as the intermediate joints: since there are only two rods to be connected, instead of three, both of the internal mechanisms in the standard five-plate joint, for a fixed deployment angle θ, are undesired.

Recently, the use of 'long-stroke space actuators', i.e. deployable masts which can apply fairly large axial forces while deploying, particularly towards the end of their deployment, has been proposed for a large wire mesh antenna [8]. A PDM would be most suitable for applications of this sort. Although in this paper we have considered a straight version of the PDM, a curved PDM could be obtained simply by moving, in each beam pair, the internal pin to a non-central position. Here, there is a link with the work by Zanardo [9] on curved pantographs, and Escrig and co-workers [10] on deployable space grids to cover grandstands, swimming pools, etc. The use of some active and passive cables could enhance the performance of these systems as well. Generally, we believe that our use of active and passive cables as key elements of a deployable structure can lead to new, more efficient concepts. So far, the only related proposal which we have encountered is the deployable hyperboloidal antenna by Waters [11], which relies on passive cables to arrest deployment but does not become pretensioned.

ACKNOWLEDGMENTS

We would like to thank Professor C.R. Calladine and Professor K. Miura for helpful comments and continuing interest in this work. Support from the SERC and the MoD (Royal Aircraft Establishment, Farnborough) is gratefully acknowledged.

REFERENCES

1. Eiden, M., Brunner, O. and Stavrinidis, C. Deployment analysis of the Olympus Astromast and comparison with test measurements, Proc. 4th European Symp. Photovoltaic Generators in Space, ESA SP-210, pp 356-363, 1984.
2. Okasaki, K., Sato, S., Obata, A., Natori, M. and Miura, K. Design considerations on the mechanical and deployment properties of a coilable lattice mast, Proc. 15th Int. Symp. Space Technology and Science, Tokyo, Japan, 1986.
3. Rimrott, F.P.J. Self-extension of an open section tube, in Theory of Shells (Ed. Koiter, W.T. and Mikhailov, G.K.), pp 495-508, North-Holland Publ., 1980.
4. SENER, Collapsible Tube Mast - Final Report, ESA CR(X) 2642, 1988.
5. DORNIER System, Development of the Extendable and Retractable Mast (ERM), Final Report, ESA CR(P) 2481, 1987.
6. Coyner, J.V. Box truss development and its applications, Proc. Conf. Large Satellite Antennas and Space Technology, NASA-CP-2269-Pt1, pp 527-543, NASA Langley Research Center, 1985.
7. Kwan, A.S.K. and Pellegrino, S. A cable-rigidised 3D pantograph, Proc. 4th European Symp. Space Mechanisms and Tribology, ESA SP-299, pp 149-154, Cannes, France, 1989.
8. Miura, K. and Miyazaki, Y. Concept of the tension truss antenna, AIAA Journal, Vol. 28, pp 1098-1104, 1990.
9. Zanardo, A. Two-dimensional articulated systems developable on a single or double curvature surface, Meccanica, Vol. 21, pp 106-111, 1986.
10. Escrig, F. and Valcarcel, J.P. Curved expandable space grids, in Proc. Int. Conf. Design and Construction of Non-conventional Structures (Ed. Topping, B.H.V.), pp 157-166, Civil-Comp Press, Edinburgh, 1987.
11. Waters, T.J. Hyperboloidal Deployable Space Antenna, U.S. Patent No 4825225, 1989.

APPENDIX: Measurement of joint coordinates and cable tensions

To measure the x-, y- and z-coordinates of the joints of the PDM we have used a Remote Measuring System which consists of two Zeiss electronic theodolites, linked via a communications box to an IBM PC microcomputer. A program, written in Turbo Pascal, collects the data from the theodolites, performs a series of checks, and computes the coordinates of the target. To initialise the system, the position of the theodolites is calculated by pointing the theodolites to the two ends of an Invar bar. The standard target used is a 1mm diameter ball, stuck to the intersection point of the five hinge lines on each joint of the PDM. The achievable accuracy is ± 0.1mm.

To measure the force in any cable of the PDM, we have made a tensiometer which measures the force required to introduce a small central deflection (up to 1mm) in a short length of the cable. This force is applied to the cable by a thin strain-gauged, spring steel cantilever: both ends of the cable are clamped before the cantilever is released against the cable. The relationship between the strain at the root of the cantilever and the force in the cable is almost linear for small cable tensions, but - as the tension increases - the curvature in the cantilever tends to a constant value and the tensiometer becomes less accurate.

The stiffness of the cantilever has been optimised for cable tensions in the range 1-40 N: the accuracy is ± 0.5 N.

Expandable Long Span Enclosures

A.M. Wolde-Tinsae

Department of Civil Engineering, University of Maryland College Park, Maryland 20742, U.S.A.

ABSTRACT

Prebuckled sandwich arches are constructed and tested in the laboratory to study the feasibility of using this new type of structural system as a skeleton for long span enclosures. The prebuckled sandwich arch is formed by laminating the core material to one face on the sandwich panel and buckling it into shape. The other facing then is buckled into shape and laminated to the core thus forming a composite section. Sandwich arches with discrete cores spaced at various intervals are studied. The test results show extremely high strength-to-weight ratios and suggest that the prebuckled arches will perform well as a skeleton for long span enclosures.

INTRODUCTION

The need to provide relief and shelter to people displaced by natural disasters such as earthquakes, draught, etc. has created a demand for structural systems that are transportable, erectable and salvageable. Similar type of structures have found applications as enclosures for exhibition, recreational complexes and for human communities associated with exploration drilling, pipeline construction, as well as, for semi-permanent settlements in hostile environments such as the Arctic. Because of the nomadic nature of man's current Arctic lifestyle, easy mobility is a vital characteristic for successful Arctic shelters (Nagrani[1]). Thus the concept of a transportable and erectable shelter for remote locations has its attractions for both industrialized countries as well as developing nations. Some additional desirable features for a shelter in a remote location are:

- The shelter should be capable of being erected by people with a minimal amount of training;
- There should be little or no requirement for heavy machinery or equipment;

- The shelter should be tolerant to high degree of ground unevenness; and
- The shelter components need to have high degree of damage tolerance with no progressive degradation occurring after damage.

The growing need for relocatable shelters and specialized structures for terrestrial and extra terrestrial applications has, in recent years, generated the development of various types of non-rigid, self-erecting structures that can easily be erected and dismantled. One type that has come into wide use is the pneumatic structure. Pneumatic structures may be of the air-supported type made of a single layer membrane with pressure elevation inside the enclosure, with pressure maintained at slightly above ambient; or they may be of tubular or cellular construction frequently referred to as the air-inflated type, made of double layer membranes with pressurization only within the walls. Pneumatic structures with negative pressure where the pressure inside the enclosure is maintained at below ambient have not found wide application. Figures 1 and 2 show membrane-type and beam type pneumatic structures (Kawaguchi, et al.[2]). The necessary air pressure is usually low for a membrane type structure and is high for a beam type.

Problems of space subdivision and environmental control in the design of air supported single-family dwellings have led to a concept, based on the principle of "pneumatics within pneumatics," in which one or more air-supported structures are contained within a single air-supported enclosure (Montero and Pohl[3]). The outer enclosure provides weather protection and resistance against wind forces, while the inner enclosures act principally as self-supporting space dividers. The resulting shelter provides fair thermal insulation due to the existence of air space between the outer and inner membrane envelopes. By the careful placement of reflective, opaque, transluscent and transparent panels in the outer membrane, heating and cooling costs for an occupied dwelling may be held to within acceptable limits in relatively mild climates. Pneumatic single-story and multi-story building types which have been developed in recent years can be used in subdividing internal spaces.

Some of the major disadvantages of air structures include the requirement for maintaining pressure differential within the structure. This results in a need for special doors to prevent excessive loss of air. This characteristic of air structures tends to limit their general use. The need for an external source to maintain its shape, leakage and vulnerability to damage have prompted some engineers to look for other types of erectable structures. Another class of self-erecting structures is the expandable structure with rigid components that unfolds in some manner to reach its final form.

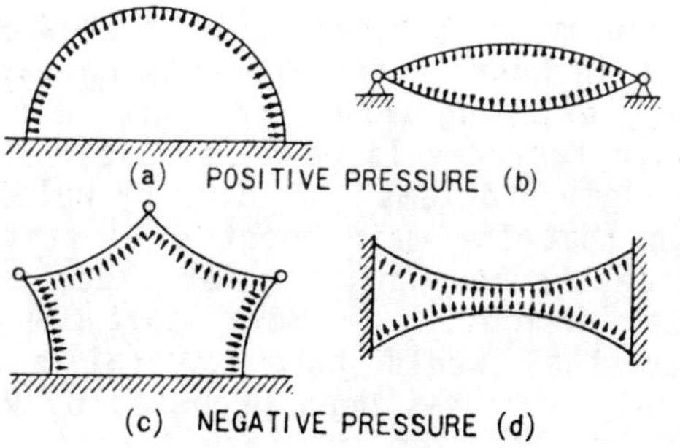

Figure 1. Membrane Type Pneumatic Structures

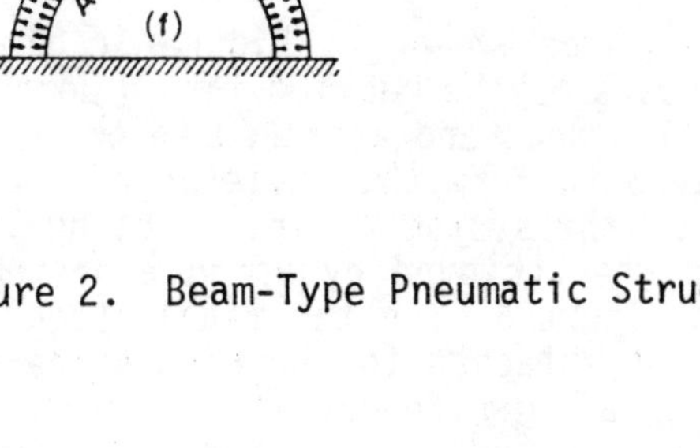

Figure 2. Beam-Type Pneumatic Structures

PREBUCKLED SANDWICH DOME

Several scientists like Fuller[4] and Zuk[5] have advocated the use of a structural system as a super-roof to create a clear-span controlled climate in hostile environments on earth as well as in space. However, existing structural forms like the geodesic dome are unsuitable for very-large-size, clear-span enclosures because of instability problems. Studies by Wolde-Tinsae[6-9] and others have shown that the main agent of instability and the primary limiting factor in span size for clear-span enclosures is structural self-weight. A more suitable new type of structural system that would have several advantages over existing structural forms has been proposed by Wolde-Tinsae[10]. The proposed structural system has been termed "The Prebuckled Sandwich Dome."

Prebuckled sandwich dome construction involves the formation of prebuckled sandwich arches and their assemblage into a dome. The prebuckled sandwich arch is formed by laminating the core material to one face of the sandwich panel and buckling it into shape. Then the other facing is buckled into shape and laminated to the core thus forming a composite section.

An erection procedure that may be used to construct the skeleton of the prebuckled sandwich dome is shown in Figure 3. The core material shown in Fig. 3a is of the corrugated type made of say aluminum (other shapes and materials may also be used; examples are shown in Figure 4). The core material is attached by means of bolts or other types of connections to the bottom skins. The skins are wide shallow members made of a ductile material like aluminum or steel. The members are then joined to a central piece to form an array of intersecting members. The central piece may take the form of a ring or the like. The top skin is then attached to the central piece but is not yet attached to the core (Figure 3b).

By applying forces as shown in Figure 3c the system of intersecting flat members may be buckled into a dome shape. The top skins are now pulled down and attached to the core by means of bolts or other means to form the skeleton of the prebuckled dome. The buckling of the system of intersecting flat members into a dome shape may be achieved by using a system of cables and jack as shown in Figure 5. In the final stage a cable may be passed through loops attached to the ends of each member or the dome may rest on a ring foundation or each leg may be attached to individual supports. The prebuckled sandwich dome skeleton may now be covered with reinforced membrane or other means, depending on intended usage.

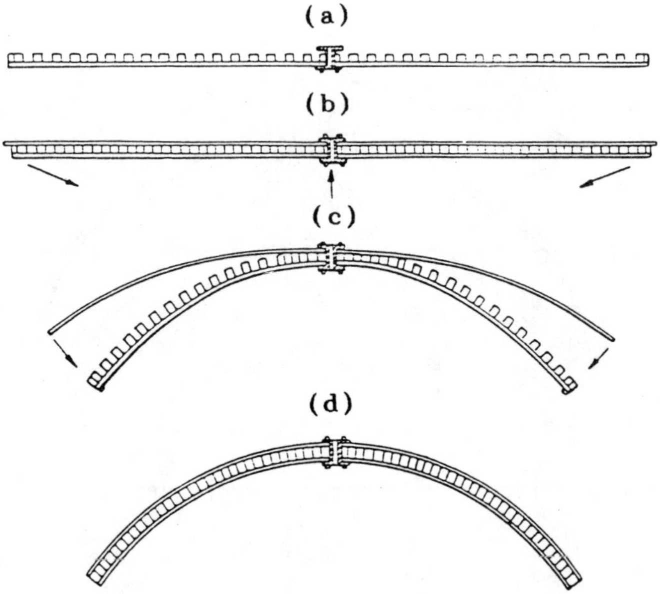

Figure 3. Stages of Construction

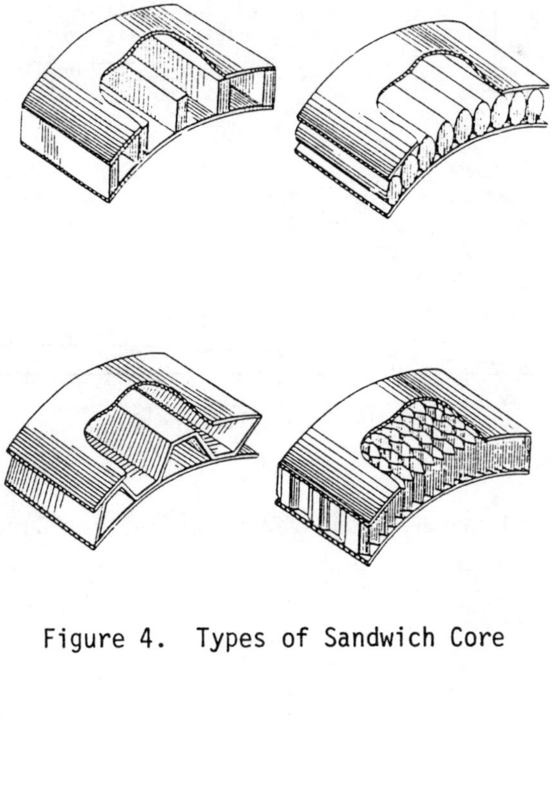

Figure 4. Types of Sandwich Core

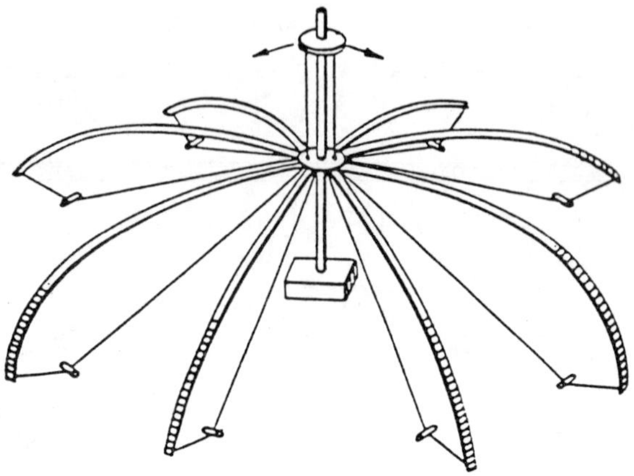

Figure 5. Erection Procedure

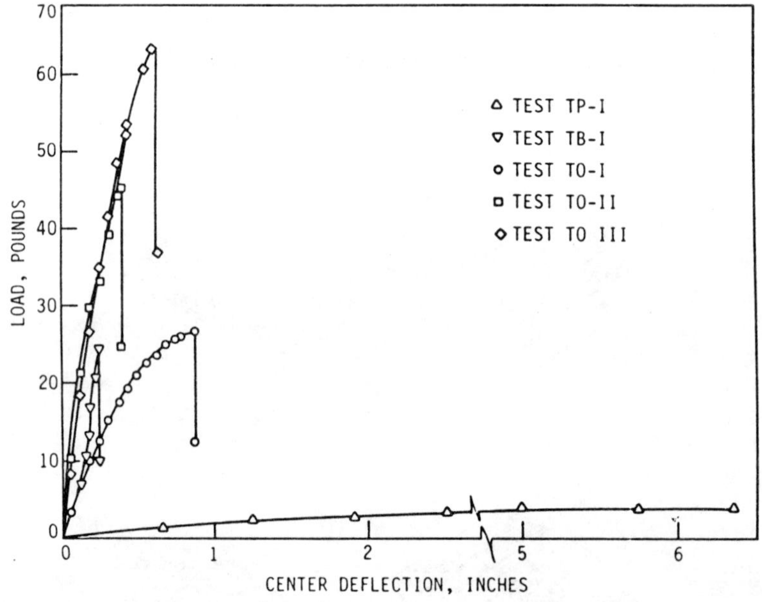

Figure 6. Load Versus Deflection Curves

SPECIMEN DESCRIPTION

The first model (TP-I), which was used as a basis for comparing the rest of the models, was constructed from a piece of 2024 T3 aluminum 0.040 inch (1.016 mm) thick, 3 inches (76.2 mm) wide and 48 inches (1219.2 mm) long. The thickness for TP-I was chosen so that it was twice as thick as each sandwich facing, thus suggesting the characteristics of an arch with a core of zero thickness.

The remaining four models utilized sandwich construction with two types of wood at different spacings. To prevent composite action during the erection procedure, the core was initially laminated to only one facing. It was decided to make the initial lamination to the bottom facing so that final lamination would be on the top surface, thereby preventing overhead work.

For the models involving sandwich construction, adhesive was used for lamination in all cases except for specimen TO-III, in which the adhesive was used in conjunction with bolts. The facing material used in all of the sandwich arches was 2024 T3 aluminum, 0.20 inch (0.508 mm) thick. A discrete core was constructed with balsa wood in test TB-I, TO-II and TO-III and spaced to various intervals. Table 1 gives a summary of materials used for the core and the facing, as well as the spacings used for the different tests. Again, strain gages were located at areas of interest.

In sandwich construction, special consideration to details must be given at locations where loads or restraints are applied. For the models tested in the laboratory, special closure clamps were made for the ends of the arch which then were attached to the hinges. The end clamps facilitate the transfer and distribution of restraint loads. Where these clamps attach to the sandwich arch, the core was spaced at much closer intervals to relieve the stress concentrations resulting at these locations. At the center, where the load was to be applied, the core also was spaced at much closer intervals to alleviate stress concentrations and also to prevent local buckling due to the application of the load.

GENERAL TEST SETUP

A six foot square plywood platform was constructed on an existing frame in the laboratory for the purpose of testing the arches. The platform was supported on two sides by I beams, which were part of the frame. An opening in the platform allowed for the crown of the arch to be moved lower than the supports. The supports for the arch consisted of hinges which were located at the edge of the opening. This allowed for rotation at the support but at the same time restricted horizontal and vertical displacement.

Table 1. Specimen Description

Test (1)	Facing (2)	Core (3)	Adhesive (4)	Core Spacing (5)	Variations (6)
TP-I	0.020 in. 2024T3 Aluminum	None	None	--	Total thickness of arch = 0.040 in.
TB-I	0.020 in. 2024T3 Aluminum	Balsa	Commercial Epoxy (Scotch)	0.625 in. O.C.	
TO-II	0.020 in. 2024T3 Aluminum	Oak	AE15	1.875 in. O.C.	
TO-II	0.020 in. 2024T Aluminum	Oak	AE15	Linear	Span 1, 0.625 in. center to center, following spans increase by 0.0625 in.
TO-III	0.020 in. 2024T3 Aluminum	Oak	AE15	Linear	Each core element bolted at same spacing as TO-II

Note: 1 in. = 25.4 mm

The load was applied by controlling the displacement of a threaded rod which had a loading mechanism attached to its end. This rod passed through a hole in a steel plate which was secured to two channels, part of the existing frame, above the model. The height of the rod was controlled by turning a nut which had been threaded onto the rod. The rod was prevented from turning by a stabilizing mechanism.

EVALUATION OF TEST RESULTS

Peak load versus center displacements for all five tests are plotted in Figure 6 to show relative load carrying capacities of each model. The specimen with no core, TP-I, behaved with extreme flexibility, achieving its peak load only after the crown of the arch had deflected 5.75 inches (146.05 mm) downward. As expected, model TO-I was found to be much more flexible than Model TB-I since the spacing between the discrete core wooden pieces in TO-I was three times the spacing used in TB-I.

The balsa wood core arch, TB-I, had the lowest load carrying capacity (24.66 lbs or 109.74 N) of the arches tested with nonzero core. This, however, still represented a 616% increase in load capacity over TP-I. With an increase of only 2.29 pounds (10.19 N), the peak load for TO-I was not significantly greater than that of TB-I. However, since the spacing used for the core in TO-I was three times larger than that used in TB-I, this does represent a substantial increase. A 675% increase in load capacity was realized over TP-I in TO-I. The linear spacing in test TO-II removed the instabilities which developed in TO-I and led to a substantial increase in its load carrying capabilities, 45.61 pounds (202.96 N) for a 1113% increase over TP-I. The failure in TO-II was a bond failure indicating that a larger load could be carried if a better bond between the core and the facings was developed. To this end, in the final test performed, TO-III, the same spacing was used as in TO-II but the oak was bolted to the facings to insure that there would be no bond failure. This resulted in a local buckling failure near the north support. Before this buckling occured, the load carried 63.74 pounds (283.64 N), approximately 1575% greater than that of TP-I. Comparing test TP-I and TO-III shows that for a 67% increase in weight, an increase of 909% in the strength-to-weight ratio is obtained.

Reviewing the shear test results for the four sandwich arches showed that the modulus of shear has a significant effect on the load carrying capacity of the arch. TO-I was able to carry a slightly greater load than TB-I even though the pieces of core material were spaced three times as far apart. For TO-III where linear spacing was used with bolts connecting the core to the facing, the facing of the arch buckled indicating that if closer spacings were used, it would be capable of attaining a load even larger than the 63.74 pounds (283.64 N) reached in TO-

III. Core spacing which will give optimum results will partly depend on load distribution on the arch.

SUMMARY AND CONCLUSIONS

Five prebuckled arches, four sandwich and one plain prebuckled, were constructed for testing in the laboratory. The plain arch was used as a basis for evaluating the performance of the four sandwich arches. All four of the sandwich prebuckled arches had discrete cores with the spacing, type of wood used and any special techniques used in construction shown in Table 1. These arches then were tested to determine maximum load carrying capacity, failure mode and strains at various locations along the legs of the arch. Theoretical results for the arch with zero core also were included.

To facilitate the testing of the arches, a testing platform and loading mechanism were constructed and mounted on an existing frame in a laboratory. The platform was constructed such that the crown of the arch could be moved below the level of the platform. The loading mechanism was capable of applying a uniform line load across the crown of the arch, while also fixing the crown of the arch against lateral translation. This setup performed well in all of the tests performed.

Shear tests were performed on the two types of wood cores used. From the results of these tests, the shear modulus of the core and of the overall sandwich panel was determined. These values then were used in comparing the load carrying capacity of the arches.

Test results on the prebuckled sandwich arches showed that very significant increases in strength-to-weight ratios can be realized. Comparing the plain prebuckled arch to one of the prebuckled sandwich arches tested, it was found that for a 67% increase in weight, an increase of 909% in strength-to-weight ratio was achieved. These results combined with the relative ease with which the structure was capable of being assembled siggest that this new type of structural system will perform well as the skeletal system for long span enclosures.

REFERENCES

1. Nagrani, A.K., "Composite Relocatable Shelters," ASCE Spring Convention, American Society of Civil Engineers, Pittsburgh, Pennsylvania, Preprint 3193, April 24-28, 1978.

2. Kawaguchi, M. and et al., "Engineering Problems of Pneumatic Structures," Proceedings, 1971 IASS Pacific Symposium Part II on Tension Structures and Space Frames, Tokyo and Kyoto, Paper No. 5-5, 1972, pp. 449-460.

3. Montero, J. and Pohl, J., "The Multi-enclosure Air-supported Dwelling," IASS World Congress on Space Enclosures, Building Research Centre, Concordia University, Montreal, July 1976.

4. Fuller, R., Ideas and Integrities, Prentice-Hall, Englewood Cliffs, New Jersey, 1963.

5. Zuk, W., Super-roof, Progressive Architecture, 1963.

6. Wolde-Tinsae, A.M. and Huddleston, J.V., "Three-Dimensional Stability of Prestressed Arches," Journal of the Engineering Mechanics Division, ASCE, Vol. 103, No. EM5, paper No. 13281, October 1977, pp. 855-867.

7. Wolde-Tinsae, A.M. and Huddleston, J.V., "Non-linear Three-dimensional Analysis of Prebuckled Domes," International Journal of Computers and Structures, Vol. 14, No. 314, December 1981, pp. 215-224.

8. Wolde-Tinsae, A.M. and Assaad, M., "Non-linear Stability of Prebuckled Tapered Arches," Journal of Engineering Mechanics, ASCE, Vol. 110, No. 1, Paper No. 18443, January 1984, pp. 84-94.

9. Wolde-Tinsae, A.M. and Chini, A., "Centrifuge Testing and Numerical Stability Analysis of Prestressed Domes," Proceedings of the Third International Conference on Computational Methods and Experimental Measurements, Porto Caras, Greece, September 2-5, 1986, pp. 539-548.

10. Wolde-Tinsae, A.M., Method of Forming a Skeletal Dome Structure In Situ, United States Patent 4400927, 1983.

Deployable Structures

M. Hernandez, H. Carlos, in collaboration with W. Zalewski

Instituto Experimental de la Construccion, Facultad de Arquitectura, Universidad Central de Venezuela, apdo.47.169. Caracas 1041- A Venezuela

INTRODUCTION

There are many examples of ideas (concepts) of structural assemblies possessing foldable, extensible, capabilities, so that they may be extended, deployed when desired and folded up to a compact form for storage and or transportation. The common characteristic of foldable structures is the abilitv these structures, whose points are lying on the surface of revolution, have to assume two different configurations; first the collapsible (compact) one, suitable for storage and transportation, and second the deployed (extended) form allowing the structure to preserve (keep) its shape (usually spherical or cylindrical)(eg. Zalewski [1])

APPLICATIONS OF THE PRINCIPLE OF DEPLOYABILITY TO LARGE SCALE SITUATIONS

Deployable structures are suitable in response to the following needs :
a- A situation in which there is a need to create enclosed or protected space for a short period of time and then move that space to another location for erection or storage. Examples: -Travelling expositions,Fairs, Provisional shelters, Movable Hospitals.
b- Difficult access places, and/or lack of labor. Examples: -Remote retransmissions units , Antenna masts, Meteorological or research stations, Military installations, Emergencies in distant localities, requiring Shelters, Bridges, Hospitals, etc.
c- Special applications equipment and shelters for special equipment which can not be transported in full open size and needs to be erected in a very quick way. Examples: - Portable radars and antennas , Portable hangars , Portable bridges, Protection and camouflage of military equipment.
d- Need to enclose space due to variable weather conditions.Examples: Stadium covers, Cover-plazas.
e- Situations of high risk with elevated labor costs, hostile environments, costly transportation.Examples: Earth exploration, Space exploration, Space stations.
f- Construction aid. Example: Reusable scaffolding for complex forms. One of the problems of building shell structures is the high cost of the forms needed; this has inhibited the use of these efficient structures. Deployable structures can provide a reusable, easy to erect scaffolding, for example inflatable forms are now being used for the construction of concrete domes and vault like structures.
g- As a construction method. The traditional view of building has been one of simple accretion. Stone is put on stone, brick on brick, steel on steel, and so

forth until the final form takes shape. In modern construction, accretion is often by larger modules in the form of walls, floors, and even entire rooms.(eg. Zut [2]) A new method consists of bringing the complete structure to the site in some compact configuration and deploying it there for permanent use.

METHODS OF DEPLOYABILITY

There are many mechanisms which fall into the category of deployable structures, but we can group them into two general categories:

A) Struts Structures : scissor-hinged, tensile, and sliding mechanisms, etc.
B) Surface Structures :folded ,inflatable ,telescopic ,etc.

The general characteristic of group A is that these structures are made out of struts which commonly work as compression, tension or bending components connected by joints or hinges. This first category includes:

The Scissor-Hinged Mechanism
(This work will emphasize these mechanisms): The basic element of this system is the deformable truss shown in Fig. 1 which is a set of struts connected by nodes, and scissor-hinges (a rotational degree of freedom about the normal plane defined by two connecting struts). This assembly has the characteristic that by rotation of the struts with respect to one another, the assembly encompasses two forms; the first form is the compact state having theoretically one dimension, and the second form is the deployed state, which is a three-dimensional body.

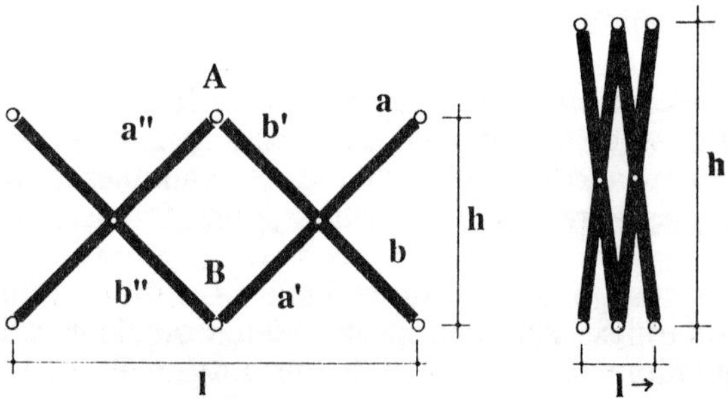

Figure 1

Sliding or Umbrella Mechanism :
This method consists of deploying the structure around a rectilinear support or guide, by sliding a cylindric or hollow joint over it, the more common example is the umbrella.

Hinged-Collapsible-Strut Mechanism :
This is another method of deployability which consists of a set of struts hinged at the ends, that allows the structure to be folded. After reaching the final open configuration the hinged joints are locked and the structure behaves as a single continuous piece.

In Group B stresses are carried by surfaces. In some, cases a continuous surface carries only tension forces like, in pressurized or inflatable construction; other structures are made out of small surfaces or planes joined together by some usually flexible means of forming a continuous structure. In this category, the more important groups are:

Inflatable or Pneumatic Structures:

These are structural forms made of a light, collapsible, very strong membrane, stabilized wholly or mainly by pressure differences of gases, liquids, foam, or material in bulk.

Folded Structures :

Another type of deployable structure can be collapsed and expanded like an accordion; it can be of two types. One made of rigid panels joined by a flexible connection along its edges, and the other is made of a continuous flexible material which allows itself to be folded.

Telescopic Structures:

These structures are based on tubular elements that can enter one inside the other, forming a compact packet which is them telescoped and stabilized to its final configuration.

DEPLOYABLE STRUCTURE COMPONENTS

Joints:

The successful behavior, duration, and reliability of a deployable structure will depend on a great way in its joints. The joints are points at which forces converge, and their ability to resist and transmit those forces will determine to an important degree the soundness of the structure the joint should meet the following criteria :

1. Should transmit the forces evenly throughout the components which arrive at that point.2. It should firmly hold all the struts which meet at that point.

3. It should give every strut enough freedom to go from the closed stage to the open one, but avoid holding the struts too loosely.

4. In the cases were reusability is required, the joint has to be designed to stand the stresses created during the erection process, minimizing fatigue in the material.

5. Friction between the moving pieces should be minimized to avoid excessive wearing and to facilitate the erection and dismantling processes.

6. As we are dealing with moving connections (usually pin connections), it is important to take into account, when designing and choosing the material, that the transference of forces between bodies which are not bonded together can occur only by the pressure exerted by one body against another; the values of compressive or tensile stresses increase at the joint. The flow of compressive forces are curving in the vicinity of the joint resulting in the need for an internal pulling action towards the center of the element; such an action always causes transverse tension to develop within the material. As a result, the element may crack and split longitudinally, if measurements are not taken. Thus, it is very important to have tension resisting materials around the connection.

Opening and Closing Mechanism:

Another fundamental aspect to consider when designing a deployable structure is the open-closing mechanism. As size and weight increase opening and closing a deployable structure become a more and more problematic factor in the effective good functioning of the concept. The structures have to be provided with a mechanism to accomplish the erection-dismantling process, these mechanisms can go from manually driven ones to fully automatic and even remote-controlled ones. But every system will depend on each structure, and in particular, on its size and weight, frequency and conditions of deployment, environmental factors, possibility of energy sources, etc.

Opening and closing mechanisms can be hydraulic systems, motor or manually-driven screws, cables and pulley mechanisms, spring-driven mechanisms, etc. But any mechanism use should provided a even movement of all the parts at a controlled rate. Since the structures usually start the deployment process running horizontally over the erecting surface, for those structures to be erected over surfaces or terrains it is convenient to mount them over wheels in order to reduce friction with the surface,thus reducing the force needed for the erection.

Covers:
There are some situations which require a protected or enclosed space, and the deployable structure has to be designed with an enclosure or cover system. There are many different solutions:

The enclosure system can be attached to the structure or free-standing; the enclosure can be added after the structure is erected, or it can be permanent part of the structure. It can also operate as a structural member, stabilizing member and/or locking member of the open configuration. Enclosures can be made out of light and flexible materials, which can act as tension carries, Such as teflon coated nylon or fiberglass fabrics, etc. Or they can be made out of rigid materials like metals or plastics.

Other Components and Materials:
The structure should be made out of materials which are light (for easier transportation) but at the same time strong enough to stand the stresses and deformations to which the structure is subjected ; in some cases flexibility is a desirable property. Materials like aluminum, laminated steel, steel, plastics, or composite materials like fiber-glass epoxy or graphite-epoxy are commonly used.

For the compression or bending members of the structure, tubular struts are a reasonable solution, due to their weight/resistance relationship and good resistance to bending; for tension components, high resistance steel cables or fabric are used.

STRUCTURAL FORMS SUITABLE FOR TRANSFORMATION.

Structural Requirements:
Clearly an essential structural requirement for all the forms with which we shall be concerned is that they shall stand and not collapse. They should carry their own weight and the live loads to which they will be subjected (wind, snow, other components attached to the structure, etc.), and under normal circumstances they should have adequate margins and stiffness in all structural elements and in their interconnections.

The basic element of the scissor-hinged mechanism, as we saw before, is the deformable truss shown in Fig.1. Due to its one degree of freedom this truss is unstable, and has no structural properties. Although if height "h" is fixed ,the truss will be able to carry loads acting on its plan. As it is not triangulated, its struts are working in flexion. By adding compression and tension members to the system, bending stresses can be reduced and the truss will become a triangulated one, where stresses are mainly transmitted by tension or compression. By using more suitable geometrical forms, and by locking the joints, the structural characteristic of the system can be improved.

The load capacity of a structure can be increased without increasing the amount of material, by choosing an adequate form. This is of special importance when designing deployable structures, because increasing the carried-capacity by adding material will make the movable joints used more expensive than they already are, and will increase the weight of all components, therefore the overall weight of the structure, reducing its efficiency and transportability.

Forms which allow the stress to be transferred mainly by compression or by tension are more efficient. The catenary and the arch are examples of those forms.There are three-dimensional forms which have the same properties of the linear funicular systems, cables and arches. They can be classified as forms with single curvature,(cylindrical or conical) or forms with double curvatures (spheres,ellipsoid, hyperbolic paraboloid, etc.).

The dome is an example of those three-dimensional forms in which, ideally, there is no bending (and therefore no variation in stress through the thickness at any point), and in which internal forces are represented by sets of principal tensions and compressions. One set of principal stresses acts radially from the crown to the base. These stresses are compressive throughout and may be likened directly to the thrusts in a large number of arches into which the shell might be split by cuts on vertical radial planes. The other set of principal stresses in the shell acts horizontally and circumferentially to prevent the upper part of the shell from falling toward the center and the lower parts from bursting outward. These stresses are compressive near the crown but become tensile nearer the base.

By a translation movement of an arch, a vault will be obtained. It works similarly to an arch, when loads acting on each transverse strip are the same. If the transverse shape of the vault is the funicular for a certain load, each strip will develop only compressive stresses in the direction of curvature and no longitudinal stresses between adjacent carried strips. If the cross section is not the funicular of the loads, bending stresses may develop in the arched strips. Other shell forms behave in a more complicated way, but in all of them the stresses are mainly tensions and compressions.

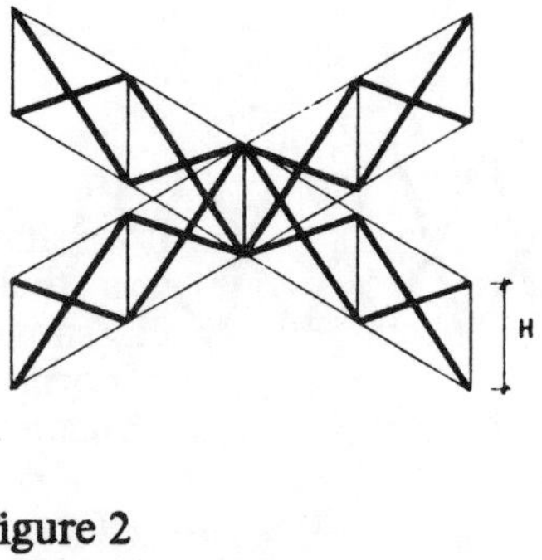

Figure 2

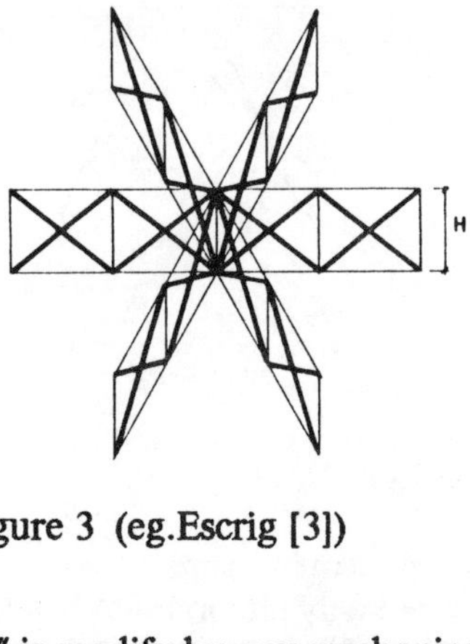

Figure 3 (eg.Escrig [3])

Kinetic Requirements:

Returning to our basic truss (Fig.1). If height "h" is modify by any mechanical mean, the "l" dimension will change with any variation of "h" according with the number of crosses and the length of the trusts. When "h" is equal to b+a the system is in its close configuration, and when "h" approach 0 height, the system will theoretical approach a straight line and "l" will be maxima.

For the system to obtain its closed configuration the only relation that have to be kept is, that around any pair of nodes A,B (a'+ b'= a"+ b"). Keeping that relationship, new combinations, with our basic system, are possible.

Until now we have only deal with flat elements, but three-dimensional combinations are possible too, and the system will closed if the same relationship used for the flat configuration is kept. By crossing two flat element perpendicular, a three-dimensional group of families is created (Fig.2); by intersecting three planes we obtain another three-dimensional configuration (Fig.3), in which the basic element is a triscissor (eg.Clarke [4]). It is a mobile assembly, of six struts, six nodes, and three scissor hinges, in which all the struts lie in one plane for a given value of the angle ß (Fig.4). The required action of a triscissor is such that, as ß increases the assembly folds rigidly out of the plane, passing through a three dimensional form and reaching a limiting one dimensional state for ß = 180 degrees. This action places constraints on the relative values of a, b, c, d. It is a necessary and sufficient condition for the mobility of a triscissor that the six struts in the plane are tangent to a circle. The triscissor is an important basic unit for some tridimensional forms like domes.

Until now we have dealt only with flat structures in which a= a' and b=b' but what happens if the center of rotation is moved out of the center ?. By making a'> a and b'< b , the basic truss starts to bend (Fig.5). The bigger the difference between a',a and b', b the smaller the radius of the curve. The radius of the curve and the number of crosses will define the length of a+a' and b+b'.

With the scissor-hinged deployable system it is possible to make flat structures, and structures with a single or double curvature, positive or negative, following the method described above.

By repeating an arch, a simple vault structure will be obtained-by rotating it, a dome, etc. The same type of geometrical forms can be obtained, by using as a basic unit the triscissor described before.

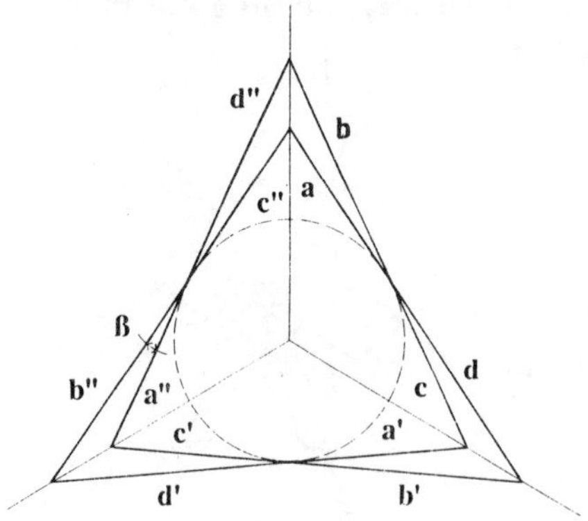

Figure 4

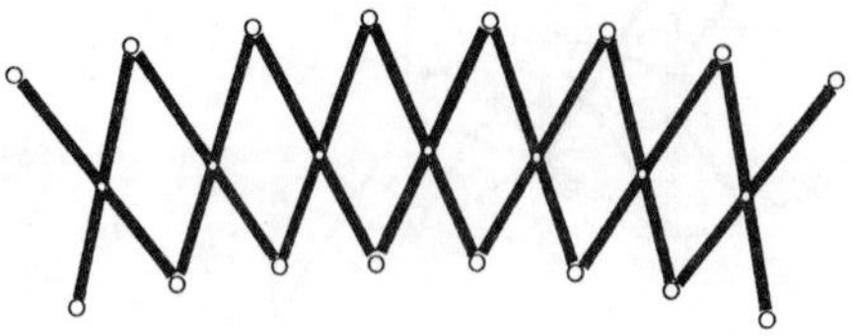

Figure 5

<u>Others Requirements:</u>

It is not only important to has a structural soundness, and the correct geometrical relationship between the parts, which allow the structure to be transform and to acquired the desire shape, but it also has to be reliable and buildable . A deployable structure has any advance, if it can not be opened, closed, or reused. As it was mention before joints are a crucial point, but the right dimensioning and positioning of all part are as important. In many of the models we have built problems during the transformation process were generally due to small mistakes in the dimension of some parts. To avoid mistakes in the dimension and position of parts, the variety of these should be as small as possible, at the same time this will facilitate the production of part, the assembly, and reduce the cost of the structure.

ESTRAN 1

Estran 1 is the result of years developing deployable structures in paper and models. It is our first scale 1/1 prototype.It was developed to be used as transportable hangar for small airplanes and it was and opportunity to test many ideas we have developed in the past years. Estran 1 is a deployable (scissor-hinged) vault structure, built by three parallel arches joined together by fourteen connections or arms and triangulated by wires in its upper plane. starting from a packet of 1x1x4.20 mts the structure deploys in two directions to an hangar of 14 mts wide, 8 mts long and 7 mts of radius.

The deployment process starts by transporting the structure horizontally as a compact packet which weight 500 Kg. after it had been stood up by a small crane (Fig.6 & Fig.7A) the deployment begin by pulling the four corners of the packet manually diagonally (Fig.7B), the structure expands over the ground surface (Fig.7C & Fig.7D) and then it begin to rise until it reaches its final position (Fig.7E - Fig.7G) The total process takes about 5 minutes to accomplish. The deployment sequence can also be achieved without a crane, by using skates, placed under the packet, to ease the expanding movement of the structure over the deployment surface, and by small winches placed between two opposite nodes in order to force the structure to arrive to its final or open configuration.Once this configuration has been reached, the structure is secured with pins and bars of reinforcement and fixed to the site where it is going to stay.

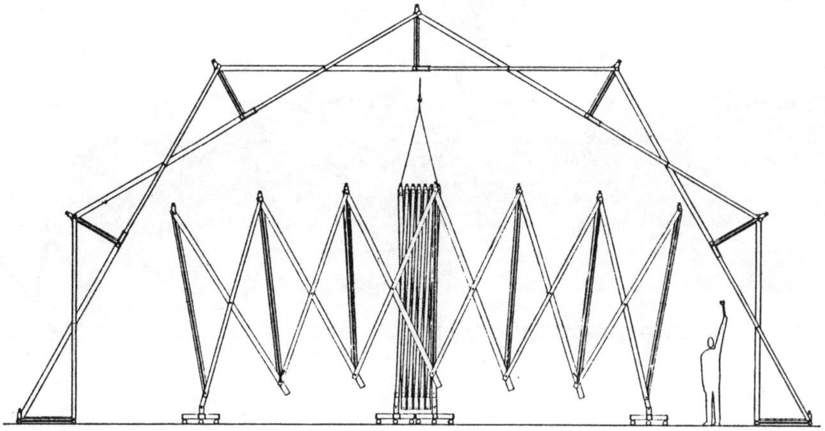

Figure 6 *Deployment process scheme for Estran 1*

The cover made out of PVC coated Nylon fabric its hanged from the structure.(weigh: 50 gr/m2, resistance:90 Kg) the cover is placed during the "erection" process, when the structure starts to rise, then when it has been fully deployed and fixed the fabric is stretched in order to reach its final form.

After nearly 50 deployment the behavior of the structure has been very satisfactory, it has not shown problems or deterioration of its parts, it has demonstrated that structures of this size and weight are easy to handled manually or with light weight equipment. It has shown the feasibility of deploying nearly instant structures from compact packets easy to transport, storage and handled.(Fig.8)

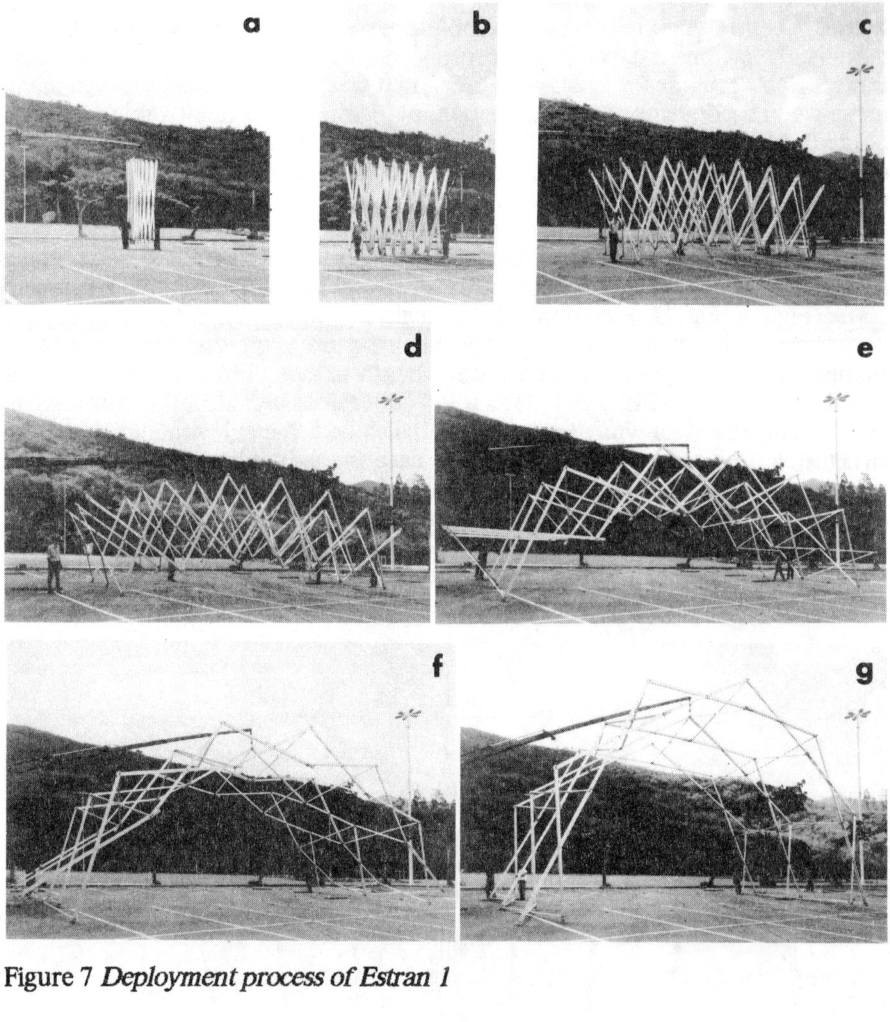

Figure 7 *Deployment process of Estran 1*

ESTRAN 2

This structure is being developed to be used for exposition and temporary events. Estran 2 is a double curvature structure formed by two series of parallel arches, (Fig.9) perpendicular between each other. The structure is supported in four corners and has a projection on the horizontal plane of 14 by 14 mts. each face has a maximum height of three meters, and the structure center is six meters high.

Figure 8

The deployment process is similar to the one in Estran 1 . Once the structure is in its open configuration, a rigid element triangulate each of the four corners and gives stability to the whole system, then a PVC coated Nylon fabric is hung from the structure as a cover.

The basic structural unit of 14x14 mts can be combined with other similar units placing them face to face, this gives a lot of possibilities and flexibility to create spaces with this deployable structure.(Fig.10)

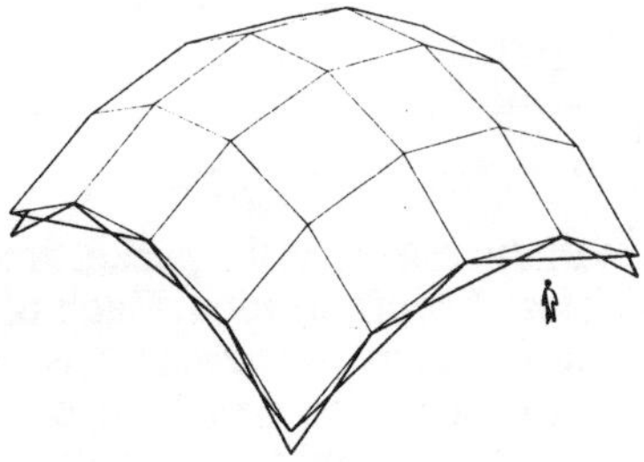

Figure 9

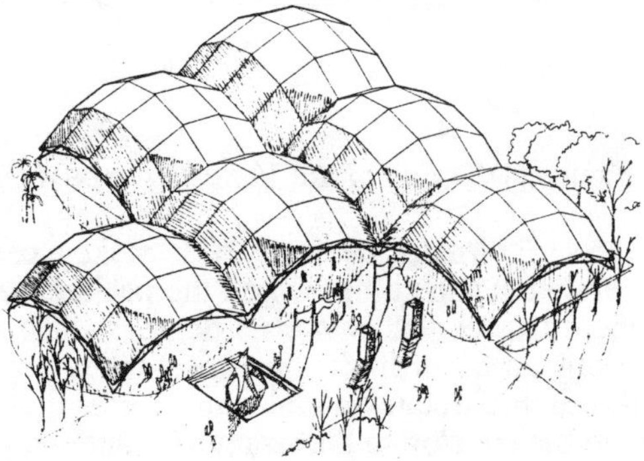

Figure 10

STRUCTURE FOR THE VENEZUELAN PAVILLION

The Venezuelan pavilion at the world fair in Servile (Fig.11), will be constructed under time limitations; transport, and labor cost. The structure is going to be built in Venezuela (building cost is many times much higher in Spain), transported to Servile , erected there, used for six months and then sent it back to Venezuela to be re-erected .

As a response to those constraints two aluminum deployable structures were proposed as the main structure of the Venezuelan pavilion . The first structure is a space structure deployable in two directions that once deployed conform a cubic space net, each cube is triangulated on the upper plane by two turnbuckle and on the lower one by a rigid roof panel.

The second structure (chose at the end for its simplicity) is a group of trusses of 18 mts and 14 mts long connected between then by hinges which allow the whole group to be folded in one direction.(Fig.13)

The trusses are made with tubular aluminum (alloy 6261) elements connected together by extruded aluminum joints (20 cm long), each joint has a hinge that connects the trusses within themselves.(Fig.13-14)

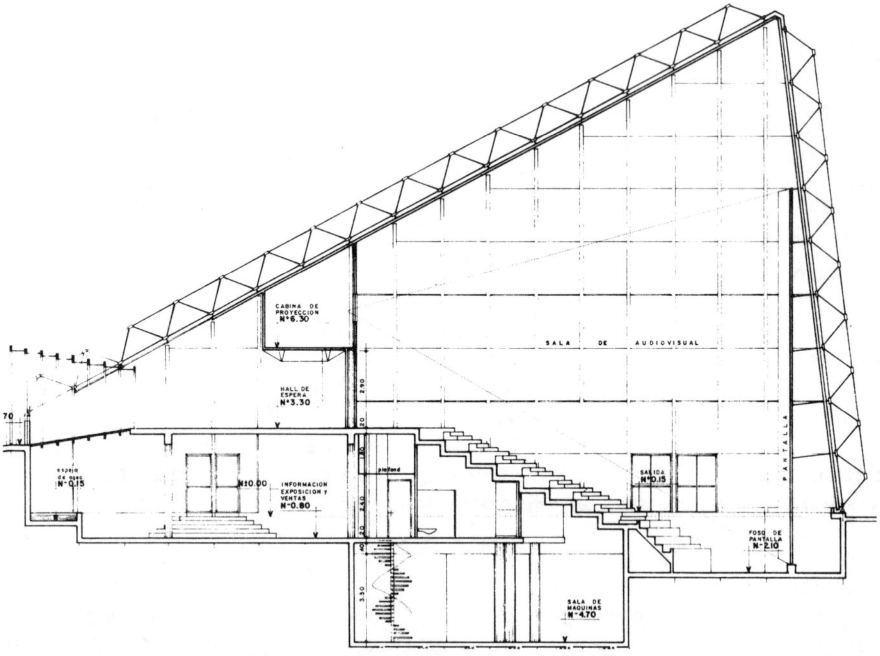

Figure 11 *Section of the Venezuelan pavillion*

22 trusses are conected to make a packet . All the trusses in the packet are parallel to each other, when the packet is are in closed configuration. Each of these trusses will rotate 45 degrees during the deployment process,and as a result an accordion- like structure is obtained.This structure works in one direction. To stand lateral winds forces that will be transmitted by vertical trusses and the envelop to the structure some perpendicular connections to the trusses are needed to give the assemblage rigidity.

The pavilion structure is going to be built out of two group of trusses. The first one is a packet of 22 trusses 14 mts long that once deployed will cover an area of 308 m2. The second group is formed by two packets of 22 trusses each of 18 mts long, one over the other,(Fig. 12A) and joined together in one extreme. In close configuration this group will be 3 mts high, 3 mts wide and 18.80 mts long (including packing) and will weight 8 Ton. The whole group will be lifted by a crane through the joined end (Fig.12B) and deploy in the air like a foldable door or curtain (both parallel packets at the same time)(Fig.12C). The end of each truss is connected with a mechanism which make a hinged connection with the opposite truss in the second packet and at the same time allowing 45º of rotation.

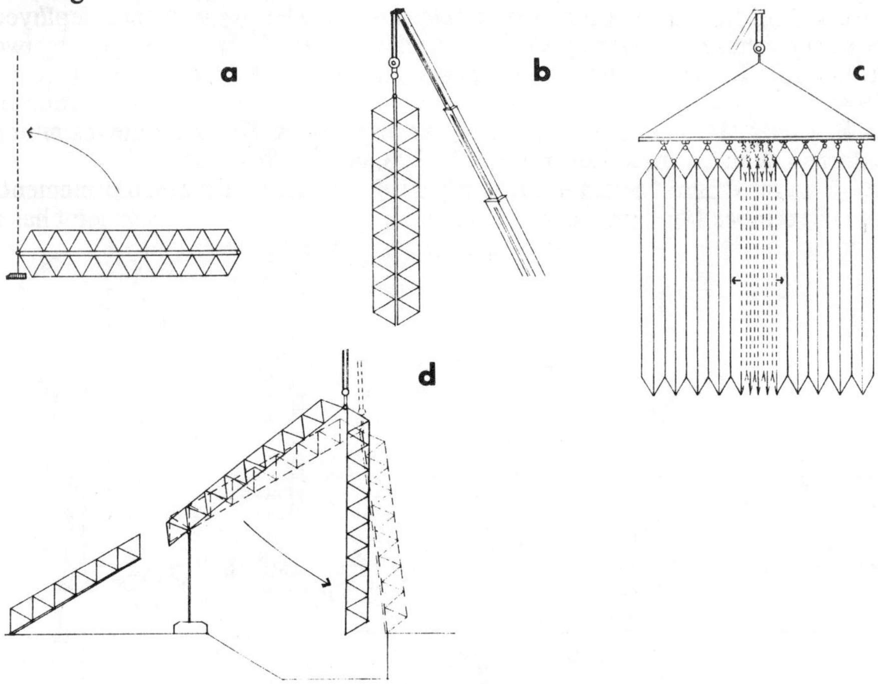

Figure 12 *Deployment process of the Venezuelan pavilion structure.*

Each of these mechanisms hangs from a pulley mounted in a roller which moves along a horizontal beam that works as a rail, over which the deployed movement is performed. The group placed in the center of the beam deploys in both opposite directions (right and left)(Fig.12C). Once this first deployment process has been achieved the structure is stabilized by placing some of the connections perpendicular to the trusses. One of the free edges of the structure is then fixed to a provisional support frame in a way that allows a rotational movement. The upper edge connected to the beam is displaced, the two parallel planes of trusses start to separate, the plane fixed to the provisional support frame will rotate until it reaches 27 degrees of inclination, the second plane will be a this point vertically and with the help of small winches it will be forced to reach the ground connections (Fig.12D) and then will be fixed. the first group of trusses, will be connected to the second group and the whole structure will have stability.

REFERENCES:

1- W Zalewski Venezuelan Patent for Estran 1, 1989.
2- William Zuk " *Kinetic Architecture*" Van Nostrand Reinhold Company, New York 1970.
3- Felix Escrig " *Estructuras Espaciales de Barra Desplegables* " Informes de la Construccion, Vol 36, #365, november 1984.
4- R. C. Clarke " *The Kinematic of a Novel Deployable Space Structure System*" Third International Conference of Space Structures, Elsevier Applied Science Publishers, New York, 1984.pp.820-822.
5- C. Henrique Hernandez " *Deployable Structures* " Thesis, MIT. 1987.

6- C. Henrique Hernandez " *Estructuras Transformables Estran 1*" Tecnologia de la Construcción, #4 , December 1988.
7- Building Systems Design Group "*Design and Development of a Building System for and Exhibition Space*" Report MIT, fall 1985.
8- Building System Design Group" *Space Station Workshop*" Report MIT, fall 1986.
9- F. Escrig, P. Valcarcel. " *Estructuras Espaciales Desplegables Curvas* " Informes de la Construcción, Vol. 39, Nº 393, january/february 1988.
10- E. Perez Piñero " *Structures Réticulées* ". L'Architecture d'aujourd'hui, Vol. 141, December 1968. pp. 76-81.

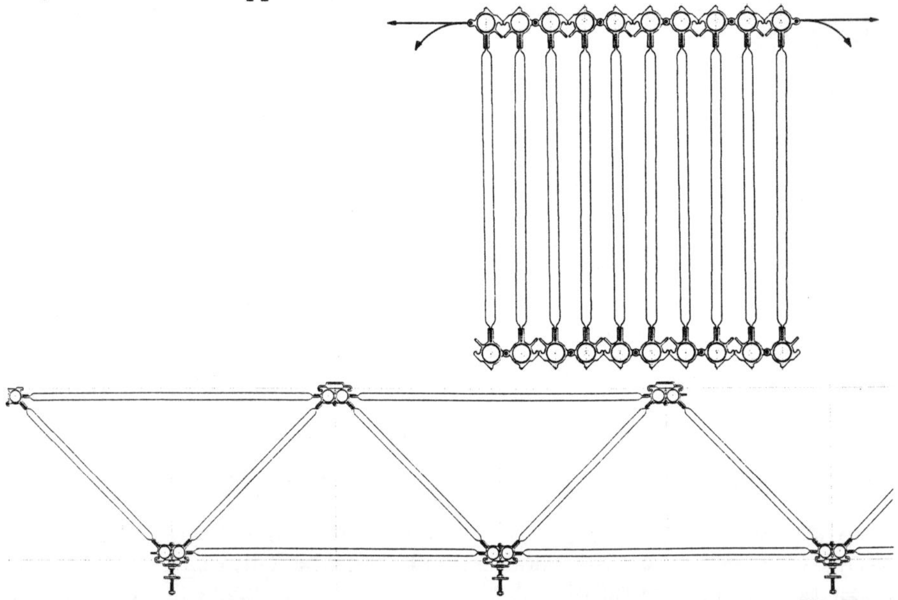

Figure 13 *View of trusses folded and unfolded with stabilizing elements.*

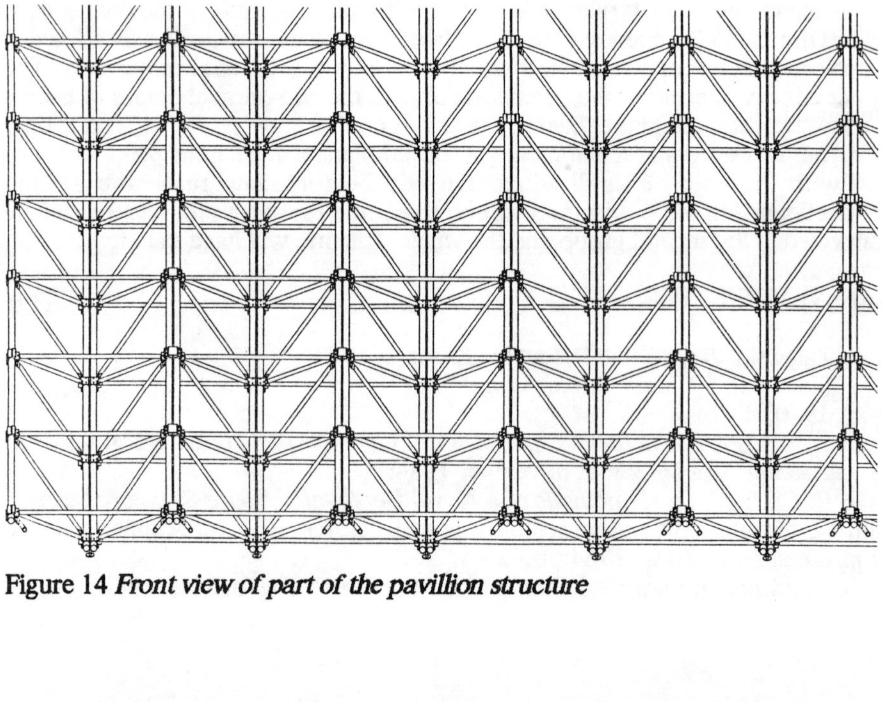

Figure 14 *Front view of part of the pavillion structure*

Geometric and Structural Design Considerations for Deployable Space Frames

C. Gantes, J.J. Connor, R.D. Logcher
Civil Engineering Department, Massachusetts Institute of Technology, Cambridge, MA 02139, U.S.A

ABSTRACT

Deployable structures that are stable and stress-free in both the deployed and folded configurations are investigated in this paper. These structures can overcome certain structural and practical disadvantages of previous designs, but show a nonlinear structural behavior during deployment. Issues related to their geometric and structural design are addressed here. The finite element method is used for the numerical modeling of the behavior in service and during deployment. The analysis results are in very good agreement with experimental results. Factors that influence the structural response are identified, and design recommendations are formulated.

INTRODUCTION

Deployable structures are prefabricated space frames consisting of straight members that can be stored in a compact bundle and can be easily deployed into large-span, load-bearing structures by articulation. An example of such a structure is illustrated in Figure 1; the folded, intermediate, and deployed configurations of a deployable semi-spherical dome are shown. Due to their ease of storage and transportation, and their speed of erection and dismantling, these structures are good candidates when portability and reusability are desired features. Potential applications include emergency shelters and bridges, lightweight camping and recreational structures, temporary buildings in remote construction sites, scaffolding systems, traveling theaters (Pinero [1]), as well as aerospace structures (Freeland [2], Rhodes [3], Cohen [4], Miura *et al.* [5]).

A detailed review of several types of deployable structures is given by Merchan [6]. Most structures that have been designed and constructed in the past suffered important drawbacks. Some of these structures required the addition of new members in order to be stable in the deployed configuration, resulting in need for temporary support and increased erection time (Pinero [1], Freeland [2],

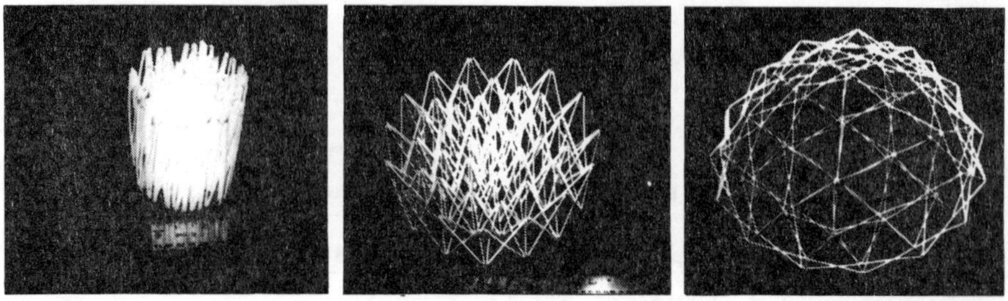

Figure 1: A Model of a Deployable Dome

Miura *et al.* [5], McNulty [7], Escrig *et al.* [8]). Others were self-stabilizing but had bent members and members with residual stresses, which reduced their load-bearing capacity (Zeigler [9], McNulty [7]).

Research at M.I.T. during the last few years has been focused on an improved version of a deployable structure that features stable and stress-free states in both the folded and the deployed configuration (Gantes [10], Rosenfeld *et al.* [11]). During deployment, however, stresses develop in the members which increase gradually, and lead to a 'snap-through' phenomenon which 'locks' the structure into a stable deployed position. The key idea to achieve this behavior is that the members should satisfy certain geometric compatibility constraints.

The present paper summarizes the work carried out to date at M.I.T. First, the geometric constraints necessary for deployability, and the limitations imposed by these constraints on the types of surfaces that are feasible, are investigated. Then, the experimental work is described. Next, modeling and structural analysis issues are addressed. Analysis in the deployed configuration under service loads is simple, since the response is assumed to be linear. The behavior during deployment however, is highly nonlinear, and requires a geometrically nonlinear finite element formulation. Parametric analyses are performed to investigate the influence of several factors on the structural response. Finally, the interaction between geometric constraints and structural behavior is studied, and the results are used to formulate design recommendations for this type of deployable structure.

GEOMETRIC CONSIDERATIONS

The basic element of the deployable structure is illustrated in Figure 2. It is a scissor-like-element, consisting of two straight bars connected to each other at an intermediate point with a pivotal connection that allows relative rotation of the bars about an axis perpendicular to their plane. At their end nodes, the two bars are hinged to end nodes of other scissor-like-elements.

The configuration of these basic units permits them to have a folded configuration in which both bars lie theoretically on a straight line. Then, the bars can unfold and form an angle between each other. During this process the scissor-like-element behaves as a mechanism and is unstable, unless some additional constraints are imposed. In manually locking structures (Pinero [1], Freeland [2],

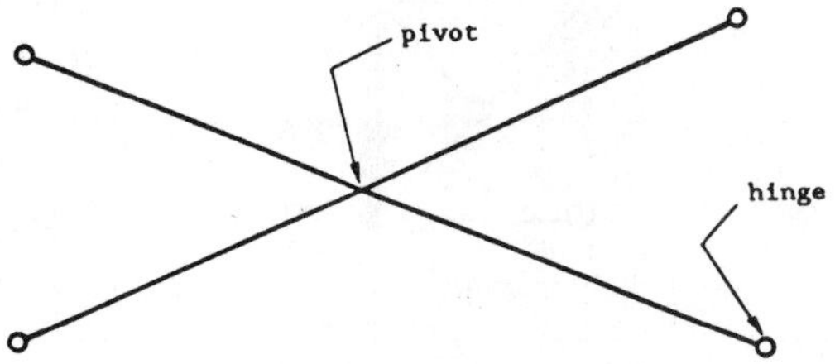

Figure 2: A Typical Scissor-Like-Element

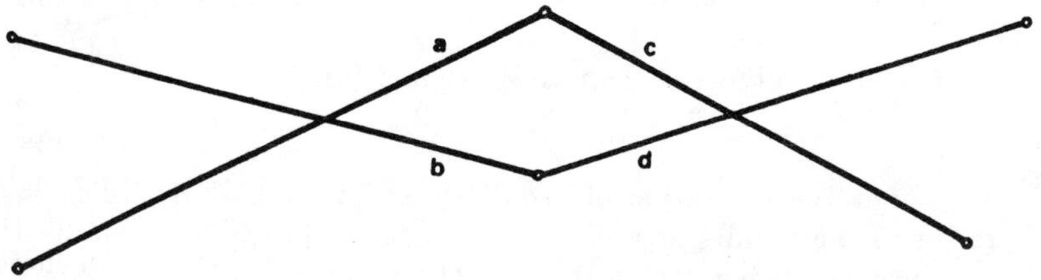

Figure 3: The Basic Deployability Constraint

Miura *et al.* [5], McNulty [7], Escrig *et al.* [8]) such constraints are imposed externally by adding members or cables that stabilize the structure in its desired deployed configuration. In self-locking structures the constraints are internal and result from geometric compatibility requirements between the lengths of the members.

In previous designs of self-locking structures however, some members remained bent in the deployed configuration. This made them more susceptible to buckling, and reduced the overall load bearing capacity of the structure. The structures we have been looking at avoid this problem by imposing some additional geometric constraints which in essence require that all bars are straight in the deployed and the folded configuration. This is illustrated in Figure 3 for a simple structure consisting of only two scissor-like-elements. The figure shows the deployed configuration of the structure in which all bars are assumed to be straight, and a,b,c,d are the lengths of the four inner parts of the bars, between pivotal connection and hinge. In order for all bars to be straight in the folded configuration as well, the following constraint must be satisfied:

$$a + b = c + d \tag{1}$$

Many scissor-like-elements can be combined to form some basic building units, and then many units can be assembled to create larger deployable structures. The first units that were investigated had a planview of a regular polygon. All sides and all diagonals of the polygon are scissor-like-elements (Figure 4). Such polygonal units can be used to assemble either flat or curved structures. In flat units, the lines connecting upper and lower corner and center nodes are parallel to each other, while in curved units these lines are concurrent (Gantes [10]).

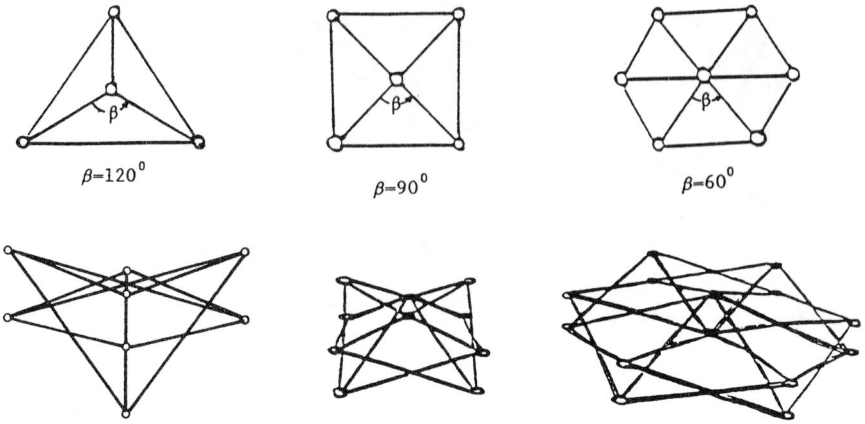

Figure 4: Typical Polygonal Units

Details about the derivation of geometric constraints, and the geometric design process of these units as well as trapezoidal units for flat structures can be found in previous papers (Gantes *et al.* [12],Gantes *et al.* [13]). Since the behavior is very sensitive to small geometric incompatibilities, discrete joint size should be taken into account in this process. The procedure used for these derivations involves looking at a geometric development on a plane of the typical scissor-like-elements of the unit in question in the deployed configuration. Then, basic geometric and trigonometric formulas are applied to relate the lengths of the members and the angles between members. In addition, the straightness constraint explained above has to be satisfied.

The geometric design process starts by choosing some dimensions as known design parameters. These parameters should ideally be external dimensions of the unit that are imposed by architectural requirements. The member lengths and the angles between the members in the deployed configuration are the unknown quantities. Since the constraint equations are, in general, complex trigonometric equations, it is not possible to solve them analytically. Certain simplifications however, can be achieved through algebraic manipulations. What is left is a system of nonlinear simultaneous equations for a number of unknowns. Iterative numerical schemes, usually Newton's method, can then be used for the solution of these equations.

A disadvantage of this type of structure is that only flat structures and curved shapes with constant curvature are currently possible. An investigation for the feasibility of other shapes is being carried out. It is based on the observation that all units that have been designed so far consist of triangles attached to each other, the sides of each triangle being scissor-like-elements. Therefore, two types of triangular units are proposed that consist of three scissor-like-elements as shown in Figure 5. One unit has the shape of a truncated triangular prism, and is intended for structures consisting of flat segments. The other unit looks like a truncated triangular pyramid and could be used for arbitrarily curved structures.

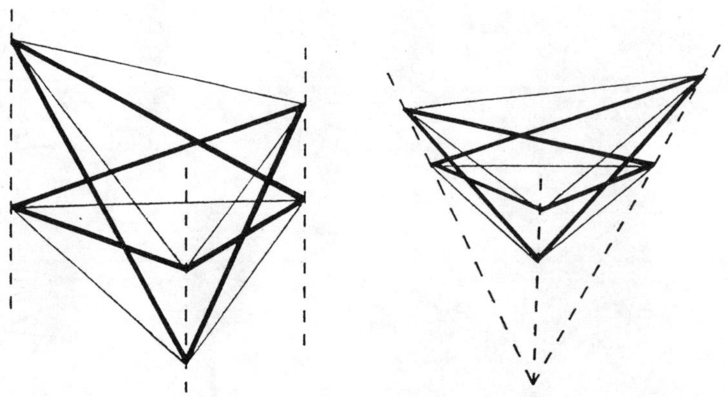

Figure 5: Generic Triangular Units

The geometric constraint equations for these structures have been derived. The geometric design process would consist of placing a unit somewhere on the desired surface, and then propagating by adding adjacent triangles, making sure that the deployability constraints are satisfied, and that the third corner of each new unit lies on the desired surface. This procedure has been tested for folded plates and parabolic arches, but difficulties have been encountered in the convergence of the iterative solution algorithm. It is currently being investigated whether this means that these shapes are not feasible, or that another numerical scheme has to be used.

PHYSICAL MODELS

An important part of our work was the design and construction of physical models. The first objective of this work was the verification of the feasibility of the concept, and of the proposed geometric design process. Several small scale models consisting of plastic coffee stirrers as members and steel wire rings as joints were built. The semi-spherical dome shown in Figure 1 is an example. These models served their purpose well, and are a powerful tool for demonstrating how such a deployable structure behaves. In addition, they convinced us of the importance of joint size for geometric design.

A second goal of the model building effort was the identification of practical engineering problems related to these structures. Several large scale models were constructed for this task. They include the shallow dome shown in Figure 6, and a full-size circular arch constructed as part of a parallel project in Israel, and described in detail by Ben-Ami [14], and Rosenfeld *et al.* [15]. A combination of aluminum and plastics has been used for their members. Their construction highlighted connection detailing as a very important factor for the feasibility, strength, ease of manufacturing, and cost of deployable structures. Other critical issues include the selection of a dismantling and deployment mechanism, the attachment of a cover on the skeletal form, and the selection of the optimum material for members and hubs.

Finally, the third objective was to obtain a basis for comparison of the numerical analysis results. This was achieved by performing a dismantling test on

Figure 6: Large Scale Model of a Shallow Dome

a structure consisting of a single pentagonal unit as illustrated in Figure 7. The unit was supported at the lower center node, while the upper lower node was attached to a load cell that recorded the load required for dismantling. This load is also shown in Figure 7 as a function of the distance between lower and upper node, starting from the deployed configuration and going towards the folded one. The nonlinearity of the response is obvious, and the illustrated snap through required a displacement controlled test. The stiffening near the folded configuration is due to friction between the bars.

STRUCTURAL ANALYSIS CONSIDERATIONS

What makes deployable structures very interesting from a structural point of view, is that they have to be designed for two completely different loading conditions (Gantes *et al.* [16], Logcher *et al.* [17]). The first one is service loads in the deployed configuration. The structural behavior in that phase is largely linear, therefore the analysis is quite straightforward, although practical problems might appear when the number of degrees of freedom increases. Response during the second phase, the deployment process, is highly nonlinear, which makes the numerical modeling of the problem very interesting. The work that has been carried out related to structural analysis in these two stages of the life of deployable structures is described in this section.

Analysis in the Deployed Configuration

Three are the potential failure modes of the structure in the deployed configuration when it is subjected to service loads. These three modes are strength, stiffness or buckling failure. The first two modes involve that the maximum stress in a member exceeds a predefined allowable stress that depends on the material, or that the maximum displacement that occurs is larger than a limit value that is specified according to the shape of the structure and the application for which

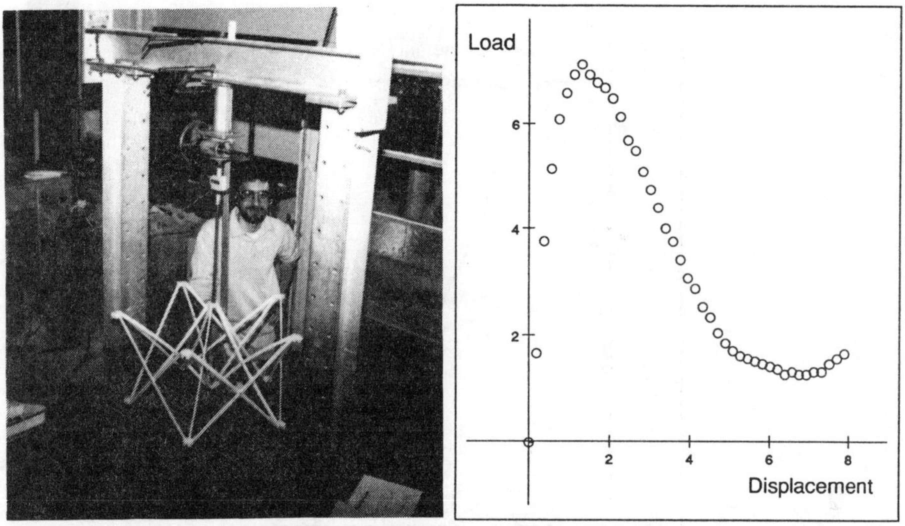

Figure 7: Experimental Setup and Load-Displacement Curve

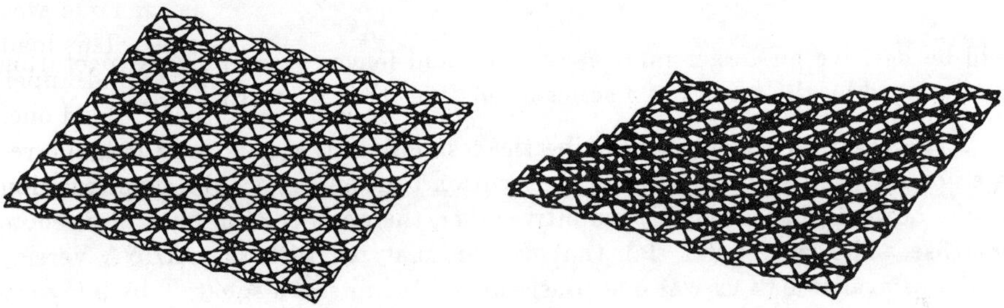

Figure 8: Deployable Flat Slab Under Service Loads

it is used. Both stresses and deflections can be obtained through a linear analysis of the structure. Figure 8 shows the original and deformed shape of a flat slab subjected to a uniformly distributed load. The slab was analyzed with the finite element program ADINA (ADINA R&D [18], [19]). Only one fourth of the structure was modeled due to symmetry considerations.

The third failure mode occurs when the load imposed on the structure exceeds the critical buckling load. This critical load can be obtained with a linearized buckling analysis with very good accuracy, since the pre-buckling behavior of the structure is characterized by small displacements.The fundamental buckling mode of flat deployable slabs consists of out-of-plane buckling of some of the scissor-like-elements in a symmetric or anti-symmetric fashion.

It is not easy to draw any general conclusions about which of the above three modes is critical, since this depends on the properties of individual members of the structure. It is reasonable to assume however, that strength and buckling will govern the design for small number of units and short spans, while stiffness

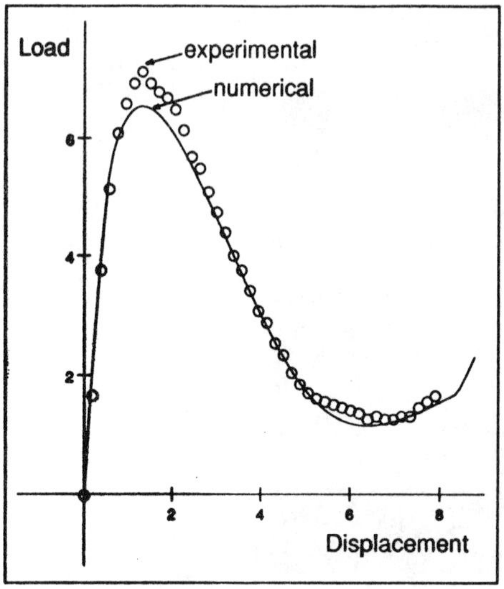

Figure 9: Comparison of Numerical and Experimental Results

will be decisive for larger number of units and longer spans. This expectation was verified by the results of a series of analyses that we carried out.

The analyses needed to describe the response in the deployed configuration are not very demanding in terms of execution time and cost. The space requirements however, can become prohibitive when the number of degrees of freedom increase. The largest flat slab that can be analyzed with the ADINA version that was available to us was one consisting of 81 units in a square 9 by 9 layout. It was shown that an equivalent continuum model that we derived can be used with satisfactory accuracy to predict the deflections of larger slabs when they are subjected to uniformly distributed loads.

Deployment Analysis

The response of the deployable structure investigated here during deployment is geometrically nonlinear, therefore, a large displacement-small strain finite element formulation implemented in ADINA was used for the numerical modeling of the problem. The members were modeled with 2-node isoparametric beam elements (Bathe [20]). A displacement controlled Newton Raphson method or an automatic load incrementation algorithm can be used to trace the load-displacement curve beyond the limit point. The effects of the discrete joint sizes and initial member imperfections were included in the numerical model (Gantes *et al.* [21]). Furthermore, friction was incorporated by using appropriate nonlinear rotational springs at all pivotal connections. Figure 9 shows the excellent agreement between numerical and experimental results for a single-unit deployable structure.

Once the accuracy of the numerical model was verified for single-unit structures, it was extended and used for the deployment analysis of structures consisting of many units (Gantes *et al.* [22]). The analyses of multi-unit structures were performed on the M.I.T. supercomputer facility CRAY-2. Figure 10 shows

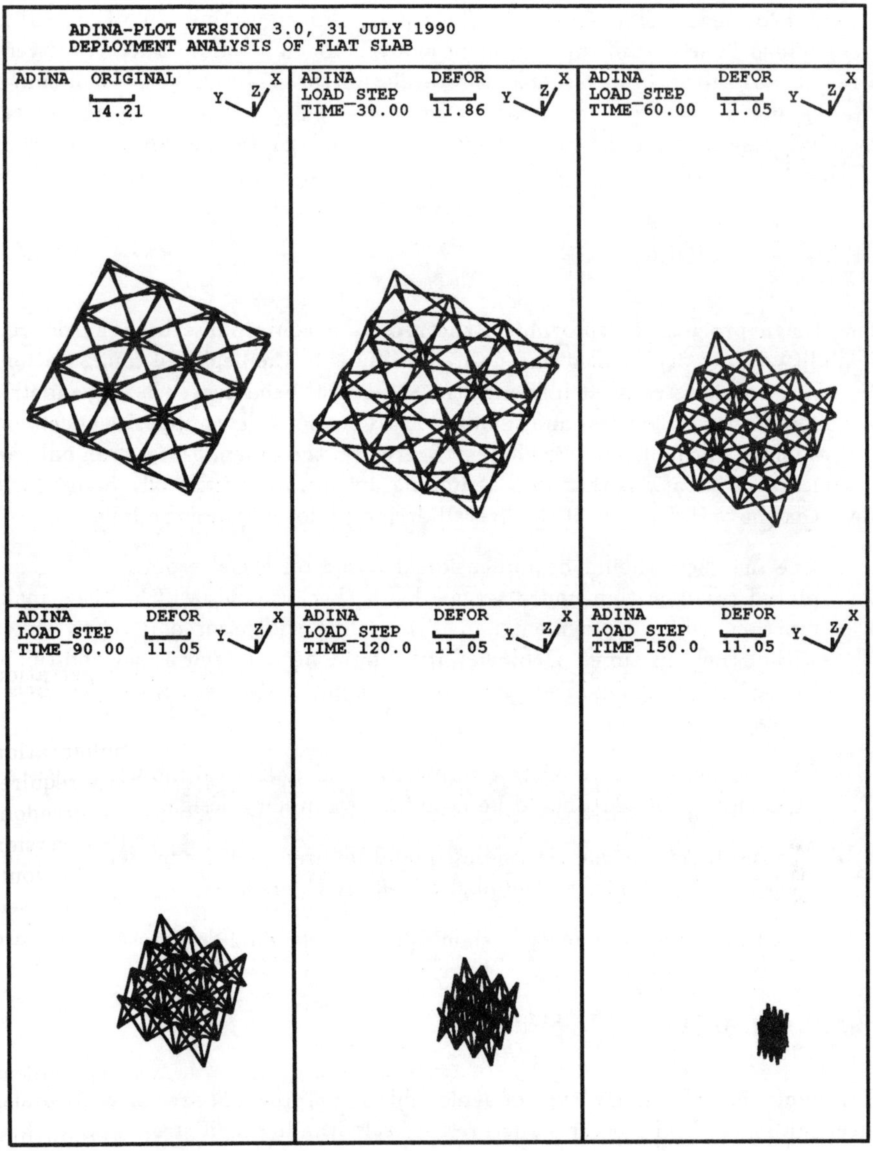

Figure 10: The Dismantling Process of a Flat Slab

several deformed configurations of a flat square slab consisting of 9 units as it is being folded. The cost of this nonlinear analysis for multi-unit structures is very high. It was shown however, that the results of the analysis, both for member stresses and for required load, can be predicted quite well by using the corresponding results from the deployment analysis of one of the units (Gantes *et al.* [23]).

DESIGN CONSIDERATIONS

The design process of deployable structures is a compromise between desired flexibility during deployment and desired stiffness in the deployed configuration. A series of parameters that influence the structural behavior, such as geometric layout, material properties, and member cross-sections, have opposite effects on the responses during these two phases. Therefore, the design process can only be a sequence of iterations that aim at finding an optimum trade-off. Some basic ideas that have to be part of the overall design philosophy are the following:

- The members should be dimensioned based on their response in the deployed configuration under service loads, because this will be their main function, while deployment and dismantling will occur only a few times within their lifetime. Deployability should be achieved as an additional 'bonus', without significant increase in weight or decrease of the load bearing capacity.
- The inexpensive approximate numerical and analytical models suggested by Gantes *et al.* [23] should be used for preliminary design.
- The results of parametric studies should be used in order to identify useful trends, and minimize the number of required iterations.
- Exact finite element analysis should be used for final design.

SUMMARY AND CONCLUSIONS

This paper describes a new type of deployable structures that are currently under investigation at MIT. These structures are self-standing and stress-free in their deployed and their folded configuration, thus overcoming some disadvantages of previous designs. The feasibility of these structures has been demonstrated both analytically, and experimentally. Furthermore, numerical models for their analysis have been proposed, and recommendations for their geometric and structural design have been formulated. The need of using an advanced computer environment as design tool has been identified, but the use of simpler models in case of limited computer resources, or for preliminary design purposes, has been suggested.

ACKNOWLEDGMENT

This work has been supported primarily by the National Science Foundation under Grants No. MSM-8718134 and No. MSM-8919287, and the Army Research Office as part of the Program for Advanced Construction Technology. This support is gratefully acknowledged. The first author would also like to acknowledge the partial financial support provided to him by the Alexander S. Onassis Foundation.

REFERENCES

[1] Pinero, E.P., Project for a Mobile Theater, Architectural Design, Vol. 31, No. 12, p. 570, Dec. 1961.

[2] Freeland, R.E., Survey of Deployable Antenna Concepts, N.A.S.A. Conference Publication 2269, Part 1, pp. 381-421, Large Space Antenna Systems Technology '82, Hampton, Virginia, Nov. 30-Dec. 3, 1982.

[3] Rhodes, M.D., New Concepts in Deployable Beam Structures, N.A.S.A. Conference Publication 2368, Part 1, pp. 331-348, Large Space Antenna Systems Technology '84, Hampton, Virginia, Dec. 4-6, 1984.

[4] Cohen. M.M., Lightweight Structures in Space Station Configurations, Proceedings of the First International Conference on Lightweight Structures in Architecture, Sydney, Australia, 24-29 August 1986, Vol. 1, pp. 507-541.

[5] Miura, K., and Furuya, H., Adaptive Structure Concept for Future Space Applications, Journal of the American Institute of Aeronautics and Astronautics, Vol. 26, No. 8, pp. 995-1002, 1988.

[6] Merchan, C.H.H., Deployable Structures, Thesis Submitted to the Department of Architecture of the Massachusetts Institute of Technology in Partial Fulfillment of the Requirements for the Degree of Master of Science in Architecture, June 1987.

[7] McNulty, O., Foldable Space Structures, Proceedings of IASS Symposium on Shells, Membranes, and Space Frames, in Osaka, Japan, Vol. 3, edited by K. Heki, Elsevier Science Publishers B.V., pp. 277-284, 1986.

[8] Escrig, F., Valcarcel, J.P., and Gil Delgado, O., Design of Expandable Spherical Grids, Proceedings of the XXX IASS Congress, Madrid, Spain, 1989.

[9] Zeigler, T.R., Collapsable Self-Supporting Structures, U.S. Patent No. 4,437,275, 1984 (Filed Aug. 7, 1981).

[10] Gantes, C., New Concepts and Preliminary Design Procedures for Deployable Structures, Thesis Submitted to the Civil Engineering Department of the Massachusetts Institute of Technology in Partial Fulfillment of the Requirements for the Degree of Master of Science, Sept. 1988.

[11] Rosenfeld, Y., and Logcher, R.D., New Concepts for Deployable-Collapsable Structures, International Journal of Space Structures, Vol. 3, No. 1, pp. 20-32, 1988.

[12] Gantes, C., Logcher, R.D., Connor, J.J., and Rosenfeld, Y., Deployability Conditions for Curved and Flat, Polygonal and Trapezoidal Deployable Structures, International Journal of Space Structures, (to be published).

[13] Gantes, C., Logcher, R.D., Connor, J.J., and Rosenfeld, Y., Geometric Design of Deployable Structures with Discrete Joint Size, International Journal of Space Structures, (to be published).

[14] Ben-Ami, Y., Developing a Light Foldable Structure for Quick Sheltering, Thesis Submitted to the Senate of the Technion-Israel Institute of Technology in Partial Fulfillment of the Requirements for the Degree of Master of Science in Civil Engineering, May 1989.

[15] Rosenfeld, Y., Ben-Ami, Y., and Logcher, R.D., A Prototype Clicking Scissor-Link Deployable Structure, (submitted for publication).

[16] Gantes, C., Connor, J.J., Logcher, R.D., and Rosenfeld, Y., Structural Analysis and Design of Deployable Structures, Computers & Structures, Vol. 32, No. 3/4, pp. 661-669, 1989, also presented at the 7th Conference on Nonlinear Finite Element Analysis and ADINA, M.I.T., August 2-4, 1989.

[17] Logcher, R.D., Connor, J.J., Rosenfeld, Y., and Gantes, C., Preliminary Design Considerations for Flat Deployable Structures, Proceedings of the Structures Congress 89, A.S.C.E., San Fransisco, May 2-5, 1989.

[18] ADINA R&D, Inc., ADINA-IN: User's Manual, Report ARD 87-4, December 1987.

[19] ADINA R&D, Inc., ADINA Theory and Modeling Guide, Report ARD 87-8, December 1987.

[20] Bathe, K-J., Finite Element Procedures in Engineering Analysis, Prentice-Hall, Englewood Cliffs, New Jersey, 1982.

[21] Gantes, C., Connor, J.J., and Logcher, R.D., Finite Element Analysis of Movable, Deployable Roofs and Bridges, Proceedings of the Third Biennial Symposium on Heavy Movable Structures and Movable Bridges, St. Petersburg, Florida, November 12-15, 1990.

[22] Gantes, C., Connor, J.J., and Logcher, R.D., Simulation of the Deployment Process of Multi-Unit Deployable Structures, Cray Channels, (accepted for publication).

[23] Gantes, C., Connor, J.J., and Logcher, R.D., Combining Numerical Analysis and Engineering Judgment to Design Deployable Structures, 8th Conference on Nonlinear Finite Element Analysis and ADINA, Cambridge, Massachusetts, July 17-19, 1991, (accepted for presentation).

Expandable, Rapidly Assembled Sandwich Domes

A. Mirmiran, A.M. Wolde-Tinsae

Civil Engineering Department, University of Maryland College Park, Maryland 20742, U.S.A

ABSTRACT

A new type of self-erecting, rapidly assembled structure is proposed for large-size, clear-span enclosures. The prestressed sandwich dome falls in the category of expandable systems. It consists of a series of arches, each formed from a three-layer sandwich composite whose layers are individually buckled into an elastica shape. The prestressed sandwich arch has been evaluated using a nonlinear finite element method. Numerical results show that prestressed sandwich arches have a better stability behavior than their homogeneous counterparts.

INTRODUCTION

Studies have shown that for large-span systems, the self-weight of the structure has a pronounced effect on its stability (Huddleston[1]). Therefore, a logical consideration for this type of structures is to reduce structural self-weight, while maintaining a high strength-to-weight ratio. Light-weight self-erecting systems have been the subject of interest for the past two decades.

Tensile systems, such as cable roof structures, membrane systems and pneumatic structures have been widely used for large-span enclosures (Otto[2]). Pneumatic systems may be either air-supported (Ikoma[3]) with a covering membrane that is kept in tension by internal air pressure; or air-inflated (Tang[4]) which consists of self-enclosed membranes stiffened with air pressure. The cost of maintaining the air pressure is a major drawback for this type of structure, (Herzog[5]).

Another type of self-erecting structures is the expandable system, which can be easily and rapidly transported, erected or dismantled. Geodesic domes, as primarily architectured by Fuller[6], respond to these requirements. The geodesic dome consists of a number of tetrahedrons joined together to form an interlocked multi-tetrahedral surface. Another class of expandable systems was proposed by Escrig[7]. The basic idea is to use an assembly of bars pinned together to form a mechanism. A mechanism is unstable in its nature, and can not carry any load. The assembly is therefore deformable and movable by pushing or pulling from the ends. Once the desired shape is achieved, the system is made stable by means of a few additional bars or supports.

In the same category of foldable structures, yet another class of transportable systems may be defined by using a triangular panel as the basic element (Foster[8]). For the structure to fold into a flat and compact form, each panel must have one angle of at least 90° degrees. Each module may consist of four or more panels with continuous hinges. Different modules can be used in one structure. Again, the geometry allows the system to fold for transportation purposes, while it yields a rigid system when erected.

Other forms of expandable systems are made available by using air pressure for construction. Meyers[9] presented such a method for assembling plastic sandwich shell structures, in which the majority of construction is accomplished in a flat configuration and then the system is pneumatically deformed into the desired domical shape.

Wolde-Tinsae and Huddleston[10-12] offered yet another class of self-erecting structures, prestressed domes. The system consists of a series of arches assembled into an arch-framework or dome. Each arch is made by elastically buckling of an initially flat thin strut. Figure 1 shows the basic elements of a prestressed dome. A cable passes through loops attached to the ends of each strut. As the cables are pulled by the reeling mechanism, the struts buckle and the dome erects itself. Once the desired height is achieved, the members are connected to supports; and then reinforcing rings, horizontal spars and covering membranes may be added.

The stability of prestressed arches and domes was further investigated by Chini[13]. Chini and Wolde-Tinsae[14] conducted a series of centrifuge experiments on models of prestressed arches. They[15-16] also developed a finite element model to predict the stability behavior of this type of structure. Their studies revealed that prestressed domes would buckle at a lower critical load as compared to rigid domes of the same shape by a margin of at least 25 percent. Therefore, prestressing technique, in the form of buckling, could serve as a construction method for expandable systems. However, it does not improve load-carrying capacity of an arch system.

SANDWICH CONSTRUCTION

The logical solution to circumvent the drawbacks of prestressing technique is to combine its benefits with a new type of material,

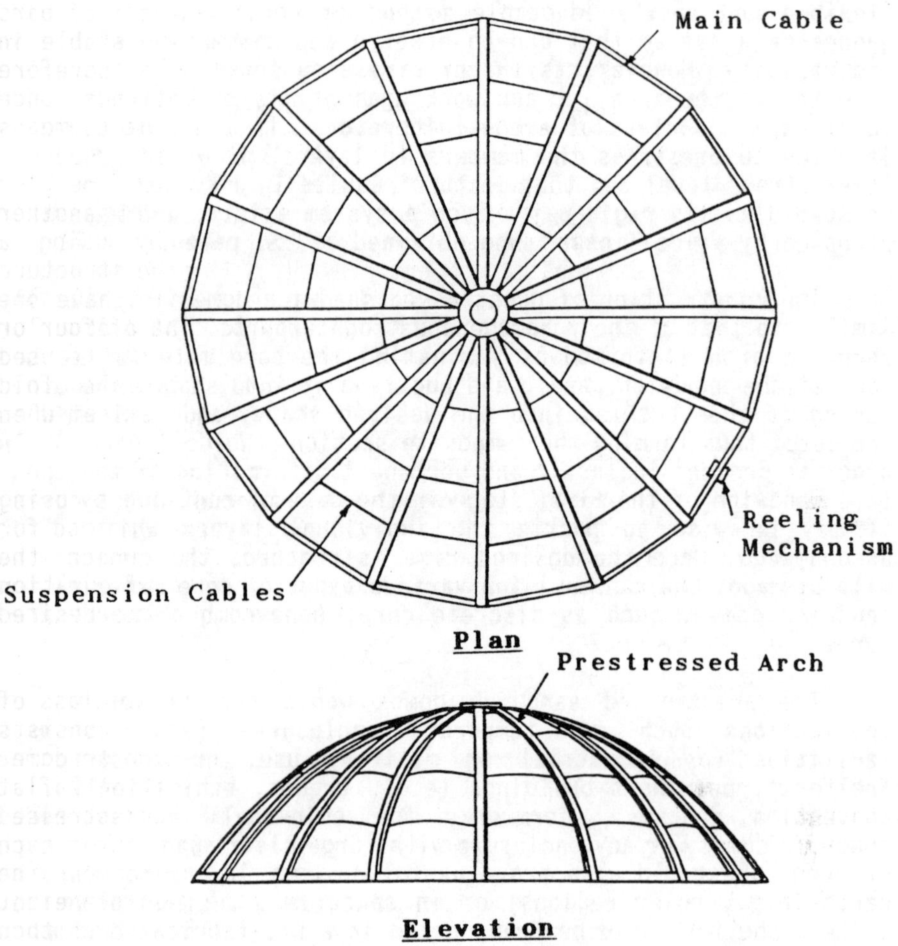

Figure 1. Prestressed Dome

namely sandwich composites. The prestressing technique offers a flexible and easily adaptable method of construction. It also generates a system that can be erected and dismantled with ease and rapidity. However, as the prestressing force helps save the geometry of the arch, it can work against its stability for the same shape and class of arches. Therefore, if a method could be invented to prestress the members in layers, it would produce a lower stress level and that in turn results in a system , perhaps as stable as its rigid pair. Such a system exists, and was first proposed by Wolde-Tinsae[17] who obtained a U.S. patent for it.

The construction of prestressed sandwich domes is basically similar to that of their homogeneous counterparts. The prebuckled sandwich arch is formed by laminating the core material to one face of the sandwich panel, and buckling it into shape. The other facing is then buckled into the desired shape, and laminated to the core; thus forming the composite section (Wolde-Tinsae[18]). In order to prevent inelastic and permanent deformation of the open-face sandwich of the first stage in the case of continuous cores, it may be wise to buckle the individual layers while still unconnected. Once the desired rise is reached, the connections will be made. One can envision various types of core material for sandwich domes, such as discrete core, honeycomb or corrugated core.

The prestressed sandwich domes would have a variety of applications such as; temporary enclosures (e.g., storage facilities for industrial or military use, or construction shelters), permanent buildings (e.g., arenas, exhibition halls, convention centers), formworks for thin-shell hemispherical concrete domes, or any enclosure with large clear spans. This type of structure could also prove useful in hostile environments on earth (e.g., polar regions) or in space (e.g., other planets), where a shelter could be transported in a pre-fabricated compact form. Because of the relative ease of erection, there would be a minimal exposure of humans to the environment during erection.

The proposed structure has the basic advantages of self-erecting systems and sandwich composites:

1. Easily erectable and dismantable;
2. High strength-to-weight ratio;
3. Good stability behavior;
4. Good thermal insulation;
5. Good acoustical insulation;
6. Good fatigue properties;
7. High life expectancy; and
8. Low maintenance cost, as it does not require a constant power supply for air pressure.

EVALUATION OF THE PROPOSED SYSTEM

Prestressed homogeneous arches are different from the conventional rigid ones in that:

1. Compressive stresses exist at each section of the arch, even without any external load;
2. Once released from the supports, they will return to their original flat shapes; and
3. The buckling load is at least 25 percent less than that of the rigid arch.

Prestressed sandwich arches share with their homogeneous pairs, the presence of internal compressive forces; however once the connections are made, part of these forces will be taken by the shear at the layer interfaces. Therefore, as one previous study shows (Nelsen[19]); upon removing the arch from its supports it will not rebound to its flat configuration. The amount of partial rebound is based on the weight of the arch and the amount of prestressing force.

The effects of prestressing force on arches have been investigated by several authors. Clifton[20] studied the effect of an initial thrust on the buckling of shallow circular arches with fixed ends. Thompson and Hunt[21] analyzed the buckling of prestressed shallow sinusoidal arches with hinged supports, and subjected to a concentrated load. However, the actual shape of a prestressed arch , the elastica, is neither a circle nor a half-sine wave (Timoshenko and Gere[22]).

Wolde-Tinsae and Huddleston[10] developed a method to uncouple a prestressed dome into its individual arches, and then re-assembled the results. Chini[13], and Chini and Wolde-Tinsae[15-16] applied the finite element technique to the stability problem of prestressed domes. In 1979, Nelsen[19] conducted an experimental feasibility study on prestressed sandwich arches with discrete cores. He reported a very high strength-to-weight ratio for the proposed structure. For example for a sandwich arch made of aluminum facings and oak wood discrete core, a strength-to-weight ratio that was 9.09 times greater than its equivalent prestressed homogeneous arch was achieved.

Stability of a ribbed dome can be most efficiently investigated by the study of the behavior of one of its members. Here, buckling and postbuckling behavior of prestressed sandwich arch is carried out by using a nonlinear finite element model. A corotational formulation of a two-node, three-layer sandwich beam element is adopted. Shear deformations of the core and facings are both considered. The discontinuity in shear strains at the layer interface is accounted for by introducing a shear warpage angle. A bilinear constitutive law is used which assumes a linear strain hardening behavior, once the elastic limit is reached for the corresponding layer. Geometric nonlinearities due to large displacements and large rotations are both considered. Postbuckling analysis is performed by using the augmentation technique. A special algorithm is devised to model the erection procedure for the prestressed sandwich arch.

NUMERICAL RESULTS

The overall stability of prestressed sandwich arches depends on several factors including the geometric and mechanical properties such as thickness and material of the core layer, and rise-to-span ratio of the arch. Here, for comparison a common face layer is used for various types of sandwich arches. The facings are assumed to be 0.1 inch thick, and made of 2024-T3 aluminum alloy. The core layer is assumed to be made of aluminum honeycomb material. The arch section is taken of unit width for simplicity. A homogeneous arch made of 2024-T3 aluminum with a thickness of 0.2 inch is used as a basis for comparison, because it is twice as thick as each sandwich facing. To perform a parametric study, three types of arches are defined based on the thickness ratio of the core layer in the composite section :

1. Homogeneous arch, with no core layer, i.e., $c/t = 0$.
2. Medium thick core sandwich arch, with a thickness ratio of $c/t \leq 60\%$.
3. Thick core sandwich arch, with a thickness ratio of $c/t \geq 80\%$.

where c is the thickness of the core layer, and t is the overall thickness of the arch section. The results are obtained in terms of displacements and internal forces as a function of incremental loads. The load parameter is taken as PL^2/EI_o for a concentrated load, and wL^3/EI_o for a uniform load; where L is the length of the members prior to buckling, and EI_o is the flexural rigidity of the composite section. The vertical displacements of the crown is plotted against the load parameter in Figure 2 for a concentrated load at the crown. The normal force at the crown of the arch is drawn against the load parameter in Figure 3. In order to compare the load-carrying capacity of the proposed structure with homogeneous arch, the critical loads as well as load parameters are shown in Tables 1 and 2. In the tables, d is the center to center distance between the facings. As shown in the tables, by increasing the core thickness, the limit and bifurcation loads both increase. However, the non-dimensional load parameter decreases due to the sharp increase of flexural rigidity of the composite section. Therefore, prestressed sandwich arches are more stable than their homogeneous counterparts. For example, a medium thick core sandwich section has a limit load of 8.27 times its equivalent homogeneous arch, under a concentrated load. This ratio sharply increases to 31.63 for a thick core sandwich arch. The results of this study (Mirmiran[23]) also indicated that thickness and shear modulus of the core not only affect the critical load of the arch, but may also change the governing mode of buckling of the arch. Furthermore, the effect of prestressing on stability of sandwich arches is negligible comparing to that of homogeneous arches, and as the thickness of the core increases these effects practically vanish. This indicates a good buckling behavior for

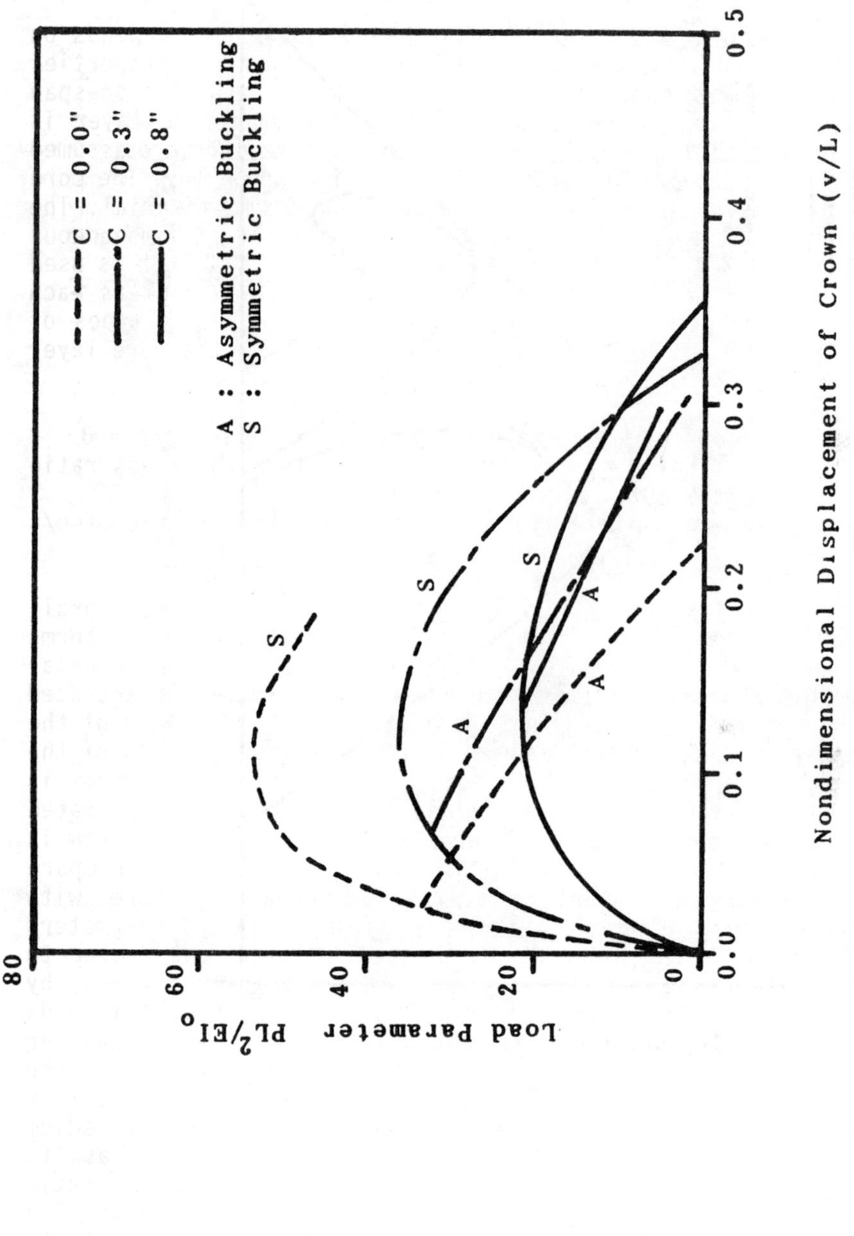

Figure 2. Load Deflection Curve for Prestressed Arches

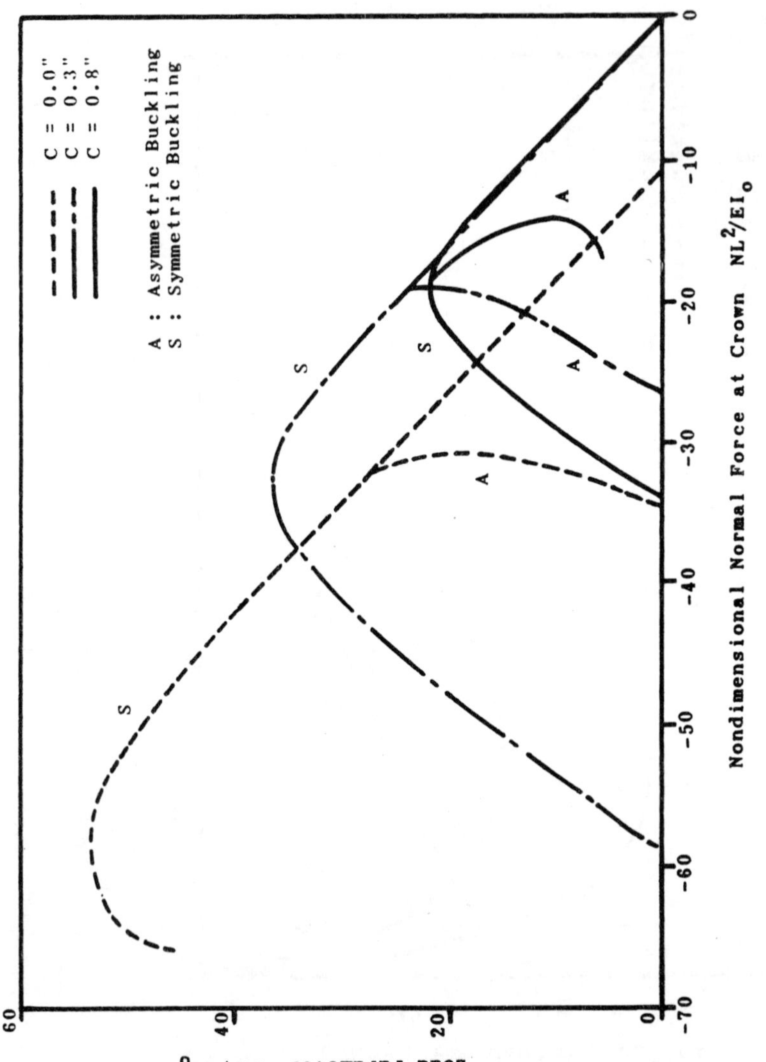

Figure 3. Normal Forces at Crown for Prestressed Arches

Table 1 Effect of Core Thickness on Stability of Sandwich Arches (1)
(Concentrated Load, Symm. Buckling Mode)

					Symm. Buckling			
Case	Core Thickness (in.)	c/t (%)	c/d (%)	EI_o (k-in^2)	Limit Load, P (kips)	Relative Value	PL^2/EI_o	Relative Value
1 (2)	0.0	0.0	0.0	6.67	0.70	1.00	53.70	1.000
2	0.1	33.3	50.0	21.67	2.09	3.01	49.66	0.925
3	0.2	50.0	66.7	46.67	3.80	5.47	41.93	0.781
4 (3)	0.3	60.0	75.0	81.67	5.76	8.27	36.26	0.675
5	0.4	66.7	80.0	126.67	7.87	11.30	31.94	0.595
6	0.5	71.4	83.3	181.67	10.08	14.48	28.54	0.531
7	0.6	75.0	85.7	246.67	12.37	17.77	25.79	0.480
8	0.7	77.8	87.5	321.67	14.71	21.14	23.53	0.438
9 (4)	0.8	80.0	88.9	406.67	17.11	24.58	21.64	0.403
10	1.0	83.3	90.9	606.67	22.01	31.63	18.66	0.347

Notes:

(1) For all cases :
Core: E_c = 0 ksi, G_c = 40 ksi
Facings: f = 0.1 in., E_f = 10000 ksi, G_f = 4000 ksi
Span/Rise (h/l) = 0.25, L = 22.68 in.

(2) c = 0.0 in.: Homogeneous Section.

(3) c = 0.3 in.: Medium Thick Core Sandwich Section.

(4) c = 0.8 in.: Thick Core Sandwich Section.

Table 2 Effect of Core Thickness on Stability of Sandwich Arches (1) (Concentrated Load, Asym. Buckling Mode)

					Asym. Buckling			
Case	Core Thickness (in.)	c/t (%)	c/d (%)	EI_o ($k\text{-}in^2$)	Bifurcation Load, P (kips)	Relative Value	PL^2/EI_o	Relative Value
1 (2)	0.0	0.0	0.0	6.67	0.44	1.00	33.74	1.000
2	0.1	33.3	50.0	21.67	1.59	3.64	37.82	1.121
3	0.2	50.0	66.7	46.67	3.19	7.28	35.11	1.041
4 (3)	0.3	60.0	75.0	81.67	5.16	11.80	32.49	0.963
5	0.4	66.7	80.0	126.67	7.34	16.79	29.82	0.884
6	0.5	71.4	83.3	181.67	9.69	22.15	27.43	0.813
7	0.6	75.0	85.7	246.67	12.11	27.70	25.26	0.749
8	0.7	77.8	87.5	321.67	14.60	33.39	23.35	0.692
9 (4)	0.8	80.0	88.9	406.67	17.05	38.98	21.56	0.639
10	1.0	83.3	90.9	606.67	22.00	50.30	18.65	0.553

Notes:

(1) For all cases :
Core: E_c = 0 ksi, G_c = 40 ksi
Facings: f = 0.1 in., E_f = 10000 ksi, G_f = 4000 ksi
Span/Rise (h/l) = 0.25, L = 22.68 in.

(2) c = 0.0 in.: Homogeneous Section.

(3) c = 0.3 in.: Medium Thick Core Sandwich Section.

(4) c = 0.8 in.: Thick Core Sandwich Section.

prestressed sandwich arches and domes.

CONCLUSIONS

This study confirms the feasibility of the prestressed sandwich arch and arch framework (dome) as an expandable, rapidly assembled structure with good buckling behavior. The buckling and post-buckling analysis of the structure is performed by using an augmentation technique. The prestressed sandwich arches demonstrate a better stability behavior and a higher critical load than their homogeneous counterparts.

REFERENCES

1. Huddleston,J.V. Dead-Load Effects on a Proposed Very-Large-Size Dome, Proc. of the IASS Symposium on Shell Structures and Climatic Influences, pp. 395-404, Calgary, Canada, 1972.

2. Otto,F.(Ed.) Tensile Structures, Vols 1 and 2, MIT Press, Cambridge, Mass., 1973.

3. Ikoma,T. Structural Properties of Air-Supported Structures, International Journal of Space Structures, Vol. 2, No. 4, pp. 195-203, 1987.

4. Tang,M.C. Lenticular Air Inflated Bubble Roof Structures, Proc. of ASCE Symposium on Long Span Roof Structures, New York, 1981.

5. Herzog,T. Pneumatic Structures - A Handbook of Inflatable Architecture, Oxford University Press, New York, 1976.

6. Fuller,R.B. Synergetics, Exploration in the Geometry of Thinking, MacMillan Publishing Co. , New York, 1975.

7. Escrig,F. Expandable Space Structures, International Journal of Space Structures, Vol. 1, No. 2, PP. 79-91, 1985.

8. Foster,C.G., and Krishnakumar,S. A Class of Transportable Demountable Structures, International Journal of Space Structures, Vol. 2, No. 3, pp. 129-137, 1986.

9. Meyers,V.J. Pneumatically Formed Sandwich Shell Roof, Journal of the Structural Division, ASCE, Vol. 104, No. ST3, pp. 381-389, 1978.

10. Wolde-Tinsae,A.M., and Huddleston,J.V. Nonlinear Three-Dimensional Analysis of Prebuckled Domes, Computers and Structures, Vol. 14, No. 3-4, pp. 215-224, 1981.

11. Wolde-Tinsae,A.M., and Huddleston,J.V. Nonlinear Analysis of Self-Erecting Domes, Proc. 7th Canadian Congress of Applied Mech., Sherbrooke, 1979.

12. Wolde-Tinsae,A.M., and Huddleston,J.V. Three-Dimensional Stability of Prestressed Arches, Journal of the Engineering Mechanics Division, ASCE, Vol. 103, No. EM5, pp. 855-867, 1977.

13. Chini,S.A. Nonlinear Stability Analysis of Prestressed Domes Using the Finite Element Method, Ph.D. Thesis, University of Maryland at College Park, MD., 1986.

14. Chini,S.A., and Wolde-Tinsae,A.M. Buckling Test of Prestressed Arches in Centrifuge, Journal of Engineering Mechanics, ASCE, Vol. 114, No. 6, pp. 1063-1075, 1988.

15. Chini,S.A., and Wolde-Tinsae,A.M. Critical Load and Postbuckling of Arch Frameworks, Journal of Engineering Mechanics, ASCE, Vol. 114, No. 9, pp. 1435-1453, 1988.

16. Chini,S.A., and Wolde-Tinsae,A.M. Effect of Prestressing on Elastica Arches, Journal of Engineering Mechanics, ASCE, Vol. 114, No. 10, pp. 1791-1800, 1988.

17. Wolde-Tinsae,A.M. Method of Forming a Skeletal Dome Structure In Situ, United States Patent 4400927, 1983.

18. Wolde-Tinsae,A.M. Prebuckled Sandwich Domes and Super Domes, Proc. of the IASS Symposium on Shells, Membranes and Space Frames, pp. 345-352, Osaka, 1986.

19. Nelsen,D. A Feasibility Study on the Use of Sandwich Construction in Prebuckled Arches, Masters Thesis, Iowa State University, Ames, 1979.

20. Clifton,R.J. Snap-Buckling of Shallow Prestressed Arches, Journal of the Engineering Mechanics Division, ASCE, Vol. 96, No. EM6, pp. 1157-1170, 1970.

21. Thompson,J.M.T., and Hunt,G.W. On the Buckling and Imperfection-Sensitivity of Arches With and Without Prestress, International Journal of Solids and Structures, Vol. 10, No. 5, pp. 445-459, 1983.

22. Timoshenko,S.P., and Gere,J.M. Theory of Elastic Stability, McGraw-Hill Book Co., New York, 1961.

23. Mirmiran,A. Nonlinear Stability Analysis of Prestressed Sandwich Arches Using the Finite Element Method, Ph.D. Thesis, University of Maryland at College Park, MD., 1991.

To Cover A Swimming Pool with an Expandable Structure

F. Escrig(*), J.P. Valcárcel(**)
(*) School of Architecture of Seville (Spain)
(**) School of Architecture of La Coruña (Spain)

ABSTRACT

Expandable Space Structures based on X-frames will be of great use in solving special problems like movable Architecture.

In this case we must provide a cover on a swimming-pool for bioclimatic profiting. This means that we need to eliminate the cover for variable periods of time as well as extend it. The area covered is of 3.000 sqm. with a section of sphere that runs hanging below a spatial arch and folds at the top.

1. INTRODUCTION

The question to be solved is how to cover a swimmimg pool of great size (50 × 21 sqm) with a structure able to be removed in the summer and erected temporarily during other seasons to capture solar energy to heat the air and the water.

To achieve this we provide an Expandable structure of bars based on movable X-frames in accordance with the model developed by authors [Ref. 1].

Expandable X-frames are mechanisms capable of showing different geometrical configurations from minima when they are folded to great surfaces and when they are expanded. In both limit states they can be immobilized by means of the blockage of some elements to guarantee the global stiffness of the whole and we can achieve structures useful to building transformable architecture.

Figure 1 shows how a set of 3-X-frames works while Figure 2 shows the foldability of a multiple X-frame.

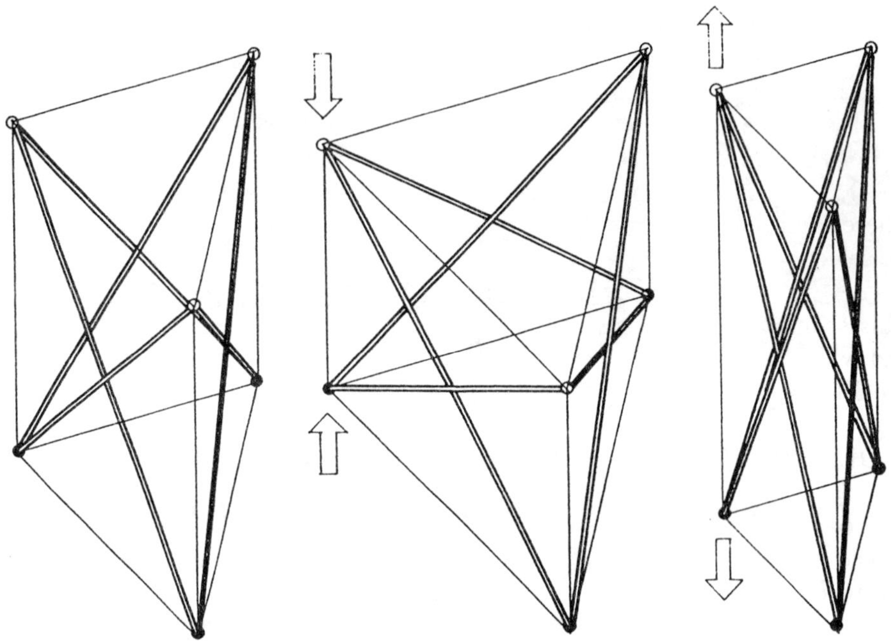

Figure 1

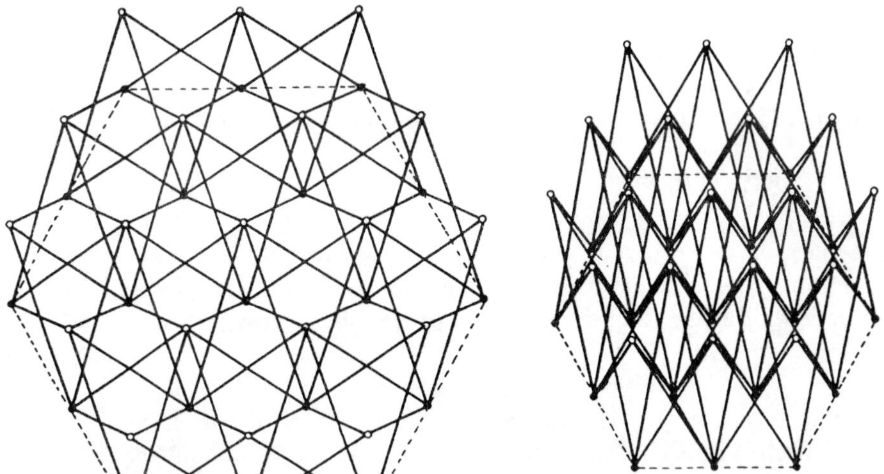

Figure 2

2. FORMAL PROPOSAL

To satisfy our need for a completely removable cover we propose a spherical shell section 60 meters in diameter and 15 meters in height. This solution has some important advantages:

a) The ratio between surface enclosure and covered area is minimum.

b) Roof and walls are included into the same geometry.

c) It admits a well studied expandable X-frame.

There are a lot of spherical expandable X-frames to be used, out of which we select a triangular spherical grid according to the authors in [Ref. 2].

This spacial frame will be supported along its boundaries when expanded and will be hung from a central point of a main arch when folded or during the folding process.

3. DESIGN AND GEOMETRY OF GLOBAL X-FRAME

The basic grid will be obtained from a triangular three way plane mesh placed at ground level by projecting it on the sphere surface from its farthest pole (Figure 3).

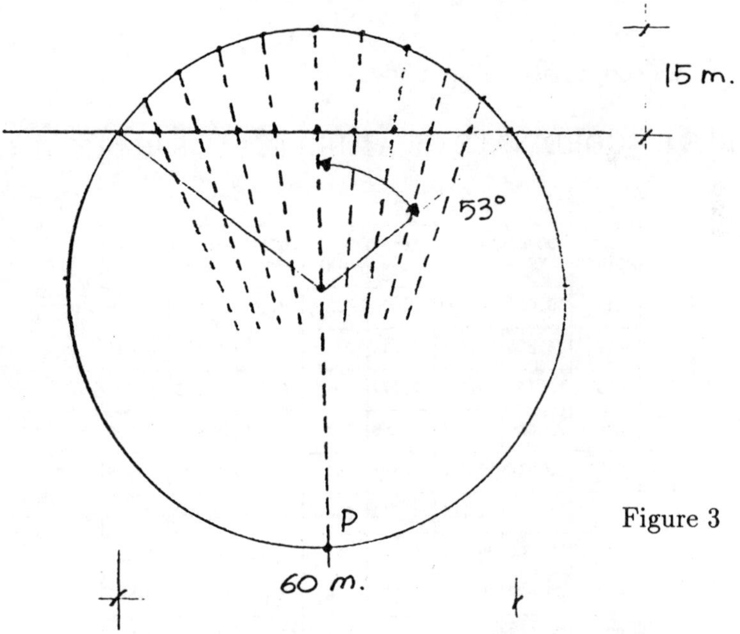

Figure 3

Figure 4 shows the grid in this way.

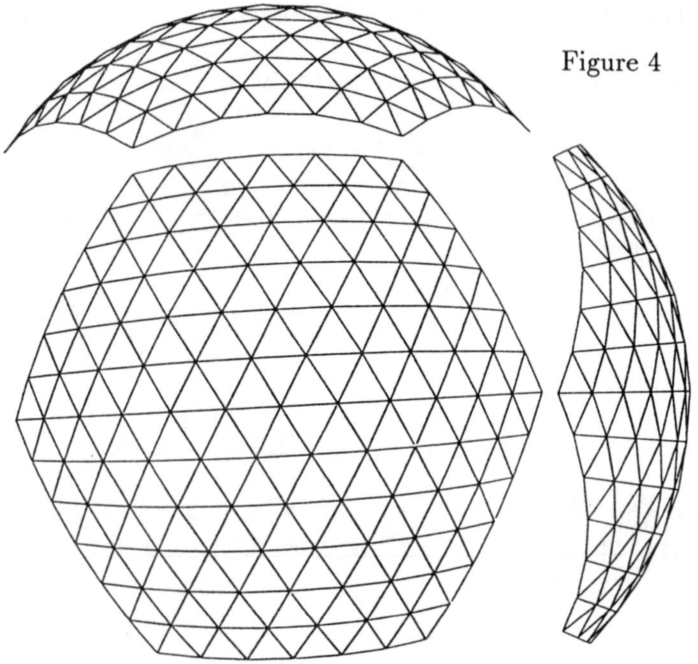

Figure 4

Table 1 shows coordinates of its nodes.

TABLE I: COORDINATE OF A THREE WAY SPHERICAL GRID

POINTS	COORDINATES (meters) X	Y	Z	POINTS	COORDINATES (meters) X	Y	Z
P11	30.000	0.000	0.000	P32	20.064	4.964	8.790
P12	28.286	4.454	1.714	P33	17.308	9.992	9.231
P13	26.163	9.063	2.791	P34	14.331	14.893	8.790
P14	23.684	13.674	3.158	P35	11.250	19.485	7.500
P15	20.930	18.126	2.791	P41	17.647	0.000	10.588
P16	18.000	22.269	1.715	P42	14.901	5,162	11.523
P17	15.000	25.980	0.000	P43	11.921	10.323	11.523
P21	26.627	0.000	3.905	P44	8.824	15.282	10.588
P22	24.545	4.724	5.455	P51	12.162	0.000	12.973
P23	22.086	9.563	6.258	P52	9.184	5.302	13.469
P24	19.325	14.345	6.258	P53	6.081	10.532	12.973
P25	16.364	18.895	5.455	P61	6.207	0.000	14.483
P26	13.314	23.059	3.906	P62	3.103	5.375	14.483
P31	22.500	0.000	7.500	P71	0.000	0.000	15.000

Into this spherical grid (D) we must now superimpose an expandable frame. To this we must solve the next system of equations (Figure 5) where α_i are known for each line of triangular spherical three way grid.

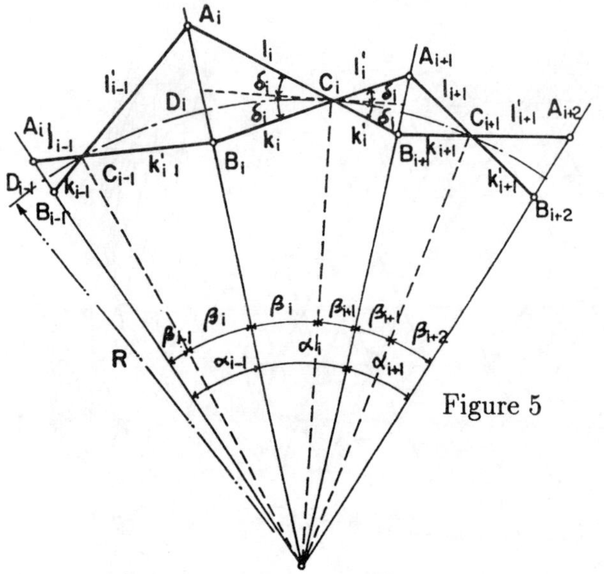

Figure 5

....

$$\alpha_{i-2}=\beta_{i-2}+\alpha_{i-1}$$

$$\alpha_{i-1}=\beta_{i-1}+\beta_i$$

$$\alpha_i=\beta_i+\beta_{i-1}$$

$$\alpha_{i+1}=\beta_{i+1}+\beta_{i+2}$$

....

In general this system is not compatible because it contains more equations than variables. Nevertheless we can optimize the solution to obtain acceptable tolerances to make the work of the structure possible.

Grids projected from a pole are the best for our purpose.

By fixing now the angle "δ" for maximum expandability we can determine the frame (AB) that achieves the necessary conditions to guarantee the kinematic compatibility in the folded and completely expanded positions.

Unfortunately the compatibility in intermediate positions is not assured and in most cases we must force the structure to pass through. Nevertheless with little imposed deflections it can work in the complete range.

Figure 6 shows the whole grid obtained with a δ= 0'15. Table II shows lengths of the bars. Figures 7 and 8 give information on criteria used to identify each bar position.

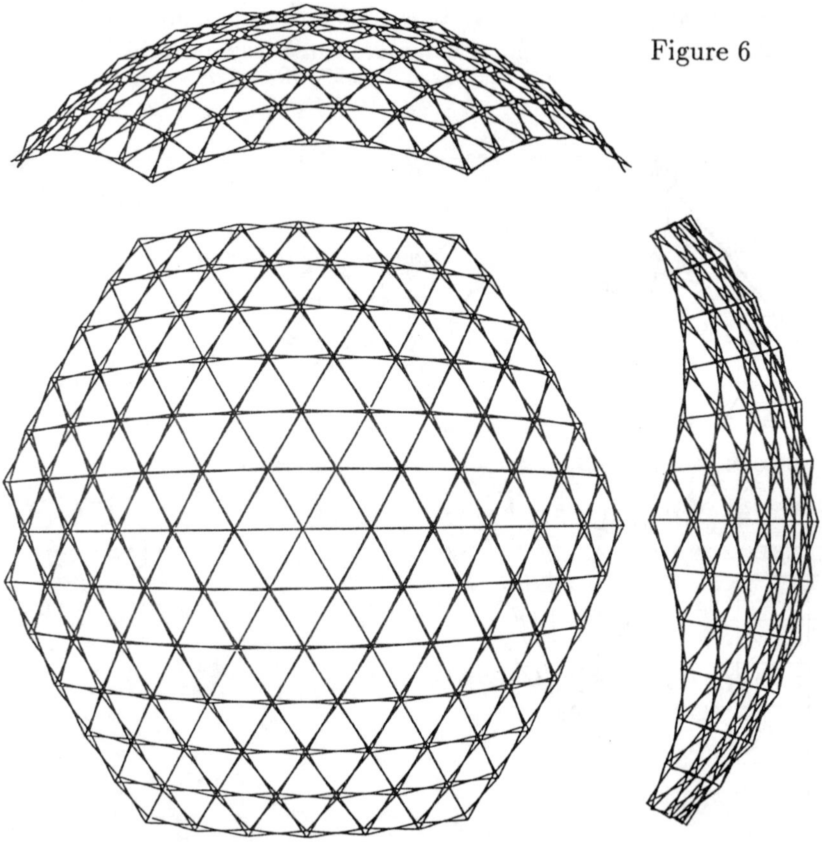

Figure 6

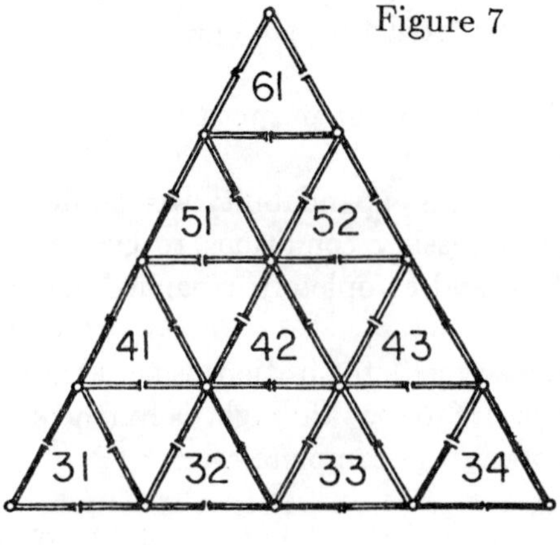

Figure 7

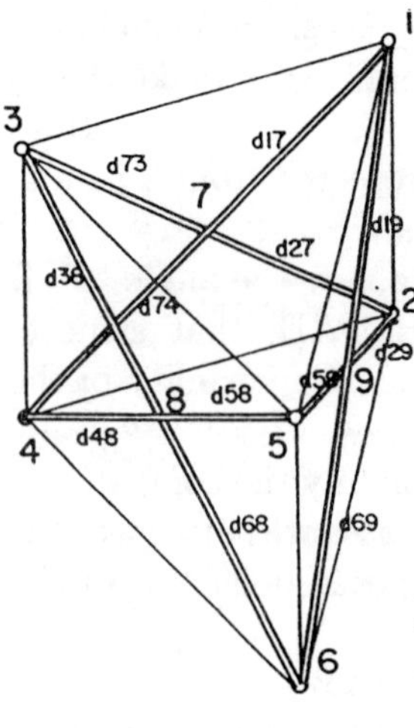

Figure 8

TABLE II: LENGTHS OF A THREE WAY SPHERICAL X-FRAME

ELEM	D17	D27	D73	D74	D38	D48	D85	D86	D19	D29	D95	D96
E61	3.2121	3.1319	3.1895	3.1105	3.1895	3.1104	3.1895	3.1104	3.2120	3.1318	3.1895	3.1104
E51	3.1892	3.1102	3.1236	3.0477	3.1238	3.0479	3.1479	3.0679	3.1894	3.1103	3.1451	3.0681
E52	3.1895	3.1104	3.1452	3.0682	3.1449	3.0679	3.1238	3.0480	3.1892	3.1101	3.1235	3.0478
E41	3.1230	3.0471	3.0195	2.9485	3.0201	2.9492	3.0597	2.9868	3.1235	3.0477	3.0601	2.9872
E42	3.1447	3.0677	3.0604	2.9875	3.0603	2.9875	3.0603	2.9875	3.1447	3.0677	3.0603	2.9875
E43	3.1236	3.0478	3.0601	2.9873	3.0597	2.9868	3.0201	2.9493	3.1229	3.0471	3.0195	2.9487
E31	3.0186	2.9477	2.8849	2.8200	2.8860	2.8211	2.9404	2.8730	3.0197	2.9487	2.9408	2.8734
E32	3.0593	2.9865	2.9414	2.8740	2.9423	2.8749	2.9594	2.8912	3.0603	2.9874	2.9597	2.8915
E33	3.0603	2.9875	2.9598	2.9815	2.9594	2.8912	2.9423	2.8750	3.0593	2.9864	2.9413	2.8740
E34	3.1097	2.9489	2.9409	2.8736	2.9404	2.8731	2.8860	2.8214	3.0186	2.9478	2.8849	2.8203
E21	2.8839	2.8191	2.7285	2.6704	2.7299	2.6718	2.7952	2.7342	2.8854	2.8205	2.7956	2.7346
E22	2.9399	2.8725	2.7962	2.7352	2.7976	2.7366	2.8299	2.7674	2.9413	2.8739	2.8302	2.7678
E23	2.9592	2.9809	2.8313	2.7688	2.8312	2.7687	2.8312	2.7687	2.9591	2.8908	2.8312	2.7687
E24	2.9414	2.8741	2.8303	2.7678	2.8299	2.7674	2.7976	2.7368	2.9399	2.8726	2.7962	2.7353
E25	2.8855	2.8209	2.7957	2.7348	2.7952	2.7344	2.7299	2.6722	2.8839	2.8193	2.7288	2.6707
E11	2.7277	2.6696	2.5588	2.5076	2.5604	2.5092	2.6331	2.5789	2.7294	2.6713	2.6330	2.5788
E12	2.7948	2.7338	2.6335	2.5793	2.6351	2.5809	2.6796	2.6235	2.7965	2.7355	2.6795	2.6234
E13	2.8296	2.7671	2.6806	2.6245	2.6822	2.6261	2.6952	2.6384	2.8312	2.7687	2.6951	2.6383
E14	2.8312	2.7687	2.6952	2.6384	2.6952	2.6384	2.6822	2.6262	2.8295	2.7670	2.6806	2.6245
E15	2.7976	2.7357	2.6796	2.6235	2.6796	2.6235	2.6352	2.5812	2.7948	2.7340	2.6335	2.5795
E16	2.7295	2.6718	2.6331	2.5792	2.6331	2.5792	2.5604	2.5097	2.7277	2.6700	2.5587	2.5080

4. KINEMATICS OF THE X-FRAME GRID

To fold and expand the grid, it will be supported by its highest node to a framed arch that moreover will contain the rod to facilitate the displacement of the other two nodes. These nodes guarantee the stability of the whole.

Figure 9 shows the arch with the mechanism of folding. The expanding is achieved by forces due to its own weight.

Figure 9

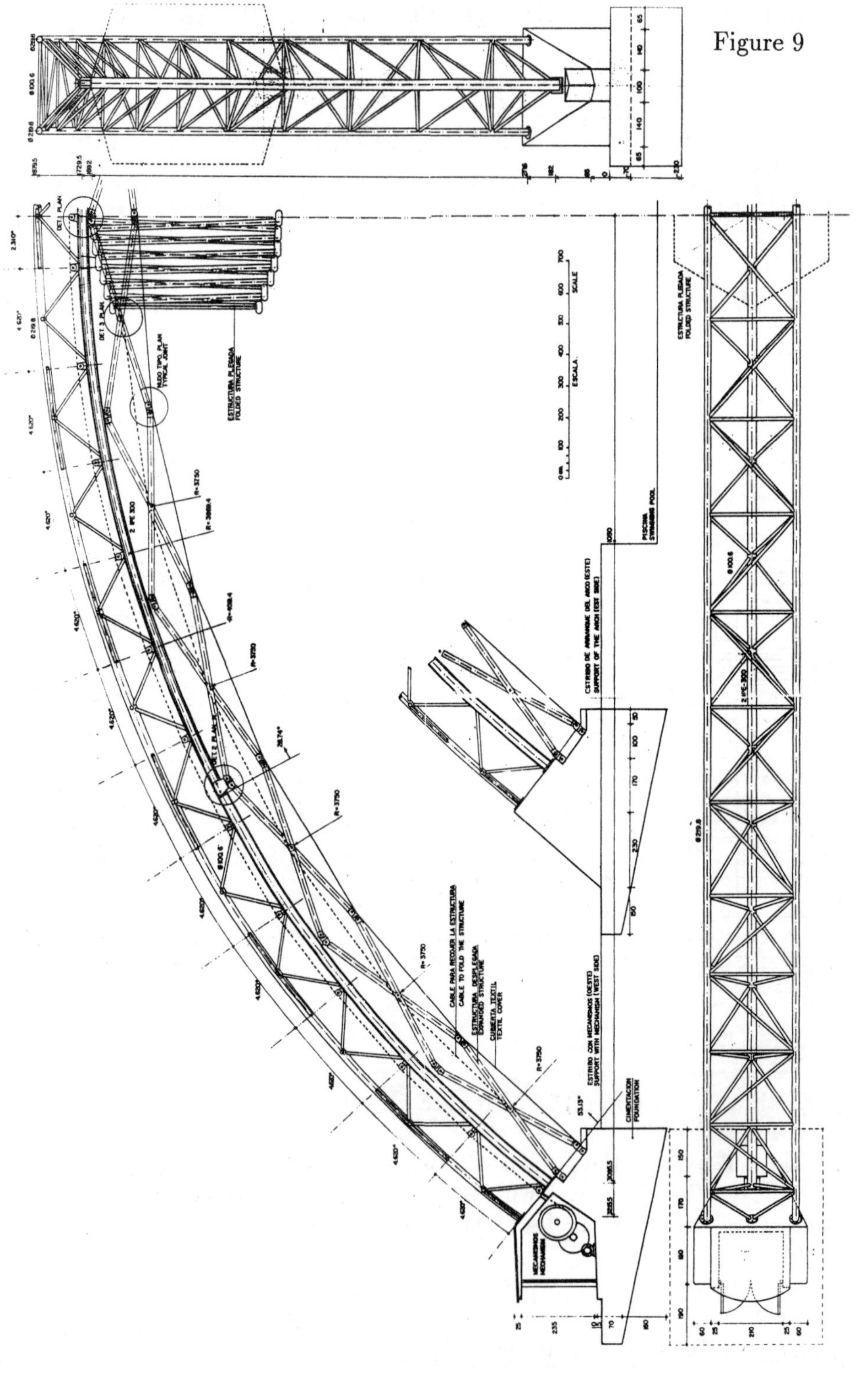
ESTRUCTURA PLEGADA
FOLDED STRUCTURE
ESCALA
SCALE
PISCINA
SWIMMING POOL
CABLE PARA RECOGER LA ESTRUCTURA
CABLE TO FOLD THE STRUCTURE
ESTRUCTURA DESPLEGADA
EXPANDED STRUCTURE
CUBIERTA TEXTIL
TEXTIL COVER
CIMENTACION
FOUNDATION
MECANISMOS
MECHANISM

Figure 10 shows details designed for joints and special pieces.

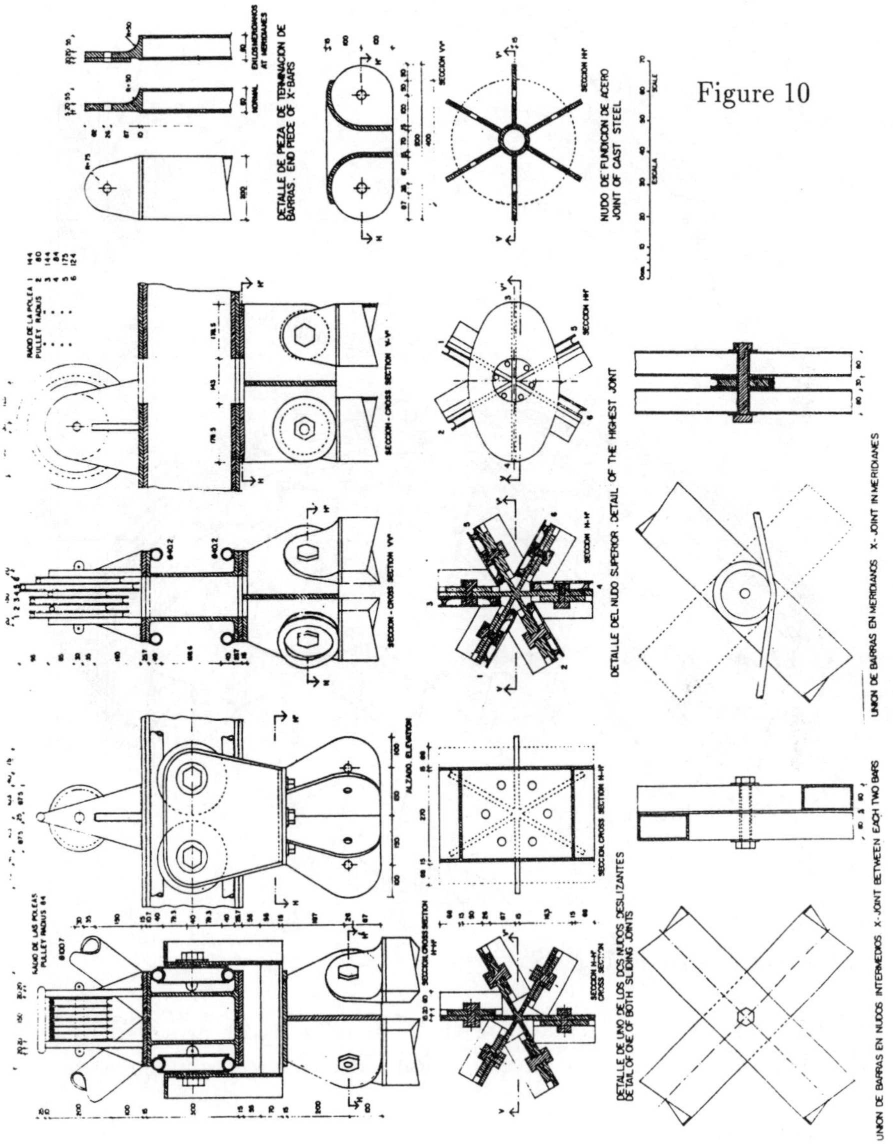

Figure 10

Figure 11 shows the complete structure folded while Figure 12 shows it expanded.

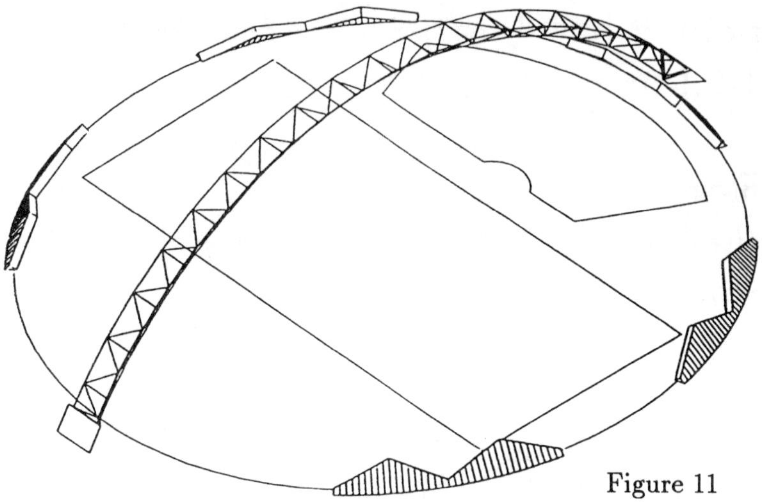

Figure 11

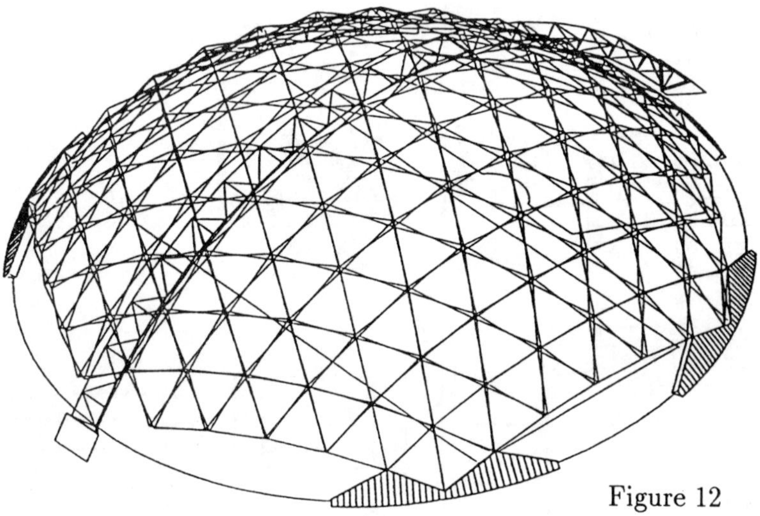

Figure 12

5. ACKNOWLEDGEMENTS

This research has been achieved with financial support provided by DGICYT (Dirección General de Investigación Científica y Tecnológica).

REFERENCES

(1) ESCRIG, F. "Expandable Space Structures", Space Structures Int. Journ. Vol. 1 num. 2., Elsevier pp 79-91.

(2) ESCRIG, F., P. VALCARCEL and J.R. GIL DELGADO "Design of Expandable Spherical Grids", IASS Simp. 1989, Madrid, 16pp.

Expandable Space Structures with Self-folding Textile Cover

J.P. Valcárcel(*), F. Escrig(**), E. Martín(*), M. Freire(*)

(*) E.T.S. Architecture. Castro de Elviña s/n. La Coruña, Spain

(**) E.T.S. Architecture. Reina Mercedes s/n. Sevilla, Spain

ABSTRACT

This paper presents a new modulus based on the bundle modulus for expandable structures with a central bar which allows stiffening of the structure, stretches the textile cover out in deployed position, and creases the cover in folding position. The paper will contain a solution of non-linear analysis of curved expandable space bar structures with the effects of this bar. Matricial methods are used, combined with the effects of stability functions and geometric changes.

INTRODUCTION

The expandable space structures consist of units of bars, articulated at their ends and at a central point, which are capable of folding into compact packs so as to allow easy transportation and unfolding wherever necessary. Although their use is not yet widespread, the potential applications are promising. They can be used as shelters, military quarters of provisional dwellings in areas where an emergency situation has arisen. Their usefulness as auxiliary elements such as masts, bridges, or antenna is remarkable as well. However, the most extended application is, perhaps, in the covering of large areas, thus defining an architectural space of great aesthetic value.

The expandable structures are made of two kinds of modulus: Scissor modulus made by three or four scissors united by their ends, or bundle modulus made by a bundle of three or four bars intersecting at an inner point around a pivot with three or four screws. These modulus are based on a triangular prism for triangular scissor modulus, or antiprism for a triangular bundle modulus or a squared prism for rectangular ones. For curved spacial grids such as cylindrical vaults or domes, these modulus may be curved in space.

These frames have a very good structural behaviour as we could verify by computer simulations. However, many of them are not stiff and they may be stiffened by wires, bars or even the textile cover.

Another important problem which these structures pose lies in the difficulty to define the adequate length of the bars to allow for the free movement of folding and unfolding in all intermediate positions. Generally speaking, our research has proven that, with the exception of flat trusses, in other structures, and namely in domes, the geometrical compatibility in starting and end positions can be guaranteed, but not so in intermediate positions. Therefore, an additional energy must be applied during the unfolding process so as to force incompatibility. Quite often such an energy is sent back by the structure, completing in this way its expansion.

Once the structure is unfolded, it needs to be covered in order to be used. The roofing material can consist of stiff panels placed on top of the frame, as shown in fig. 1, a pavilion designed by E. Pérez Piñero in 1962, the first expandable structure actually built.

SELF-FOLDING SYSTEM WITH A FIXED CENTRAL BAR.

Another possibility for roofing consists of a self-folding textile cover, an umbrella being the simplest example of such a device.

The use of the self-folding textile cover presents serious difficulties in these structures and not much attention has been devoted to these, with the consequent lack of effective solutions.

This paper presents a new modulus based on the bundle one but with a central bar which allows the structure to be stiffened, stretches the textile cover out in deployed position, and creases the cover in folding position. Two possibilities are examined here, corresponding to both a central fixed bar and a central sliding bar. Some guidelines are also presented concerning the study of geometric conditions that allow for the self-folding process.

The proposed modulus is based on a bundle of four bars, articulated on a central piece (figure 2). What is new about this system is the use of a bar welded onto the said piece (fig. 3) which pulls the textile cover height in the unfolded position, as well as forcing the structure to fold up (fig. 4).

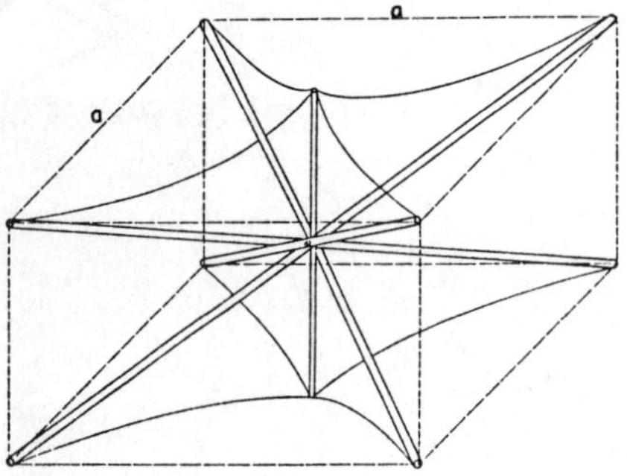

Figure 2

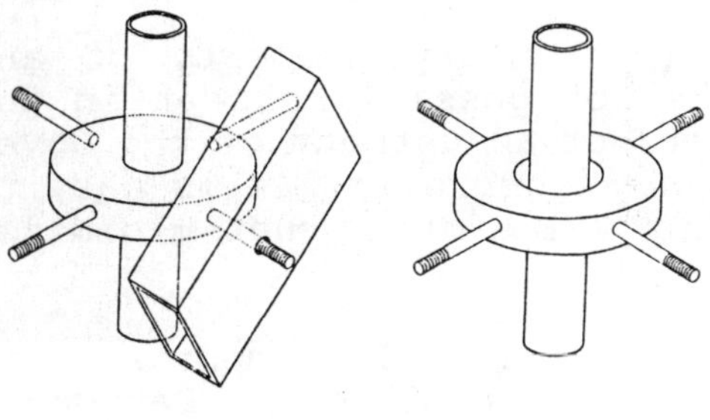

Figure 3

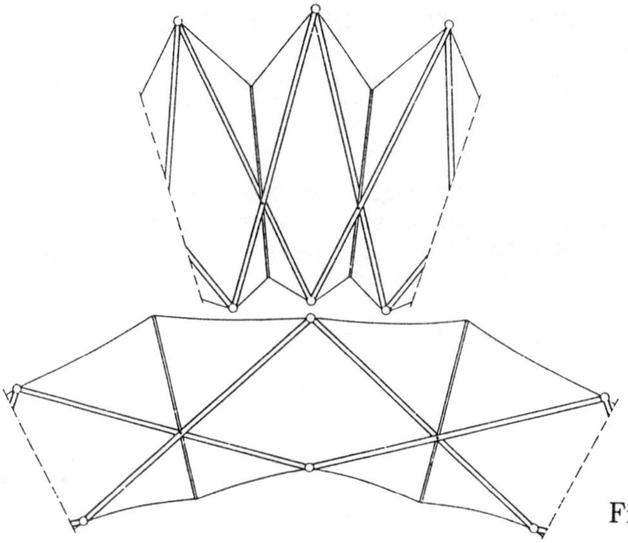

Figure 4

In order to study the conditions of geometric compatibility, we are going to consider the movement of the diagonal of the parallelepiped (fig. 5).

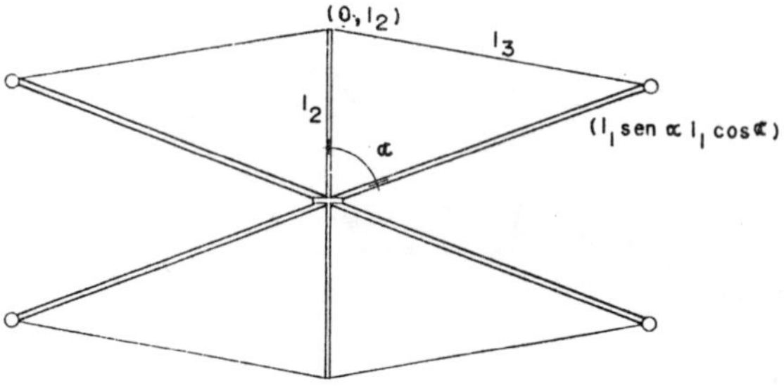

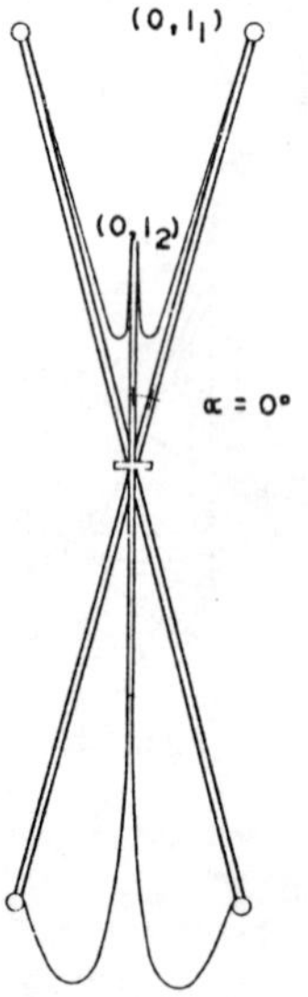

Figure 5

Since $l_3 > l_1 - l_2$ (elementary property of any triangle) height folding is not possible except in the insignificant case of a perfect adjustment of the cover to the bar, such as that of an umbrella. It is easy to prove that there is not any value of l_2 that minimizes $l_3 - (l_1 - l_2)$.

In this case it is especially important to determine which position makes the distance between B and B' (fig. 6) unchangeable during deploying, so a simple cable can produce a correct folding. One can easily prove that such a distance corresponds to an angle α = 70º 31' 43".

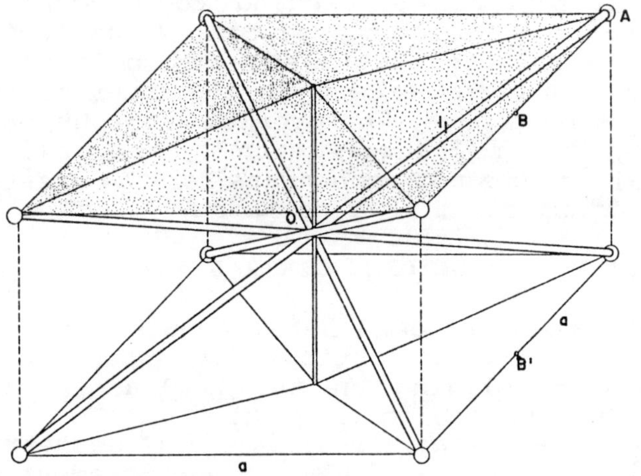

Figure 6

Also considered is the case of a textile cover which initially has the dimensions to form a flat surface and which is pretensed by displacing the central point a distance **d**. (Fig. 7).

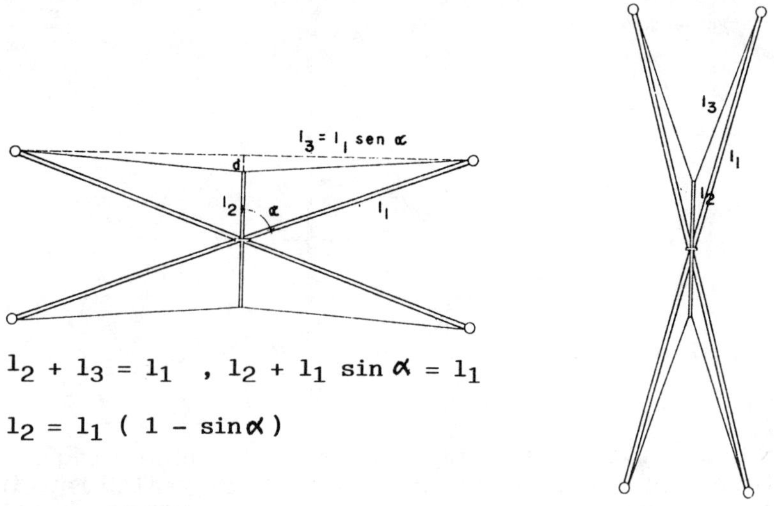

$l_2 + l_3 = l_1$, $l_2 + l_1 \sin\alpha = l_1$

$l_2 = l_1 (1 - \sin\alpha)$

Fig 7

A central bar of this length allows for folding across the diagonal, providing at the same time a much needed pretension in order to stiffen the truss. The study of the geometric conditions shows that for a given angle α = 45º, a maximum of the required pretension takes place, so it is advisable to achieve more open angles. Our studies show that best results are obtained from angles ranging between 60º and 70º.

We have also studied the possibility that an adequate initial diagonal length of the cover material could minimize pretensing. Proceeding to a complete study of the problem of minimum produced, we can prove that, although the solution seems somewhat better, no minimal value is achieved, except in the case of the "umbrella" already mentioned. The problem remains open but the first results suggest that the slight improvement obtained does not compensate for the greater difficulty in cutting new fabric patterns.

SELF-FOLDING SYSTEM WITH SLIDING CENTRAL BAR.

The second system proposed here, consists of the use of a central sliding bar articulated at connecting rods and capable of sliding on the ball-and-socket joint, gathering up the fabric (fig. 8).

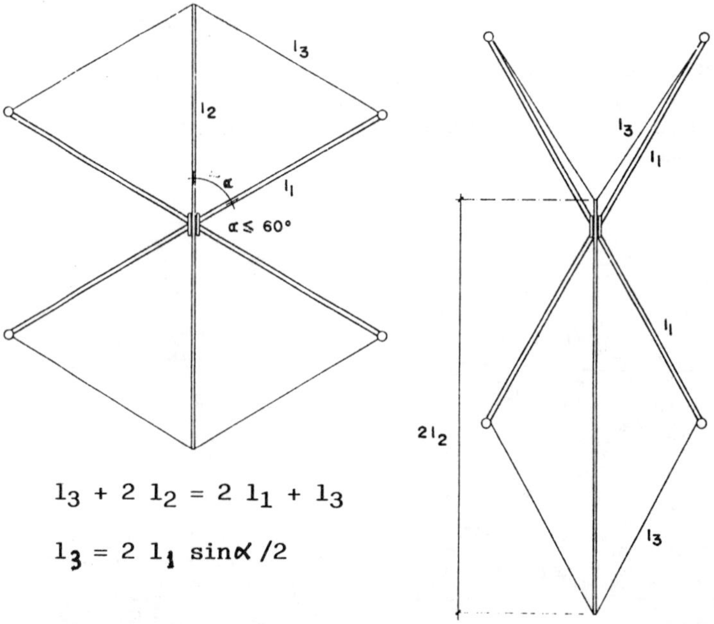

$$l_3 + 2\,l_2 = 2\,l_1 + l_3$$

$$l_3 = 2\,l_1 \sin\alpha/2$$

Fig. 8

To begin with, we study the case of connecting rods articulated at the ends of the lower bars, which implies a folding angle of less than 60º, often rather inadequate. The opening angle may be wider if the connecting rods are articulated at a point other than the end of the bars. The system can also function when the diagonal of the fabric has a different length than the bar (Fig. 9).

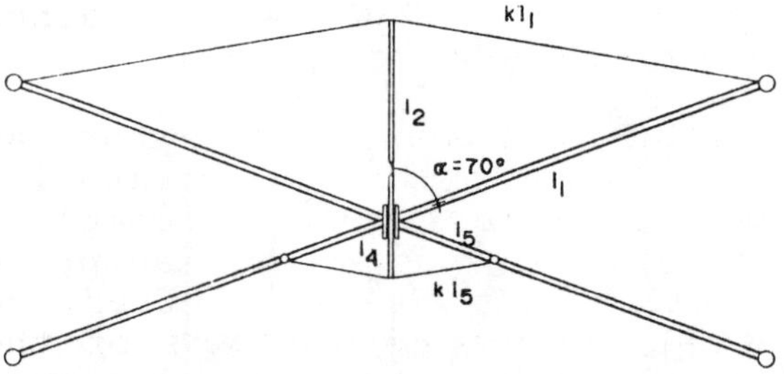

$$k^2 l_1^2 = l_1^2 + l_2^2 - 2 l_1 l_2 \cos\alpha$$

$$l_2 = l_1 \left(\cos\alpha + \sqrt{\cos^2\alpha - (1 - k^2)} \right) = \varphi l_1$$

$$l_5 = (1 - k - \varphi) / (\varphi - 1 - k)\, l_1$$

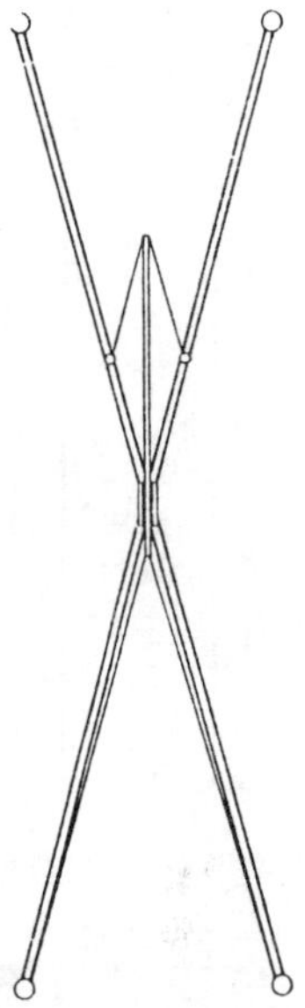

Figure 9

ANALYSIS OF THESE STRUCTURES.
BAR WITH FIXED INNER JOINT.

For the bars with three joints the equilibrium is completely defined by the system of forces P_1, P_2, N_1, N_2 (Fig 10). The reference 2 explain the linear calculus of these structures. However the effect of inner articulation make a bending on the bars and axial forces modifies the value of bending moment at the inner point. The figure shows the equilibrated system of forces and reactions with non-linear effects of axial forces.

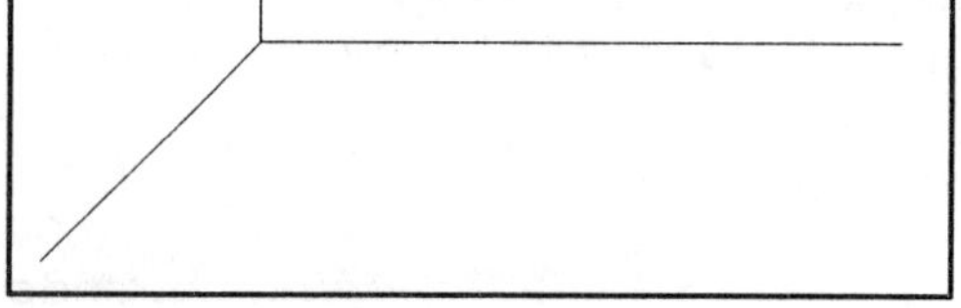

$$R_{11} = \frac{P_1 l_2}{l} + \frac{N_2 - N_1}{l} \delta_1$$

$$R_{12} = \frac{P_1 l_1}{l} - \frac{N_2 - N_1}{l} \delta_1$$

$$R_{21} = \frac{P_2 l_2}{l} + \frac{N_2 - N_1}{l} \delta_2$$

$$R_{22} = \frac{P_2 l_2}{l} + \frac{N_2 - N_1}{l} \delta_2$$

Figure 10

Being δ_1 and δ_2 the deflections at the inner point.

STIFFNESS MATRIX IN LOCAL COORDINATES. $\tilde{P} = \tilde{K} \cdot \tilde{Z}$

$$N_1 = \frac{EA}{l_1} u_1 \qquad N_2 = \frac{EA}{l_2} u_2$$

$$P_1 = \frac{3\ E\ I\ l}{\phi_1 l_1^2 l_2^2}\ \delta_1 \qquad\qquad P_2 = \frac{3\ E\ I\ l}{\phi_2 l_1^2 l_2^2}\ \delta_2$$

$$\begin{vmatrix} N_1 \\ N_2 \\ P_1 \\ P_2 \end{vmatrix} = \begin{vmatrix} \frac{EA}{l_1} & 0 & 0 & 0 \\ 0 & \frac{EA}{l_2} & 0 & 0 \\ 0 & 0 & \frac{3\ E\ I\ l}{\phi_1 l_1^2 l_2^2} & 0 \\ 0 & 0 & 0 & \frac{3\ E\ I\ l}{\phi_2 l_1^2 l_2^2} \end{vmatrix} \cdot \begin{vmatrix} u_1 \\ u_2 \\ \delta_1 \\ \delta_2 \end{vmatrix}$$

COMPATIBILITY MATRIX

x axis → ($\cos\alpha_1$, $\cos\beta_1$, $\cos\Gamma_1$) Direction of bar.
y axis → ($\cos\alpha_2$, $\cos\beta_2$, $\cos\Gamma_2$) Direction of normal to plane of scissor. We can obtain by vectorial product of vectors of bars.
z axis → ($\cos\alpha_3$, $\cos\beta_3$, $\cos\Gamma_3$) Direction perpendicular of x and y axis.

After defining local axis, the relationship between local and global systems are

$$u_1 = |u'_2 - u'_1|$$
$$u_2 = |u'_3 - u'_2|$$
$$v = \delta_2 - \delta_1 \cdot l_2/l - \delta_3 \cdot l_1/l$$
$$w = \phi_2 - \phi_1 \cdot l_2/l - \phi_3 \cdot l_1/l$$

In matrix form $\tilde{Z} = \tilde{A} \cdot \tilde{X}$

$$\begin{vmatrix} u_1 \\ u_2 \\ v \\ w \end{vmatrix} = \tilde{A} \begin{vmatrix} x_1 \\ y_1 \\ z_1 \\ x_2 \\ y_2 \\ z_2 \\ x_3 \\ y_3 \\ z_3 \end{vmatrix}$$

Being of compatibility matrix $\tilde{A}$

$$\begin{vmatrix} -\cos\alpha_1 & -\cos\beta_1 & -\cos\Gamma_1 & \cos\alpha_1 & \cos\beta_1 & \cos\Gamma_1 & 0 & 0 & 0 \\ 0 & 0 & 0 & -\cos\alpha_1 & -\cos\beta_1 & -\cos\Gamma_1 & \cos\alpha_1 & \cos\beta_1 & \cos\Gamma_1 \\ \frac{-l_2\cos\alpha_2}{l} & \frac{-l_2\cos\beta_2}{l} & \frac{-l_2\cos\Gamma_2}{l} & \cos\alpha_2 & \cos\beta_2 & \cos\Gamma_2 & \frac{-l_1\cos\alpha_2}{l} & \frac{-l_1\cos\beta_2}{l} & \frac{-l_1\cos\Gamma_2}{l} \\ \frac{-l_2\cos\alpha_3}{l} & \frac{-l_2\cos\beta_3}{l} & \frac{-l_2\cos\Gamma_3}{l} & \cos\alpha_3 & \cos\beta_3 & \cos\Gamma_3 & \frac{-l_1\cos\alpha_3}{l} & \frac{-l_1\cos\beta_3}{l} & \frac{-l_1\cos\Gamma_3}{l} \end{vmatrix}$$

STIFFNESS MATRIX IN GLOBAL COORDINATES.

Matrix equation is $\underset{\sim}{L} = \underset{\sim}{A}^t \cdot \underset{\sim}{K} \cdot \underset{\sim}{A} \cdot \underset{\sim}{X} = \underset{\sim}{S} \cdot \underset{\sim}{X}$

The references 2, 4 shows the mathematical development of stiffness matrix in global coordinates. In our last programs we prefered to make the matrix product of the single matrix. This method improves the structuration of computer program, but it is a little bit more slow.

STRESS ANALYSIS.

Bending Moments $$M_1 = \frac{P_1\, l_1\, l_2}{l} - \frac{N_1\, l_2}{l}\,\delta_1 - \frac{N_2\, l_1}{l}\,\delta_1$$

$$M_2 = \frac{P_2\, l_1\, l_2}{l} - \frac{N_1\, l_2}{l}\,\delta_2 - \frac{N_2\, l_1}{l}\,\delta_2$$

Shear stresses

$$V_{11} = \frac{P_1\, l_2}{l} + \frac{N_1 - N_2}{l}\,\delta_1 \qquad V_{21} = \frac{P_2\, l_2}{l} + \frac{N_1 - N_2}{l}\,\delta_2$$

$$V_{12} = \frac{P_1\, l_1}{l} - \frac{N_1 - N_2}{l}\,\delta_1 \qquad V_{22} = \frac{P_2\, l_1}{l} - \frac{N_1 - N_2}{l}\,\delta_2$$

BAR WITH SLIDED INNER JOINT.

In several expandable structures we can use bars with a slided inner joint in axial direction. The movement is only restrained in transverse directions, but it causes bending moments on the bar. For these bars the equilibrium is completely defined by the system of forces P_1, P_2, N, being the axial force the same in both spans' structures.

STIFFNESS MATRIX IN LOCAL COORDINATES.

$$N = \frac{E\,A}{l}\,u \qquad P_1 = \frac{3\,E\,I\,l}{\phi_1 l_1^2 l_2^2}\,\delta_1 \qquad P_2 = \frac{3\,E\,I\,l}{\phi_2 l_1^2 l_2^2}\,\delta_2$$

$$\begin{vmatrix} N \\ P_1 \\ P_2 \end{vmatrix} = \begin{vmatrix} \frac{EA}{l_1} & 0 & 0 \\ 0 & \frac{3\,E\,I\,l}{\phi_1 l_1^2 l_2^2} & 0 \\ 0 & 0 & \frac{3\,E\,I\,l}{\phi_2 l_1^2 l_2^2} \end{vmatrix} \cdot \begin{vmatrix} u_1 \\ \delta_1 \\ \delta_2 \end{vmatrix}$$

COMPATIBILITY MATRIX. $\tilde{Z} = \tilde{A} \cdot \tilde{X}$

The formulation is the same as the bar with fixed inner joint.

$$\begin{vmatrix} u \\ \\ \\ v \\ \\ w \end{vmatrix} = \tilde{A} \begin{vmatrix} x_1 \\ y_1 \\ z_1 \\ x_2 \\ y_2 \\ z_2 \\ x_3 \\ y_3 \\ z_3 \end{vmatrix}$$

The compatibility matrix $\tilde{A}$ is

$$\begin{vmatrix} -\cos\alpha_1 & -\cos\beta_1 & -\cos\Gamma_1 & 0 & 0 & 0 & \cos\alpha_1 & \cos\beta_1 & \cos\Gamma_1 \\ \frac{-l_2\cos\alpha_2}{l} & \frac{-l_2\cos\beta_2}{l} & \frac{-l_2\cos\Gamma_2}{l} & \cos\alpha_2 & \cos\beta_2 & \cos\Gamma_2 & \frac{-l_1\cos\alpha_2}{l} & \frac{-l_1\cos\beta_2}{l} & \frac{-l_1\cos\Gamma_2}{l} \\ \frac{-l_2\cos\alpha_3}{l} & \frac{-l_2\cos\beta_3}{l} & \frac{-l_2\cos\Gamma_3}{l} & \cos\alpha_3 & \cos\beta_3 & \cos\Gamma_3 & \frac{-l_1\cos\alpha_3}{l} & \frac{-l_1\cos\beta_3}{l} & \frac{-l_1\cos\Gamma_3}{l} \end{vmatrix}$$

Matrix equation is $\tilde{L} = \tilde{A}^t \cdot \tilde{K} \cdot \tilde{A} \cdot \tilde{X} = \tilde{S} \cdot \tilde{X}$

Bending Moments $M_1 = \dfrac{P_1\ l_1\ l_2}{l} - \dfrac{N\ (l_2 + l_1)}{l}\ \delta_1$

$$M_2 = \frac{P_2\ l_1\ l_2}{l} - \frac{N\ (l_2 + l_1)}{l}\ \delta_2$$

Shear stresses

$$V_{11} = \frac{P_1\ l_2}{l} \qquad V_{21} = \frac{P_2\ l_2}{l}$$

$$V_{12} = \frac{P_1\ l_1}{l} \qquad V_{22} = \frac{P_2\ l_1}{l}$$

CALCULUS OF STABILITY FUNCTIONS.

The process of calculus explained above needs the formulation of stability functions because the effect of inner articulation means that, in general, axial forces are different in two spans. The way chosen is the integration of differential equations of bar in two spans and apply the boundary conditions. Axial forces can be compression, traction or null. The reference 4 explains in detail the mathematical developments. In this paper we are going to support only the case of bar with sliding inner joint until undeveloped.

N<0

$$\phi = 3\,\frac{\left(\frac{1}{p\,l_1}+\frac{1}{p\,l_2}\right)-\frac{1}{tg\,p_1 l_1}-\frac{1}{tg\,p_2 l_2}}{l_1 l_2\left[\frac{1}{tg\,p_1 l_1}+\frac{1}{tg\,p_2 l_2}\right]}$$

N<0

$$\phi = 3\,\frac{-\left(\frac{1}{p\,l_1}+\frac{1}{p\,l_2}\right)+\frac{1}{th\,p_1 l_1}+\frac{1}{th\,p_2 l_2}}{l_1 l_2\left[\frac{1}{th\,p_1 l_1}+\frac{1}{th\,p_2 l_2}\right]}$$

D.- EXAMPLES.

One example of an application of the system proposed here is the author's work on the Asturias Pavillion at the EXPO'92 in Seville, wich was awarded fourth prize in national open competition. One of the major challenges was to design a low-cost system to minimize the effects of the strong sun radiation, common in Seville during spring and summer. The proposed solution consists of three expandable structures, covering the pavilion as sunshades and ready to be folded at night, when not in use. The modulus chosen was that described here, which we considered appropriate in this case.

Another challenge was the oblique deployment of the sunshade. As shown in the graph, the solution lies in using a fixed articulation placed on a corbel and an upper articulation attached to a ring which slides down the mast. A set of cables driven through the inside of the mast and a set of pulleys at its upper part, allow for a simple and economic folding system. At the same time, these cables work as a system of braces in the position of total aperture of the truss.

His structure has been calculated with a computer program based on the planning developed in this paper, and assuming that each bar has the strictly necessary dimensions, the total weight of the structure is of 8.3 kg/m^2 and the maximum section necessary would be 80.60.4mm. The following table shows the stress results obtained in some of the bars in the linear and non-linear hypotheses and with the effect of large deflections. It is worth remarking that for the dimensions used, the non-linear effects show a small, but far from negligible effect.

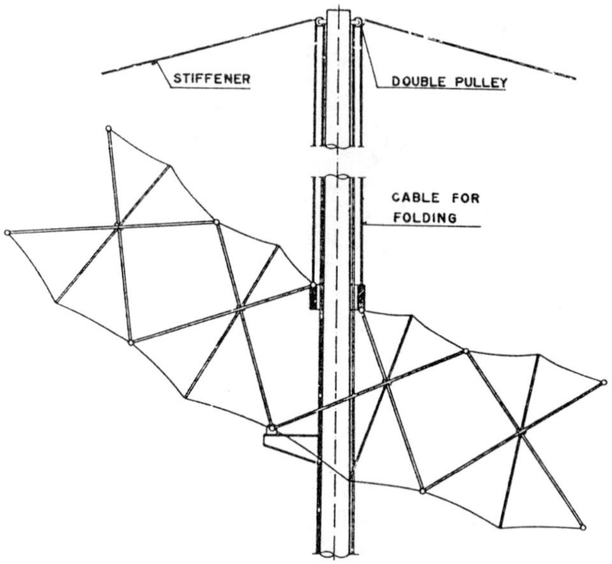
STIFFENER
DOUBLE PULLEY
CABLE FOR
FOLDING

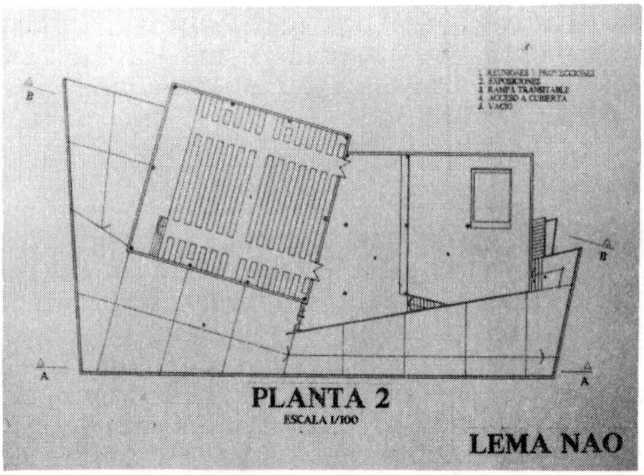
PLANTA 2
ESCALA 1/100
LEMA NAO

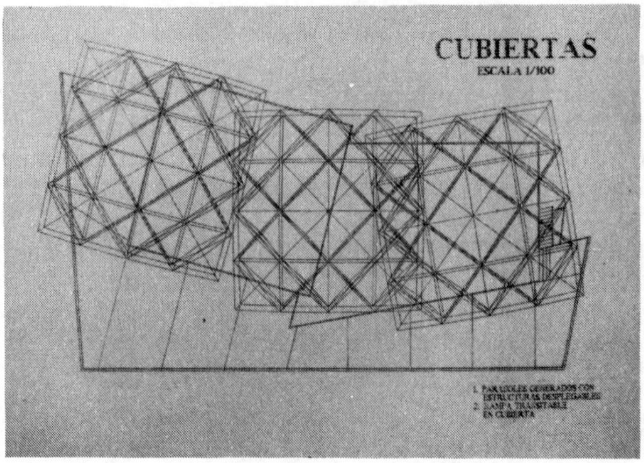
CUBIERTAS
ESCALA 1/100

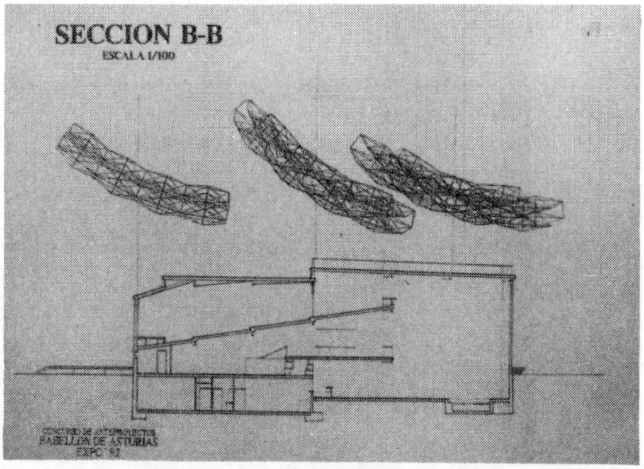

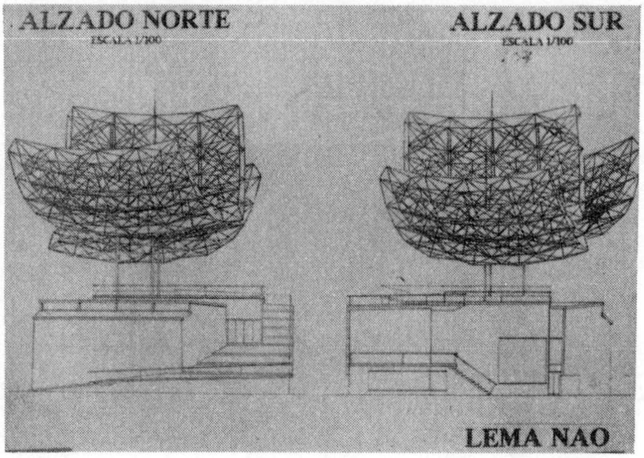

ACNOWLEDGEMENTS.

This research has been achieved with financial support provided by DGICYT (Dirección General de Investigación Científica y Técnica).

REFERENCES.

1.- Pérez Piñero, E."Estructures reticulées".L'Architecture d'Aujourd'hui. Vol.141.Dec.1968.pp 76-81.
2.- Escrig, F.; P.Valcárcel, J.B. "Analysis of Expandable Space Bar Structures". Int. Symposium on Membrane Structures and Space Frames. IASS. Osaka.1986.
3.- Escrig, F.; P.Valcárcel, J.B. "Curved Expandable Space grids". Non-conventional Structures '87. London 1987.
4.- P.Valcárcel, J.B.; Escrig, F. "Analysis of curved expandable space bar structures" Int. Symposium on 10 Years of Progress in Shell and Spatial Structures. IASS. Madrid 1989.

Supabeam- Large Lightweight Platforms for Construction and Maintenance Access

P.B. Tindall

Technical Director P.M.L.

1. INTRODUCTION

The provision of a safe and stable working platform for gaining access to the underside of a large roof or the soffit of a bridge has traditionally been achieved by building up a forest of tubes and fittings (scaffolding) from below. There are, however, many situations where this is either impractical or undesirable, and the preferred solution may be to suspend the working platform from the structure itself. The major problem encountered when one considers suspending a platform is that very few structures have been designed for this possibility; frequently there are very few suitable members for attaching fixings, and there are usually fairly severe limitations on the loading that can be applied to the structure.

Any proposed solution has therefore to be able to span a considerable distance to accommodate whatever fixings are available and to have a very low self weight.

The "Supabeam" platform is a modular aluminium alloy system designed to satisfy these criteria and, in addition, to be capable of very rapid assembly by relatively unskilled operatives.

2. BACKGROUND TO THE ORIGINAL CONCEPT

During the period 1976 to 1986 the Author and colleagues were directly involved in a number of contracts for British Rail for the major refurbishment of various railway station roofs; including Newcastle Central Station, Kings Cross Station, Victoria Station, St. Pancras Station and Hull Paragon Station. In all cases there were severe limitations with regard to obstructions at ground level as it was necessary to maintain a fully operational railway system with minimal disturbance to the travelling public. Construction of any temporary works was generally limited to short periods usually in the early hours of the morning.

The decision was therefore made to adopt a platform system suspended from a series of temporary runway beams attached to the roof structure. Initially the working platforms used were constructed from standard scaffold type materials available at the time. These, whilst serving their purpose well, were both heavy and time consuming to construct, which resulted in a decision to develop an alternative form of construction which would be lighter and faster to build.

The ensuing system (known as a Monkspeed platform) was used for creating the working platforms used for the renovation of the St. Pancras Station roof, and consisted of a series of trussed aluminium beams supporting large flooring panels made of plywood fixed to aluminium bearers. The beams were capable of spanning 24 ft (7.3m) and the overall system weight was approximately 6 $lb.ft^2$ (27 Kg/m^2).

One great advantage of installing an unusual access system in a large public space such as a railway station is that it attracts attention and generates interest from other potential users. It soon became apparent from the enquiries received that a span capability of 7.3 metres was insufficient and that a figure of 20 metres would be required to service the majority of applications. Thus was born the requirement for the Supabeam platform.

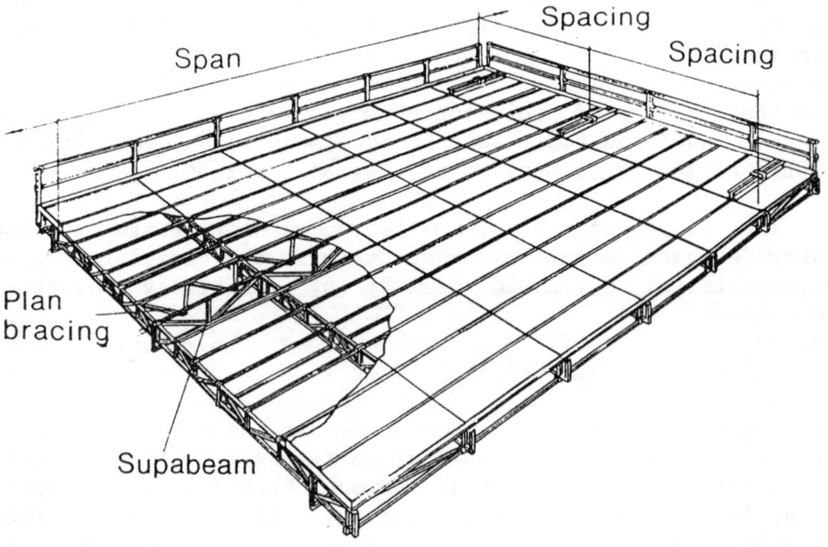

FIGURE 1

Supabeam Platform Perspective View

3. DESIGN PARAMETERS FOR THE SUPABEAM SYSTEM - LOADING

Having established that spans of 20 metres were required, the next fundamental decision to be made concerned the safe working loads that the system should be able to accommodate.

The nearest appropriate British Standard is applicable to scaffold structures and has various categories of loading starting at 15 lb/ft^2 (0.75 kN/m^2). Experience with the Monkspeed system, including careful monitoring of actual loads applied to the working platforms, had shown that for large platforms the overall loading rarely approached half this figure, although locally the loading could reach a figure of 30 lb/ft^2 (1.5 kN/m^2) over an area of up to 64 square feet (6 m^2) when construction materials were stacked ready for use, and point loads of a quarter of a ton (250 Kg) were frequently encountered. In addition, impact loads arising from materials being dropped onto the platform were encountered. The decision was therefore made to design all of the elements to withstand the worst case arising from the point load, impact load load or distributed load of 30 lb/ft^2 (1.5 kN/m^2) but to make the system adaptable so that a greater or small number of primary beams (Supabeams) could be installed to suit the global level of loading chosen for the particular application. Thus for a platform spanning 20 metres it is possible to configure the beams to give a safe working load of 30 lb/ft^2 (1.5 kN/m^2), over the whole platform, and for shorter spans the loading can be as high as 100 lb/ft^2 (5 kN/m^2), this latter capability largely arising from the local strength required for point loads and impact loads. It is readily apparent though that configuring the platform for these high loadings results in a large number of primary beams, which increases both the cost and weight of the system. It is therefore more appropriate to configure the platform for the load that will actually be encountered.

4. DESIGN PARAMETERS - PHYSICAL

The original concept which prompted the design of the Monkspeed system was based on the requirement easily to move components in the middle of the night into a railway station and rapidly assemble them into a working platform. To this end, and to comply with Health and Safety regulations already in force in some potential export markets as well as probable future legislation in the U.K., a maximum component weight of 25Kg was established with a requirement that each component could be easily handled by one man and then assembled without any special tools or skills. Threaded fasteners which could become damaged or loose and cause delay during construction, or collapse in use were avoided.

These principles had shown themselves to be absolutely sound and were embodied in the design of the Supabeam system.

The experience with the Monkspeed system had also prompted several other considerations.

Two of the components were very similar in appearance and length, (one a horizontal brace, the other a diagonal brace) and it was found to be difficult to distinguish between them. Attempts at colour coding the components were only partially successful in resolving the problem. The solution adopted for the Supabeam system was to make any components of similar length from material with different cross section thus readily enabling identification irrespective of lighting conditions or problems of colour blindness. The other consideration involved the number of different types of component in the system and it was considered essential to keep this to the absolute minimum commensurate with overall system efficiency.

5. DESIGN CONSIDERATIONS - CONNECTIONS

The main structural elements of the Supabeam system are fabricated from hollow aluminium extrusions for maximum structural efficiency and minimum weight. Experience with the earlier system which had used spigot connections with multiple pin and spring clip fastenings had shown the advantages of the pin and spring clip fastenings, but exposed the limitations of long spiggot connections for site assembly. The main connections have therefore been designed on the jaw plate principle, with a lower landing plate attached to the female jaw to aid alignment for rapid assembly. Pin and spring clip fastenings have been retained for all site assembled parts with the exception of the floor panels which are retained by a simple clip fit interlocking extrusion.

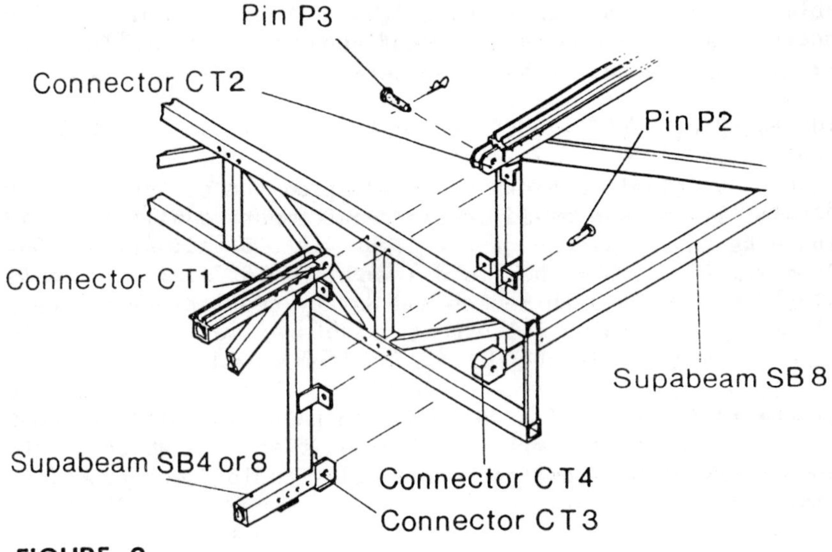

FIGURE 2

Supabeam to Supabeam Connection

6. DESIGN CONSIDERATIONS - STRUCTURAL

Design of the individual components of the system rapidly ran into two major problems, the first being that the relevant standard (British Standard Code of Practice CP 118 "The Structural Use of Aluminium") was in many instances not appropriate for the required details, the second problem being to realistically assess the behaviour of the platform under impact loads.

The problems on both fronts were resolved by fabricating prototype assemblies which were then subject to destructive testing to verify the design.

6.1 Limitations of CP 118

Two major limitations were encountered when attempting to calculate the strength of components to CP 118. The first concerned the rules applicable to pinned or bolted connections. The jaw plates used for the main Supabeam connections are designed to be bolted to the beams at fabrication stage. Consideration of bearing strength and shear strength dictate minimum bolt diameters, whilst the rules of CP 118 dictate edge distances and separation between fasteners. The comparatively thin walled extrusions used indicate that a large number of small diameter fastening should be used if the rules for maximum spacing between fasteners are to be observed, whilst the strength and geometrical requirements dictate that a small number of large diameter fastenings should be used. The design is therefore based on satisfying the strength requirements rather than CP 118 spacing rules, and an extensive test program (subsequently verified by years of actual use) has shown no resulting problems. The other problem relating to the use of CP 118 concerned the design of the welded joints within the beam itself. The standard gives clear rules on allowable stresses for welds subject to shear parallel or perpendicular to the weld axis, but gives no guidance on weld groups subject to moments as well as axial forces. In reality each weld run is subject to shears in several different directions. In addition the code does not address the possible reduction in joint strength from considerations of local buckling and punching shear when welding hollow sections. Welded joint strength was therefore established by destructive load testing.

6.2 Impact loading

Any analytical considerations for impact loading must be able to assess the energy absorption of the items concerned and also the ductility of the components in question. Whilst the former can be calculated, the wide disparity between the ductility of aluminium alloy extrusions and weld metal makes the latter somewhat problematical.

A comprehensive test programme was therefore evolved. This entailed building a complete platform which was then subject to impact loads at various positions. The loads were progressively increased until a failure occurred, the failed member replaced and loads applied to other areas of the platform. Static loads were applied to the platform such that the member subject to the impact load was stressed to its maximum working stress at the time of impact. For comparison further tests were carried out without these static imposed loads. It was notable that the level of initial static load on the platform made little difference to the impact failure level.

7. APPLICATIONS

At the time the system was designed the expected end use was predominantly for maintenance and refurbishment operations to bridge soffits and industrial building roof spaces. Whilst the system has mainly been used in these locations, the application has tended to be more often for initial construction work, and a number of other interesting applications have also ensued. In the majority of instances the platform system has been used in conjunction with a runway beam system so that the platform can be moved along the structure as work proceeds. Often the runways have been of a temporary nature, only extending part of the length of the structure to be serviced, and the runways have then been moved along the structure using the patented "Monkradle system."

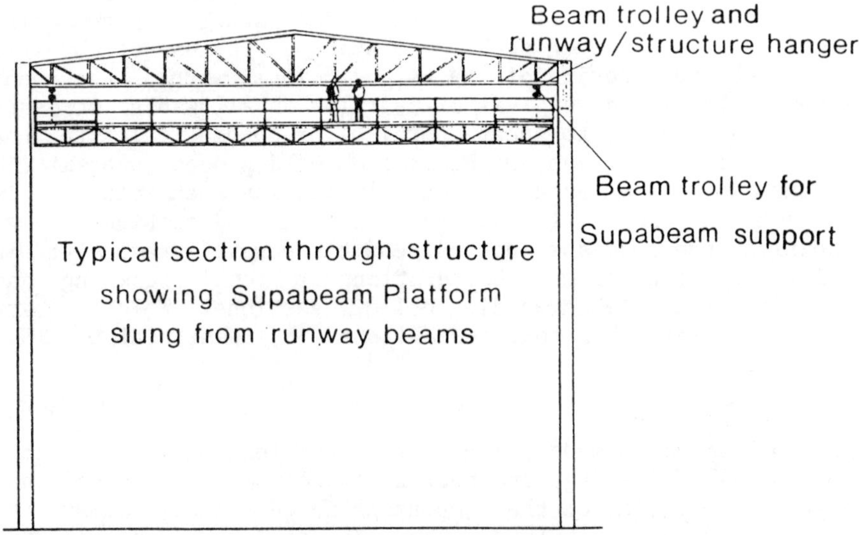

FIGURE 3
PML 'Monkradle System'

Other applications for the system have included the provision of; mobile platforms to the underdeck of North Sea oil platforms for maintenance operations; large work stations for maintenance operations to the insides of power station boilers; working platforms, both static and mobile, to the top of a large "chimney" at BNFL Sellafield; and creation of a travelling atrium cleaning gantry to an office building in the United States of America.

In nearly all cases the light weight of the system combined with the large spans available and ease of installation have been the motivating factor for our client's choice of the system. For the last two applications quoted the only practical means of transporting the components to their assembly location was via a passenger lift and in both instances the system was chosen by our clients after substantial world wide investigations into systems available.

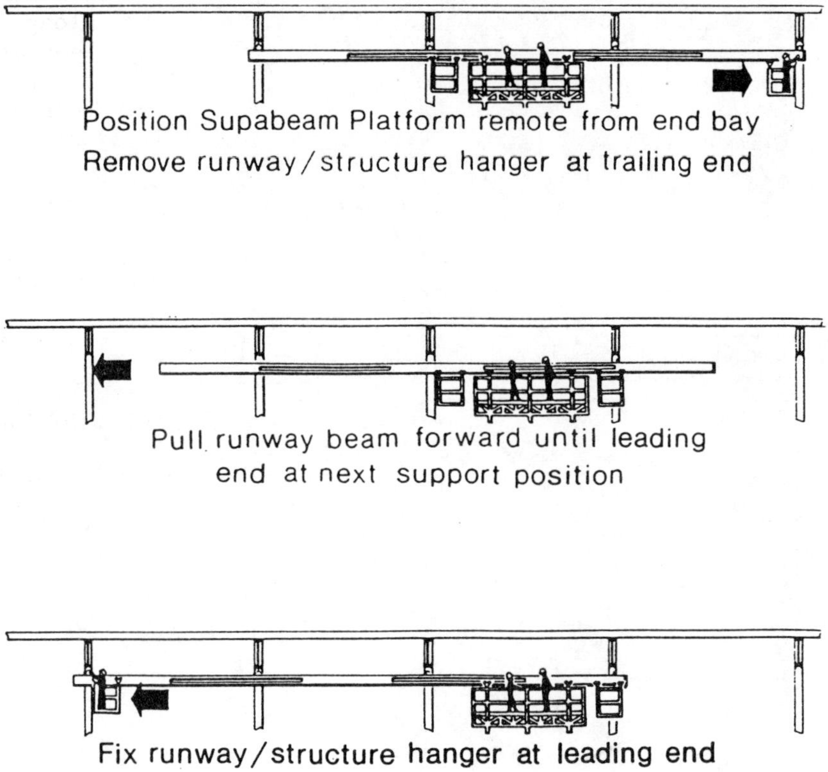

FIGURE 4

PML 'Monkradle System'

SEQUENCE FOR ADVANCING SYSTEM ALONG STRUCTURE Patent No. 1 589 515

8. FUTURE DEVELOPMENTS

Notwithstanding the success of the system to date, the Author is ever mindful that future success is not achieved by standing still. Invariably there are potential applications where larger spans or higher safe working loads are desirable, and any savings in construction time can only be beneficial. Current studies are concentrating on improving the shear resistance of the Supabeam itself (as load capacity tends in the majority of cases to be governed by shear rather than bending moment or deflection) and on developing a hemaphrodite connector for the beams.